Lecture Notes in Computer Science 14319

Founding Editors

Gerhard Goos
Juris Hartmanis

Editorial Board Members

The series Lecture Notes in Computer Science (LNCS), including its subseries Lecture Notes in Artificial Intelligence (LNAI) and Lecture Notes in Bioinformatics (LNBI), has established itself as a medium for the publication of new developments in computer science and information technology research, teaching, and education.

LNCS enjoys close cooperation with the computer science R & D community, the series counts many renowned academics among its volume editors and paper authors, and collaborates with prestigious societies. Its mission is to serve this international community by providing an invaluable service, mainly focused on the publication of conference and workshop proceedings and postproceedings. LNCS commenced publication in 1973.

Tiago Prince Sales · João Araújo ·
José Borbinha · Giancarlo Guizzardi
Editors

Advances in Conceptual Modeling

ER 2023 Workshops, CMLS, CMOMM4FAIR, EmpER
JUSMOD, OntoCom, QUAMES, and SmartFood
Lisbon, Portugal, November 6–9, 2023
Proceedings

 Springer

Editors
Tiago Prince Sales 🅾
University of Twente
Enschede, The Netherlands

João Araújo 🅾
Universidade Nova de Lisboa
Caparica, Portugal

José Borbinha 🅾
Universidade de Lisboa
Lisbon, Portugal

Giancarlo Guizzardi 🅾
University of Twente
Enschede, The Netherlands

ISSN 0302-9743   ISSN 1611-3349 (electronic)
Lecture Notes in Computer Science
ISBN 978-3-031-47111-7   ISBN 978-3-031-47112-4 (eBook)
https://doi.org/10.1007/978-3-031-47112-4

This Springer imprint is published by the registered company Springer Nature Switzerland AG
The registered company address is: Gewerbestrasse 11, 6330 Cham, Switzerland

Paper in this product is recyclable.

Preface

The International Conference on Conceptual Modeling is the main international forum for discussing the state of the art, emerging issues, and future challenges in research and practice in conceptual modeling. Topics of interest span the entire spectrum of conceptual modeling, including research and practice in areas such as theories of concepts and ontologies, techniques for transforming conceptual models into effective implementations, and methods and tools for developing and communicating conceptual models.

ER 2023 was held at the Congress Center of the Instituto Superior Técnico, which is located in the heart of Lisbon, Portugal. In addition to its main program, the conference hosted several satellite events, including seven workshops. ER workshops are forums for exchanging emergent ideas on conceptual modeling. They provide opportunities for presenting initial and ongoing work and allow researchers and practitioners to interact and collaborate. This volume contains the proceedings of the seven workshops held in conjunction with ER 2023. The chairs and program committees of the workshops were responsible for reviewing and selecting the papers. Each paper was single-blindly reviewed by at least three members of the program committee of the workshop to which it was submitted. Out of 53 submissions overall, 28 full papers and 2 short papers were accepted for publication in these proceedings.

Anna Bernasconi, Arif Canakoglu, Alberto García, and José Fabián Reyes Román organized the 4th International Workshop on Conceptual Modeling for Life Sciences (CMLS 2023). The CMLS workshop aims to bring together researchers from three different areas, namely information systems, conceptual modeling, and data management, to share their work on problems in healthcare and life sciences, with a focus on genomic data management and precision medicine. This volume contains the four full papers presented at CMLS 2023.

João Moreira, Luiz Olavo Bonino, and Pedro Paulo Favato Barcelos organized the 3rd edition of the workshop on Conceptual Modeling, Ontologies and (Meta)Data Management for Findable, Accessible, Interoperable, and Reusable Data (CMOMM4FAIR). The goal of the workshop is to investigate, discuss, and improve conceptual modeling practices towards improved FAIRness. This volume contains the one full paper presented at CMOMM4FAIR 2023.

Dominik Bork, Miguel Goulão, and Sotirios Liaskos organized the 6th International Workshop on Empirical Methods in Conceptual Modeling (EmpER 2023). EmpER brings together researchers and practitioners interested in the empirical investigation of conceptual modeling systems and practices. This volume contains the five full papers presented at EmpER 2023.

Silvana Castano, Mattia Falduti, Cristine Griffo, and Stefano Montanelli organized the 2nd International Workshop on Digital Justice, Digital Law, and Conceptual Modeling (JUSMOD 2023). JUSMOD invites researchers to share ideas on modeling, analyzing, formalizing, and interpreting legal data and related processes. This volume contains the six full papers and two short papers presented at JUSMOD 2023.

Sergio de Cesare, Frederik Gailly, Giancarlo Guizzardi, Chris Partridge, and Oscar Pastor organized the 9th International Workshop on Ontologies and Conceptual Modeling (OntoCom). This is an academic workshop that focuses on the practical and formal application of ontologies to conceptual modeling. It intends to bring together academics, researchers, and practitioners (with a background in IS engineering and/or ontology development) in order to develop an agenda for future collaborations that combine research and industrial expertise. This volume contains the five full papers presented at OntoCom 2023. We highlight that this year, OntoCom and CMOMM4FAIR merged their program.

Beatriz Marín, Giovanni Giachetti, Estefanía Serral, and Josè Luis de la Vara organized the 4th International Workshop on Quality and Measurement of Model-Driven Software Development (QUAMES). This workshop aims to attract research on methods, procedures, techniques, and tools for measuring and evaluating the quality of conceptual models that can be used in MDD environments. Its primary goal is to enable the development of high-quality software systems by promoting quality assurance in the modeling process. This volume contains the three full papers presented at QUAMES 2023.

Renata Guizzardi, Catherine Faron, Filipi Miranda Soares, and Gayane Sedrakyan organized the 1st edition of the Workshop on Controlled Vocabularies and Data Platforms for Smart Food Systems (SmartFood). This workshop is a forum meant to bring together researchers, industry workers (e.g., logistics, health), and consumer organizations that are concerned about the future of food-related systems and processes. More specifically, SmartFood seeks to gather people who believe that semantic technologies, such as controlled vocabularies and ontologies, data platforms, and consumer behavior-based models are at the core of the solutions targeting this field. This volume contains the three full papers presented at SmartFood.

We are deeply grateful to the ER 2023 general chairs, Kamalakar Karlapalem and Manfred Jeusfeld, and PC chairs, Upendranath Chakravarthy, Mukesh Mohania, and Jolita Ralyté, for their trust and support. We thank the workshop organizers for organizing the workshops and putting together exciting programs composed of keynote speeches, presentations of technical papers, and discussions. We are grateful to the workshop authors and reviewers for ensuring the high quality of the workshop programs. Finally, we thank the team at Springer for their guidance and help in preparing these proceedings.

September 2023

Tiago Prince Sales
João Araújo
José Borbinha
Giancarlo Guizzardi

Organization

General Chairs

José Borbinha   INESC-ID, IST, Universidade de Lisboa, Portugal
Giancarlo Guizzardi   University of Twente, The Netherlands

Program Committee Chairs

João Paulo A. Almeida   Federal University of Espírito Santo, Brazil
Jelena Zdravkovic   Stockholm University, Sweden
Sebastian Link   University of Auckland, New Zealand

Workshops Chairs

João Araújo   Universidade Nova de Lisboa, Portugal
Tiago Prince Sales   University of Twente, The Netherlands

Forum Chairs

David Aveiro   Universidade da Madeira, Portugal
Sotirios Liaskos   York University, Canada

Tutorial Chairs

Pnina Soffer   University of Haifa, Israel
Maribel Yasmina Santos   University of Minho, Portugal

Posters and Demos Chairs

Sergio de Cesare   University of Westminster, UK
Miguel Mira da Silva   INOV, IST, Universidade de Lisboa, Portugal

Industrial Track Chairs

Wolfgang Maass   Saarland University, Germany
Pedro Sousa   INESC-ID, IST, Universidade de Lisboa, Portugal

Panel Chairs

Veda Storey   Georgia State University, USA
Oscar Pastor   Universidad Politécnica di Valencia, Spain

Symposium on Conceptual Modeling Education Chairs

Maria Keet   University of Cape Town, South Africa
Fernanda Baião   Pontifical Catholic University of Rio De Janeiro,
                                     Brazil
Estefania Serral   KU Leuven, Belgium

Project Exhibition Chairs

João Araújo   Universidade Nova de Lisboa, Portugal
Tiago Prince Sales   University of Twente, The Netherlands

Doctoral Consortium Chairs

Helena Sofia Pinto   INESC-ID, IST, Universidade de Lisboa, Portugal
Ladjel Bellatreche   LIAS/ENSMA, France
Simon Hacks   Stockholm University, Sweden

Publicity Chairs

Anna Bernasconi   Politecnico di Milano, Italy
Claudenir M. Fonseca   University of Twente, The Netherlands

Sponsorship Chairs

Luis Bonino   University of Twente, The Netherlands
Sérgio Guerreiro   INESC-ID, IST, Universidade de Lisboa, Portugal

Volunteers Chair

André Vasconcelos   INESC-ID, IST, Universidade de Lisboa, Portugal

Treasurer

Daniel Faria   INESC-ID, IST, Universidade de Lisboa, Portugal

Web Chair

António Higgs   INESC-ID, Portugal

Contents

JUSMOD

OntoCom

QUAMES

SmartFood

CMLS

CMLS – 4th International Workshop on Conceptual Modeling for Life Sciences

Anna Bernasconi[1] 🆔, Arif Canakoglu[2] 🆔, Alberto García S.[3] 🆔,
and José Fabián Reyes Román[3] 🆔

[1] Dipartimento di Elettronica, Informazione e Bioingegneria Politecnico di
Milano, 20133 Milan, Italy
`anna.bernasconi@polimi.it`

[2] Dipartimento di Anestesia, Rianimazione ed Emergenza-Urgenza, Fondazione
IRCCS Ca' Granda Ospedale Maggiore Policlinico, Milan, Italy
`arif.canakoglu@policlinico.mi.it`

[3] Valencian Research Institute for Artificial Intelligence (VRAIN), Universitat
Politècnica de València, Valencia, Spain
`{algarsi3,jreyes}@vrain.upv.es`

The recent advances in unraveling the secrets of human conditions and diseases have encouraged new paradigms for their prevention, diagnosis, and treatment. As information is increasing at an unprecedented rate, it directly impacts the design and future development of information and data management pipelines; thus, new ways of processing data, information, and knowledge in healthcare environments are strongly needed.

The International Workshop on Conceptual Modeling for Life Sciences (CMLS) was held in 2023 in its fourth edition. Its objective was to be both a meeting point for Information Systems (IS), Conceptual Modeling (CM), and Data Management (DM) researchers working on health care and life science problems and an opportunity to share and discuss new approaches to improve promising fields, with a special focus on *Genomic Data Management* – how to use the information from the genome to understand better biological and clinical features – and *Precision Medicine* – giving each patient an individualized treatment by understanding the peculiar aspects of the disease. From the precise ontological characterization of the components involved in complex biological systems to the modeling of the operational processes and decision support methods used in the diagnosis and prevention of disease, the joined research communities of IS, CM, and DM have an important role to play; they must help in providing feasible solutions for high-quality and efficient health care.

The COVID-19 pandemic and the recent climate disruption have attracted increasing attention to the effects of pathogens and environmental change on human health. CMLS aims to become a forum for discussing the responsibility of the conceptual modeling community in supporting the life sciences related to these new realities.

The fourth edition of CMLS attracted high-quality 7 submissions about models and processes of the life sciences domain. Four papers were selected after a single-blind review process involving at least three field experts for each submission. All of them provide significant insights related to the problem under investigation, and they confirm an interesting technical program to stimulate discussion. We expect a growing interest in this area in the coming years; this was one of the motivations for continuing our commitment to this workshop in conjunction with the ER 2023 conference.

Acknowledgments. CMLS 2023 was organized within the framework of the projects: Advanced Grant 693174 (data-driven Genomic Computing) – European Research Council; PDC2021-121243-I00, MICIN/AEI/10.13039/501100011033 and PID2021-123824OB-I00 – Spanish State Research Agency; INNEST/2021/57 – Agència Valenciana de la Innovación (AVI), and CIPROM/2021/023 – Generalitat Valenciana.

We would like to express our gratitude to Stefano Ceri and Oscar Pastor, who demonstrated continuous support to the organization of CMLS 2023. We also thank the Program Committee members for their hard work in reviewing papers, the authors for submitting their works, and the ER 2023 organizing committee for supporting our workshop. We also thank ER 2023 workshop chairs João Araújo and Tiago Prince Sales for their direction and guidance.

CMLS Organization

Workshop Chairs

Anna Bernasconi — Politecnico di Milano, Italy
Arif Canakoglu — Fondazione IRCCS Ca' Granda Ospedale Maggiore Policlinico, Italy
Alberto García S. — Universitat Politècnica de València, Spain
José F. Reyes Román — Universitat Politècnica de València, Spain

Program Committee

Raffaele Calogero — University of Turin, Italy
Stefano Ceri — Politecnico di Milano, Italy
Pietro Cinaglia — Magna Graecia University, Italy
Tommaso Dolci — Politecnico di Milano, Italy
Johann Eder — University of Klagenfurt, Germany
Jose Luis Garrido — University of Granada, Spain
Giovanni Giachetti — Universitat Politècnica de València, Spain
Giancarlo Guizzardi — University of Twente, The Netherlands
Khanh N.Q. Le — Taipei Medical University, Taiwan
Sergio Lifschitz — Pontifical Catholic University of Rio de Janeiro, Brazil
Roman Lukyanenko — HEC Montreal, Canada
Giovanni Meroni — Technical University of Denmark, Denmark
Paolo Missier — Newcastle University, UK
José Palazzo — Federal University of Rio Grande do Sul, Brazil
Ignacio Panach — University of Valencia, Spain
Oscar Pastor — Universitat Politècnica de València, Spain
Barbara Pernici — Politecnico di Milano, Italy
Pietro Pinoli — Politecnico di Milano, Italy
Rosario Michael Piro — Politecnico di Milano, Italy
Tiago Prince Sales — University of Twente, The Netherlands
Monjoy Saha — National Cancer Institute, USA
Veda Storey — Georgia State University, USA
Emanuel Weitschek — Uninettuno University, Italy

Additional Reviewers

Arturo Castellanos
Francesco Invernici
Volodymyr A. Shekhovtso

An Ontology for Breast Cancer Screening

Yasmine Anchén[1][iD], Edelweis Rohrer[2][iD], and Regina Motz[2][✉][iD]

[1] Facultad de Medicina, Universidad de la República, Av. Gral. Flores 2125,
11800 Montevideo, Uruguay
yasanmo@gmail.com
[2] Facultad de Ingeniería, Universidad de la República, Julio Herrera y Reissig 565,
11300 Montevideo, Uruguay
{erohrer,rmotz}@fing.edu.uy

Abstract. This work introduces the creation of the Breast Cancer Screening Recommendation Ontology (BCSR-Onto), aimed at delivering screening recommendations for the early detection of breast cancer in individuals with varying risk levels, following different risk models and multiple recommendation guidelines. In particular, this ontology stands out in its reasoning capabilities since it generates recommendations without the need to use the Semantic Web Rules Language (SWRL) but exclusively using the reasoning mechanisms of description logic. This work also describes the central role of this ontology within a decision support system, serving as a formal communication piece between health professionals and patients. The BCSR-Onto is developed using OWL2 and is accessible under an open license.

Keywords: Breast cancer · Screening recommendation · Ontology reasoning

1 Introduction

The significance of early breast cancer detection cannot be overstated; it stands as a cornerstone in ensuring effective treatment outcomes and enhancing patient well-being. In recent years, a perceptible shift has unfolded in breast cancer screening strategies, moving away from a standardized approach towards a more personalized one, focusing on an individual's risk profile and preferences.

A crucial component of robust breast cancer screening initiatives is the assessment of breast cancer risk. This assessment plays a pivotal role in identifying individuals who stand to benefit from early detection methods, genetic testing, and preventive therapies. It also guides the general population in making informed decisions about their screening approach, as evidenced by the research conducted by Kim et al. [11]. To quantify this risk, predictive models have been developed that estimate the likelihood of a woman being diagnosed with breast cancer within a 5 to 10-year timeframe, as well as the probability of harbouring a BRCA1 or BRCA2 mutation, as explored by Pons-Rodríguez et al. [16]. Noteworthy among these models are the Tyrer-Cuzick [22], Claus [6], BOADICEA

T. P. Sales et al. (Eds.): ER 2023 Workshops, LNCS 14319, pp. 5–14, 2023.
https://doi.org/10.1007/978-3-031-47112-4_1

[2], and Gail [8] models. These models meticulously consider a multitude of factors pertinent to breast cancer risk calculation, encompassing both genetic and non-genetic determinants. These considerations are contingent on the specific evaluation model being employed. Variations exist in the risk factors considered, as well as their respective weights, among these models. Moreover, given that each model was developed based on a distinct group of women meeting specific criteria, potential biases may emerge depending on the population being studied [11].

Numerous research studies advocate for a comprehensive assessment of breast cancer risk, entailing risk calculation across multiple models. The rationale for this is twofold: firstly, estimates can differ significantly across various models; secondly, the applicability of these estimates may vary based on the specific clinical context. Barke's article [3] is illustrative in presenting a tabulated breakdown of model recommendations tailored to specific clinical scenarios. In this approach, medium- or high-risk guidelines are applied to provide breast cancer screening test recommendations based on a person's calculated breast cancer risk percentage from the risk models. Various globally recognized organizations engaged in breast cancer research have formulated practice guidelines grounded in evidence-based breast cancer screening and expert opinion for medium- and high-risk scenarios [5,12,14].

To summarize, given the need to consider different risk calculation models, the diverse range of risk factors each model takes into account, and the presence of additional guideline recommendations for different risk levels, a practical tool is required to provide accessible recommendations tailored to individual patient profiles. In this context, ontologies emerge as powerful modelling tools that offer a structured framework to describe complex domains [19]. Utilizing the OWL2 ontology language enables the automatic validation of ontology constraints and facilitates the extraction of implicit declarations. This ensures consistency through reasoning mechanisms grounded in descriptive logic [9].

The present work significantly contributes to the development and implementation of an OWL2 ontology aimed at providing breast cancer screening recommendations that consider various risk models and guideline recommendations. The potency of our ontology lies in its ability to leverage reasoning capabilities, enabling it to infer recommendations for each patient without relying on semantic web rules language (SWRL). This approach capitalizes on the power of OWL2 constructs with the Hermit reasoner. This work is an evolution of our ontology for mammography screening recommendations MAMO-SCR-Onto [1] to encompass different risk models and guidelines. We also discuss the benefits and challenges associated with using OWL2 and reasoning techniques to ensure the ontology's consistency, coherence, and inferential capabilities.

Moreover, recognizing the importance of open ontologies, we have made our ontology available as an open resource[1] with the CC-by-SA-4.0 license.

---

[1] Accessible at https://bioportal.bioontology.org/ontologies/BCSR-ONTO.

The remainder of this paper is organized as follows. Section 2 provides a brief overview of related work. Section 3 presents the ontology's key components and knowledge representation. Section 4 offers an overview of the role of BCSR-Onto in a system. Finally, Sect. 5 gives some conclusions and future directions.

2 Related Work

For more than a decade, breast cancer researchers have been committed to developing ontology-based systems aimed at enhancing decision-making within the medical domain. These systems operate by adhering to recommendations derived from acquiring clinical data and implementing knowledge-based systems. Notable instances of these systems encompass various applications, including breast cancer classification based on a combination of Case-Based Reasoning (CBR) and ontology approach [13], ontological analysis and diagnosis of breast masses [21], and more recently, initiatives such as the Horizon-2020 Desiree Project that centre around clinical guidelines for breast cancer [4], as well as the work of Oyelade *et al.* [15], which employs ontologies for breast cancer diagnosis. However, despite these endeavours employing ontologies to enhance breast cancer knowledge, their ontological frameworks exhibit two notable limitations to our objectives. Firstly, they concentrate on different uses of the ontology. In [13], the specific ontology aims to enhance how knowledge is represented, organized, and utilized within Case-Based Reasoning systems. In [4, 21] and [15], the use of ontology concentrates on clinical guidelines for treatment and diagnosis rather than recommendations for breast cancer screening, although there exists potential for adaptation to our specific context. To our knowledge, no ontology is oriented to breast cancer screening. Furthermore, recommendations and diagnostic procedures within this domain are represented using rules based on the Semantic Web Rule Language (SWRL) instead of harnessing the reasoning mechanism intrinsic to the OWL2 ontology. Nevertheless, the primary concern lies in the lack of availability of these ontologies for open reuse, stemming from the absence of essential free access necessary for open-source utilization. In the context of early breast cancer prediction, ontology accessibility within these systems is of paramount importance, given the ongoing revisions aimed at determining the most suitable risk model for each patient.

3 Breast Cancer Screening Recommendation Ontology

BCSR-Onto represents knowledge about risk assessment and recommendations by diverse guidelines for breast cancer screening for women with varying risk levels, following different risk models and multiple recommendation guidelines.

Figure 1 illustrates the conceptual model of the BCSR-Onto ontology. In this diagram, ovals symbolize concepts or classes, arrows depict relations or properties, and bullets signify individuals. To ensure clarity, our representation includes only a select group of individuals. We opt not to employ the OntoUML language

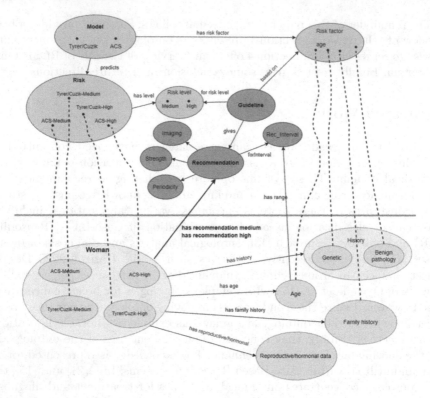

Fig. 1. Breast cancer screening guidelines ontology.

to express the conceptual model. OntoUML is based on UML, which was initially conceived for object-oriented design and primarily focuses on describing how data flows. In contrast, our notation is more straightforward; it only represents relevant concepts, relationships, and individuals. The upper segment of the figure delineates overarching concepts such as *Model*, *Risk*, and *Guideline*, which are employed by medical professionals in their interactions with patients. A model predicts the risk of breast cancer occurrence by considering distinct risk factors. Consequently, a woman may be categorized as high risk under one model while being classified as medium risk under another. Conversely, the lower part of the figure represents women (captured by the concept *Woman*) and associated properties, such as age and clinical or family history. These attributes serve as crucial determinants for risk calculation via various models. The dotted lines connecting instances of the *Risk Factor* concept and descriptions of women's properties (e.g., *Age*, *History*) denote the connection between the set of risk factors identified as instances (e.g., individual *age*) by medical professionals and the overarching concepts representing risk factors applicable to all women (e.g. the set of ages of women). Moreover, instances of the *Risk* concept are represented as subclasses of the *Woman* concept, classifying women based on the risk computed by each distinct model. Finally, the *Recommendation* concept serves as a bridge between both conceptual tiers, encapsulating recommendations provided to women by applying guidelines tailored to medium and high-risk levels.

The BCSR-Onto was implemented in OWL2 by using the ontology editor Protégé, according to the model of Fig. 1. Some terms were reused from NCI Thesaurus, in particular *Woman, History, Personal Medical History* and *Recommendation*. For clarity, Fig. 1 does not show some subproperties, inverse properties, and also a set of restrictions, that are included in the implementation. In particular, two Rbox axioms make it possible for the reasoner to infer recommendations for women, by defining the properties *hasRecommendationMedium* and *hasRecommendationHigh* as super properties of property chains. Rbox axioms expressed in description logics are as follows.

$$hasAgeMedium \ o \ hasRange \ o \ intervalOfMedium \sqsubseteq hasRecommendationMedium$$
$$hasAgeHigh \ o \ hasRange \ o \ intervalOfHigh \sqsubseteq hasRecommendationHigh$$

hasRecommendationMedium and *hasRecommendationHigh* have domain *Woman* and range the sets of recommendations given by guidelines for medium and high risk levels:
$$Recommendation \sqcap \exists gives^-.(\exists forRiskLevel.\{Medium\})$$
$$Recommendation \sqcap \exists gives^-.(\exists forRiskLevel.\{High\})$$

hasAgeMedium and *hasAgeHigh* are sub-properties of *hasAge* with range *Age* and domain the sets of women classified with risk level medium and high:
$$Woman \sqcap \exists hasRisk.(\exists hasLevel.\{Medium\})$$
$$Woman \sqcap \exists hasRisk.(\exists hasLevel.\{High\})$$

intervalOfMedium and *intervalOfHigh* are subproperties of the inverse of *forInterval* property, with domain *RecInterval* and range:
$$Recommendation \sqcap \exists gives^-.(\exists forRiskLevel.Medium)$$
$$Recommendation \sqcap \exists gives^-.(\exists forRiskLevel.High).$$

Definitions of *hasRecommendationMedium* and *hasRecommendationHigh* show that starting from a woman instance, if she has an age that fits into an interval of a recommendation for medium or high-risk level, then the woman is connected to this recommendation by the property *hasRecommendationMedium* or *hasRecommendationHigh*. Figure 2 shows the recommendation given to the 43 years old woman *Woman1* classified by ACS and Tyrer/Cuzik models as of medium risk, by applying the ACS guideline for medium risk. Figure 3 shows the recommendation given to the 56 years old woman *Woman4* classified by ACS and Tyrer/Cuzik models as of high risk, by applying the guideline for high risk. Figure 4 shows the recommendation given to the 56 years old woman *Woman2* classified by the ACS model as of medium risk and by the Tyrer/Cuzik model as of high risk, by applying both guidelines for medium and high risks.

Fig. 2. Reasoner entailments for a woman with intermediate risk.

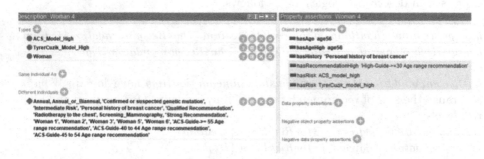

Fig. 3. Reasoner entailments for a woman with high risk.

Fig. 4. Reasoner entailments for a woman with different risks calculated by different models.

4   Ontology-Driven Breast Cancer Screening Recommendation System

The Ontology-driven Breast Cancer Screening Recommendation System is a system that assists the doctor with inferred recommendations to detect breast cancer in a healthy woman. The system is developed around the BCSR-Onto and uses various risk model calculators, electronic health records and various guidelines recommendations. In this section, we provide a concise overview of the utilization of the BCSR-Onto within the system. The comprehensive implementation of the system is currently under development by [20].

The system has an initial configuration phase in which the system administrator loads the various risk breast cancer calculators for risk models that the system will work with. Some of the potential risk calculators that can be instantiated in the system include, for example, the Gail Model Risk Calculator[2], Tyrer-Cuzick Risk Calculator[3], and BOADICEA risk calculator[4]. Furthermore, the administrator is required to load distinct recommendation guidelines intended for use as instances of the Guidelines class within the BCSR-Onto, prior to the initiation of doctor-system interaction. Illustrative instances of Guidelines encompass the medium and high-risk guidelines established by the American Association for Cancer Research [14] and [18] respectively, as well as the medium-risk guidelines, such as those recommended by [17] or [12].

Figure 5 depicts the overall structure of the system. Physicians interact through an interface that enables the collection of data and visualization of recommendation guidelines for breast cancer detection, based on patient data and different risk models, following various guidelines.

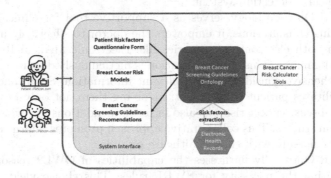

Fig. 5. Ontology-driven breast cancer screening guidelines system.

The physician can select through this interface which risk calculation models they wish to apply, being able to choose multiple models at once. For each of these models, the ontology extracts the necessary patient data from their Electronic Health Records. Data not present in the Electronic Health Record, which is necessary for the application of a certain model, is obtained from the patient through a form generated by the system. With the data integrated into the BCSR-Onto by the system, the ontology categorizes the patient into an appropriate risk group based on each risk model. This classification is accompanied by a percentage representing the likelihood of a breast cancer diagnosis within the upcoming 5 to 10 years [10]. Using this data, the system triggers the appropriate risk calculation tools for each model and fills the ontology with data that corresponds to the risk level associated with each model for the patient. By

---

[2] https://bcrisktool.cancer.gov/.

[3] https://ibis.ikonopedia.com/.

[4] https://www.canrisk.org/.

leveraging the data from the Electronic Health Record, the Patient Risk factors Questionnaire Form, and the calculated corresponding risk level for each model, the BCSR-Onto ontology employs reasoning and derives recommendations for each guideline (there may be multiple guidelines) based on the risk level identified by the models for that patient. These guideline-based recommendations are then showcased through the system interface, accessible to both the physician and the patient.

5   Conclusions and Future Work

Expanding the MAMO-SCR-Onto ontology, originally created to provide recommendations for women with average breast cancer risk, has broadened its applicability. This evolution now includes personalized recommendations for those with a high-risk level, as well as the incorporation of the essential risk assessment procedure for breast cancer. This holistic approach encompasses a crucial aspect of the early detection system.

The BCSR-Onto ontology serves as a supportive tool for clinical decision-making. During consultations, it empowers the doctor to effectively utilize data sourced from both electronic repositories like Electronic Medical Records and patient questionnaires. This data integration is meticulously documented, ensuring a comprehensive consideration of relevant information. Simultaneously, the ontology facilitates patient engagement, involving them not only in the assessment of their breast cancer risk but also in recommendations for evidence-based early detection studies. This collaborative approach promotes informed decision-making in partnership with their healthcare provider.

The BCSR-Onto fully harnesses the capabilities of OWL2 reasoning techniques, obviating the necessity for SWRL rules. This choice yields numerous advantages. Firstly, OWL2 reasoning incorporates built-in consistency checks, ensuring data integrity and averting contradictions within the ontology. Secondly, inferences derived from axioms and logical constructs in OWL2 significantly bolster the system's capacity to furnish meaningful and pertinent recommendations. Thirdly, OWL2 ontologies exhibit greater manageability, scalability, and adaptability over time when compared to SWRL rules, thanks to their standardized structure and support tools. Moreover, seamless integration with external ontologies and vocabularies is facilitated by pure OWL2 ontologies, a crucial aspect of promoting knowledge sharing in healthcare. Lastly, OWL2's adaptability effectively accommodates the evolving landscape of medical knowledge and guideline updates.

While SWRL rules have their merits, OWL2 reasoning offers a robust and comprehensive approach for creating intelligent systems that support complex decision-making processes, such as breast cancer screening recommendations, leveraging the full potential of ontological modelling and reasoning capabilities. While SWRL rules can be valuable in scenarios involving a substantial number of instances due to their capacity to optimize reasoning performance, especially when compared to OWL2, our specific case and the scale of our ontology and dataset do not justify the utilization of SWRL rules. In our ontology's use-case scenario, one aspect of the ontology is to serve as a conceptual model facilitating communication between a woman and her healthcare provider. The primary role of our ontology is to provide a conceptual framework for elucidating the complex decision-making processes within this context.

In future efforts, the ontology's scope could expand to encompass additional guidelines, such as those provided by the National Comprehensive Cancer Network (NCCN) for genetic counselling and testing decisions [7]. This would address specific criteria necessitating such counselling, where conventional risk assessment models might not be suitable. Moreover, the ontology could model optimal breast cancer risk assessment model selection based on distinct clinical scenarios, ensuring precise and suitable choices.

References

1. Acuña, C., Anchén, Y., Rohrer, E., Motz, R.: An ontology for mammography screening recommendation. In: 14th International Conference of Biomedical Ontology (ICBO 2023). CEUR-WS.org (To appear)
2. Antoniou, A.C., Pharoah, P., Smith, P., Easton, D.F.: The BOADICEA model of genetic susceptibility to breast and ovarian cancer. Br. J. Cancer **91**(8), 1580–1590 (2004)
3. Barke, L.D., Freivogel, M.E.: Breast cancer risk assessment models and high-risk screening. Radiol. Clin. **55**(3), 457–474 (2017)
4. Bouaud, J., et al.: Implementation of an ontological reasoning to support the guideline-based management of primary breast cancer patients in the DESIREE project. Artif. Intell. Med. **108**, 101922 (2020)
5. Cardoso, F., et al.: Early breast cancer: ESMO clinical practice guidelines for diagnosis, treatment and follow-up. Ann. Oncol. **30**(8), 1194–1220 (2019)
6. Claus, E.B., Risch, N., Thompson, W.D.: Autosomal dominant inheritance of early-onset breast cancer. Implications for risk prediction. Cancer **73**(3), 643–651 (1994)
7. Daly, M.B., et al.: Genetic/familial high-risk assessment: breast, ovarian, and pancreatic, version 2.2021, NCCN clinical practice guidelines in oncology. J. Natl. Compr. Cancer Netw. **19**(1), 77–102 (2021)
8. Gail, M.H., et al.: Projecting individualized probabilities of developing breast cancer for white females who are being examined annually. J. Natl Cancer Inst. **81**(24), 1879–1886 (1989)
9. Hitzler, P., Krötzsch, M., Rudolph, S.: Foundations of Semantic Web Technologies. Chapman & Hall/CRC (2009)
10. Ibarra Valencia, M.: Conocimientos y actitudes sobre prevención de cáncer de mama en adolescentes de la costa y sierra, piura 2020. ALICIA Repository (2022)
11. Kim, G., Bahl, M.: Assessing risk of breast cancer: a review of risk prediction models. J. Breast Imaging **3**(2), 144–155 (2021)
12. Klarenbach, S., et al.: Recommendations on screening for breast cancer in women aged 40–74 years who are not at increased risk for breast cancer. CMAJ **190**(49), E1441–E1451 (2018)
13. Lotfy Abdrabou, E.A.M., Salem, A.B.M.: A breast cancer classifier based on a combination of case-based reasoning and ontology approach. In: Proceedings of the International Multiconference on Computer Science and Information Technology, pp. 3–10 (2010). https://doi.org/10.1109/IMCSIT.2010.5680045

14. Monticciolo, D.L., et al.: Breast cancer screening for average-risk women: recommendations from the ACR commission on breast imaging. J. Am. Coll. Radiol. **14**(9), 1137–1143 (2017)
15. Oyelade, O.N., Aghiomesi, I.E., Najeem, O., Sambo, A.A.: A semantic web rule and ontologies based architecture for diagnosing breast cancer using select and test algorithm. Comput. Methods Prog. Biomed. Update **1**, 100034 (2021)
16. Pons-Rodriguez, A., Marzo-Castillejo, M., Cruz-Esteve, I., Galindo-Ortego, G., Hernández-Leal, M.J., Rué, M.: Avances hacia el cribado personalizado del cáncer de mama: el papel de la atención primaria. Atención Primaria **54**(5), 102288 (2022)
17. Ministerio de Salud Pública, U.: Guía práctica clínica de detección temprana del cáncer de mama - Tamizaje y diagnóstico precoz (uruguayan guide 2015) (2015). https://www.gub.uy/ministerio-salud-publica/comunicacion/publicaciones/guia-practica-clinica-deteccion-cancer-mama/
18. Saslow, D., et al.: American cancer society guidelines for breast screening with MRI as an adjunct to mammography. CA: A Cancer J. Clin. **57**(2), 75–89 (2007)
19. Staab, S., Studer, R.: Handbook on Ontologies, 2nd edn. Springer, Heidelberg (2009). https://doi.org/10.1007/978-3-540-92673-3
20. Szilagyi, B.: An ontology-drive breast cancer screening recommendation system. Engineering Computing Graduate thesis, Universidad de la República, Uruguay (To appear)
21. Trabelsi Ben Ameur, S., Sellami, D., Wendling, L., Cloppet, F.: Breast cancer diagnosis system based on semantic analysis and choquet integral feature selection for high risk subjects. Big Data Cogn. Comput. **3**(3), 41 (2019). https://doi.org/10.3390/bdcc3030041
22. Tyrer, J., Duffy, S.W., Cuzick, J.: A breast cancer prediction model incorporating familial and personal risk factors. Stat. Med. **23**(7), 1111–1130 (2004)

Integrating Nuclear Medicine and Radiopharmacy Data: A Conceptual Model for Precision Medicine and Enhanced Patient Care

Sipan Arevshatyan[1](✉) ⓘ, José Fabián Reyes Román[1](✉) ⓘ,
Elisa Caballero Calabuig[2], Mari Carmen Plancha[2], Alejandra Abella[2], Pedro Abreu[2],
and Óscar Pastor[1](✉) ⓘ

[1] Valencia Research Institute for Artificial Intelligence (VRAIN), Universitat Politècnica de València, Camino Vera s/n., 46022 Valencia, Spain
siar5@doctor.upv.es, jreyes@pros.upv.es, opastor@dsic.upv.es
[2] Nuclear Medicine Service and Radiopharmacy Unit, Hospital Universitario Doctor Peset (HUDP), Ave. Gaspar Aguilar 90, 46017 Valencia, Spain
{caballero_eli,plancha_mca,abella_ale,abreu_ped}@gva.es

Abstract. Nuclear Medicine and Radiopharmacy have emerged as pivotal branches of medical science, harnessing radioactive isotopes to diagnose and treat various diseases. The implementation of conceptual modeling in these disciplines is a paradigm shift that fosters precision, safety, and efficiency. The main objective of this research paper is to present a comprehensive Conceptual Model (CM) that integrates data from the Nuclear Medicine services and Radiopharmacy unit, laying the foundation for developing a practical application to manage the data in these different fields together. By harnessing the power of this model, the envisioned application seeks to facilitate precision medicine, thereby enhancing clinical outcomes and patient care.

Keywords: Conceptual Modeling · Nuclear Medicine · Radiopharmacy · Clinical Data · Precision Medicine

1 Introduction

In recent years, nuclear medicine and radiopharmacy have emerged as crucial and transformative medical science branches, revolutionizing how diseases are diagnosed and treated [1]. These fields harness the power of radioactive isotopes to provide invaluable insights into various medical conditions, enabling physicians to deliver targeted and effective therapies. Concurrently, the rise of conceptual modeling has opened new possibilities for advancing precision, safety, and efficiency in these disciplines.

Conceptual modeling refers to the process of creating abstract representations of complex systems to aid in understanding, analysis, and decision-making, which aids in software development [2]. By developing comprehensive Conceptual Models (CMs),

T. P. Sales et al. (Eds.): ER 2023 Workshops, LNCS 14319, pp. 15–24, 2023.
https://doi.org/10.1007/978-3-031-47112-4_2

researchers can bridge the gap between theory and practice, enabling the integration of diverse data sources into a cohesive framework. In the context of nuclear medicine (NM) and radiopharmacy (RF), conceptual modeling is poised to be a paradigm shift, offering a transformative approach to data management and analysis.

The primary objective of this research paper is to present a novel and all-encompassing CM that seamlessly integrates data from nuclear medicine and radiopharmacy (CMNMR). The proposed CMNMR aims to establish a strong foundation for developing a practical application to efficiently manage and interpret the complex datasets generated in these distinct yet interconnected fields.

The envisioned application powered by this CMNMR aspires to unlock the full potential of precision medicine. Precision medicine, an emerging approach to healthcare, emphasizes the customization of medical decisions and treatments based on individual variability in genes, environment, and lifestyle factors. By effectively managing and leveraging data from nuclear medicine and radiopharmacy, this application seeks to enhance clinical outcomes and patient care by providing tailored and personalized treatment strategies.

This paper will first explore the significance and impact of nuclear medicine and radiopharmacy in modern medicine. We will explore how these disciplines have transformed diagnostic capabilities and therapeutic interventions, demonstrating their pivotal role in improving patient outcomes. Subsequently, we will explore the principles and advantages of conceptual modeling, shedding light on how it can revolutionize data management in the context of nuclear medicine and radiopharmacy.

The core of this paper will focus on the presentation and detailed description of the proposed CMNMR. We will explore its structure, components, and the integration of data from nuclear medicine and radiopharmacy. Emphasizing its versatility and adaptability, we will highlight how the CMNMR can accommodate the rapidly evolving medical knowledge and technology landscape.

Furthermore, we will discuss the proposed practical application that will leverage the power of the CMNMR. We will delve into its potential use cases and applications, illustrating how it can streamline clinical workflows, improve decision-making processes, and contribute to the advancement of precision medicine.

To achieve our goal following this research line, in Sect. 2 we first give a brief view of the current state of the art. The CM of the domain introduced in Sect. 3. Section 4 shows the research implementation's current state and the application's development. Finally, the conclusions and future works are presented in Sect. 5.

2   State of the Art

This work is intended to illustrate the research which was done by VRAIN[1] institute and the Hospital Universitario Doctor Peset (HUDP)[2] in the scope of the project "*CalysapFénix*" [3]. The main objective was to improve data management processes in the Nuclear Medicine service and Radiopharmacy unit of the HUDP.

---

[1] https://vrain.upv.es/.
[2] https://doctorpeset.san.gva.es/es/.

Currently, the NM service and the RF unit have several computer tools to manage the data, but the applications they use do not meet the basic needs of the services.

The program called *"Calysap"* is currently being used within the units to manage the data, but this current solution does not satisfy the needs of the services, because, on the one hand, it is a system that uses Microsoft Access-based solutions, and on the other hand it does not have support or maintenance to develop care activity. Not having satisfying solutions, they have several problems when it comes to managing the data because they face data security, privacy, and data management issues from different sources.

Orion Clinic [4] is an IS for the management of clinical data created for the *Valencia Health Agency* with a clear orientation towards clinical data management, focused on improving patient care and health outcomes, for example, i) helping in the activity of health professionals, ii) increase the efficiency of the care process in as a whole, and iii) facilitate the continuity of care for citizens. The main disadvantage of this system is that it is not designed to manage radiopharmacy data when the patients are in the hospital to do explorations. There are some solutions in the market, such as *Syngo vía* [5]. Still, this application does not support data integration solutions while managing all the imaging information about the patient.

To solve the problems in the units, we use conceptual modeling first to understand the domain and then integrate the data from both units. CM serves as a method for delineating and portraying specific domains and aids in software production. By applying conceptual modeling the VRAIN research institute introduced GeIS, a framework tailored to managing disorders like usher syndrome [6], breast cancer [7], neuroblastoma [8], and alcohol sensitivity [9], among others, based on their corresponding domain conceptualizations.

In a broader context, aiming to establish a standardized lexicon and unequivocal definitions within the genomic domain, the *Conceptual Model of the Human Genome* (CMHG) [10, 11] was conceived. This model comprehensively encapsulates all pertinent information within the genomic domain. The subsequent development of a *Human Genome Database* (HGDB) [11] resulted from this model, reinforced by the utilization of *"GenesLove.Me"* [12], a well-established internal tool anticipated to contribute to genetic diagnostics significantly.

As a continuation we present the CM which defines the concepts and entities in Nuclear Medicine and Radiopharmacy to integrate the data in these units.

3   The Domain

Nuclear medicine is a medical specialty that uses small amounts of radioactive materials, known as radiopharmaceuticals, to diagnose and treat various medical conditions. On the other hand Radiopharmacy, also known as nuclear pharmacy, is a specialized branch of pharmacy that deals with the preparation, dispensing, and safe handling of radiopharmaceuticals used in nuclear medicine procedures.

In this section we are going to describe the domain showing how we were able to integrate these two branches. The CMNMR for Integrated NM and RF Data is designed to bridge the gap between nuclear medicine and radiopharmacy, facilitating seamless data exchange and fostering collaboration between these essential domains of modern health-care. This model aims to enhance treatment accuracy, patient safety, and research capabilities by harmonizing data and streamlining processes, ultimately leading to improved medical outcomes.

In order to make it easy to explain the model we have two views: the *Nuclear Medicine View* and *RadioPharmacy View*. Next, we are going to explain these views.

3.1   Nuclear Medicine View

The nuclear medicine view (Fig. 1) focuses on patient care and imaging procedures. The procedure starts when a request arrives to the Nuclear Medicine service. To manage the data of the request in the Nuclear Medicine Unit, we have a *"Medical Request"* class which is composed of attributes such as *"id_request"*, *"request_date"*, and *"observations"*. To keep track of the origins of the requests we have the classes *"Hospital"* and *"Service"*. These classes are the hospitals and the services or departments that can ask the Nuclear Medicine service for a treatment to the patient. For this purpose, the *"Patient"* class has been defined. This class contains attributes that describe the patient, such as *"id_patient"*, *"name"*, *"last_name"*, *"birthdate"*, *"address"*, *"phone"*, etc. The classes *"Province"* and *"Population"* represent the place where the patient lives currently.

Once the patient data has been reviewed, the explorations are scheduled, for this purpose we have the *"Exploration"* class. This class contains parameters like the date of the exploration, the priority, the patient's weight at that time etc. All this information is needed to give the patient the best treatment possible. When explorations are created, they are associated with the techniques by default. Hence, we can find the *"Technique"* class in the model. In the model, the *"associated_techniques"* relationship signifies that a technique can have related techniques. This implies that when a physician selects a technique with associated techniques, they are automatically implemented.

The same radioactive treatment may give different radioactivity to patients based on the administration route. Hence, we have defined the class *"Administration Route"*. Each technique has a default administration route. Nevertheless, at the end the radio-pharmacists define the final route to administer the single dose. In the *"Technique"* class we have the first integration point to the Radiopharmacy view with the *"Prescribable Radiopharmaceutical"* class which will be discussed in the Radiopharmacy View.

When the radiopharmaceutical lot is already prepared, the nurse selects the prepared lot, administers the patient and registers data about the administration. This data is reflected in the *"Administration"* class.

During the exploration the doctors assign diagnostics and findings to the patient which are specified by *"Diagnostic"* and *"Finding"* classes in the model.

After performing the exploration, the doctor is in charge of evaluating the images and coding the report to send to the applicant. This is defined by the "done_by" relationship. For this purpose, the *"Medical Report"* class has been defined and this class contains the following attributes: *"id_report"*, "date_report" and *"description"*. The doctors which can deal with the patient data are represented by the *"User"* class which should be used to log in to the application. Additionally, we have a *"validated_by"* relationship, as each report should be validated by another doctor.

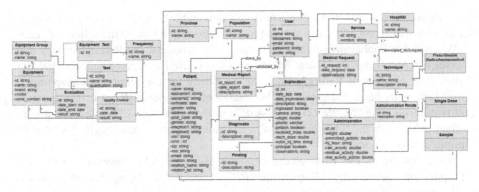

Fig. 1. The Conceptual Model of the Domain - Nuclear Medicine View

Different equipment is used to carry out the exploration. The doctors should carry out a quality control to adapt to current legislation regarding quality criteria in Nuclear Medicine. The tools represented by "*Equipment*" are grouped by a generic type of equipment defined as "*Equipment Group*" in the model. Each group of equipment should be tested periodically; that's why we have many to many relationships with the "*Test*" class. The doctors carry out the quality control of each equipment and register it in the "*Quality Control*" class. The frequency of these controls are defined by the "*Periodicity*" class. Finally, to know which tests are pending in a period of time we have an "*Evaluation*" class.

3.2 Radiopharmacy View

The Radiopharmacy View (Fig. 2) focuses on managing radiopharmaceuticals, ensuring quality control, and dispensing processes. As it was stated in the *Nuclear Medicine View* the first contact point of the integration is the "*Prescribable Radiopharmaceutical*" class. When a doctor decides to apply a technique, a prescribable radiopharmaceutical is also automatically selected. Then, the radiopharmacists evaluate whether the Radiopharmacy units have enough radiopharmaceuticals. If there is not enough radiopharmaceutical the Radiopharmacists ask for a laboratory. Hence, we have a Laboratory class and "*provides*" relationships. The "*manufacturers*" relationship indicated that the same laboratory can also manufacture products.

In the model we have four types of products to receive: "*Reactive Equipment*", "*Precursor*", "*Generator*", and "*Radiopharmaceutical Ready*". Their relationship with the "*Isotope*" class defines whether they are radioactive or not. As we see in the model only "*Reactive Equipment*" is not radioactive. The "*Generic*" class groups these products for statistical and management purposes.

Radioactivity can be defined as the property of certain atomic nuclei to spontaneously emit particles or electromagnetic radiation, resulting in the transformation of the nucleus into a more stable state. This emission of radiation occurs due to the instability of the nucleus, which seeks to achieve a more balanced and stable configuration.

The rate of radioactive decay is characterized by the half-life of the isotope which we register in the "*Isotope*" class (attribute "*half_life*"), which is the time it takes for half of the radioactive atoms in a sample to decay into stable atoms. Different radioactive isotopes have different half-lives, ranging from fractions of a second to millions of years.

When the radiopharmacist asks a laboratory for products, the Radiopharmacy unit receives batches of those products. The "*Lot*" class is used to receive and store batches of products. The lots of Radiopharmaceuticals Ready for use (*Radiopharmaceuticals Ready Lot*) as its name indicates, can be fractioned and administered if they pass the quality controls. The "*Single Dose*" class defines these fractions. The class has a 1-to-1 relationship with the "*Administration*" class meaning that each single dose can be administered only once.

Some lots from the Radiopharmaceuticals Ready batch, however, go through a preparation, defined by "*Preparation RR*" class, before being fractionated and administered.

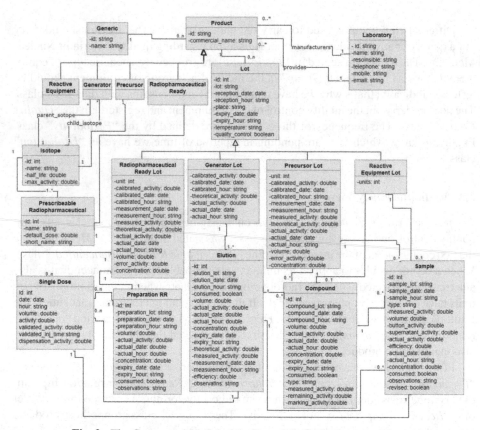

Fig. 2. The Conceptual Model of the Domain - Radiopharmacy View

Other types of lots that can be received are defined by "*Radiopharmaceutical Ready Lot*", "*Precursor Lot*", "*Generator Lot*", and "*Reactive Equipment Lot*" classes.

"*Elution*" lots are generated from the Generator batch. The elution batch can be administered directly to the patient after fractionation or may be in other preparations.

Compounds represented by the *"Compound"* class are not administered directly but are used for Autologous Samples defined by the *"Sample"* class. The user uses a Reactive Equipment Lot, an Elution Lot, or a Precursor Lot to generate a compound.

The Autologous Samples are prepared individually for each patient. To do this, the user selects a patient to validate the patient's blood. Hence, the relationship between the *"Patient"* and the *"Sample"* classes (Fig. 1). Then, depending on the type of autologous sample (Red Blood Cells, Leukocytes, or Platelets), the user selects a lot of Reactive Equipment, Elution, Precursor, or Compound.

4   Current State of the Research Implementation

We have a version of the model that has been used to develop an application to manage the data of the Nuclear Medicine and Radiopharmacy units. The CMNMR is open and demonstrates a capacity to integrate additional features that align with its foundational principles and design architecture. So, during the ongoing meetings, we may add new entities while the research continues.

During the execution of application development, the technology of the initial version was changed to develop a much more powerful, robust and scalable solution *"CalysapFénix"* based on the CMNMR that allows us to understand the domain and the needs of professionals to develop the application from a practical point of view. The CNMNR also was used to develop the database schema. In addition to this, we have used the latest technologies (React.js[3], Spring Boot[4]) to develop the application. MySQL[5] has been used as the database engine. The technologies used are open source and allow us to develop modular applications where the CMNMR plays a fundamental role with the entities and relationships described in the model.

The application manifests a modular structure, comprising six distinct modules: *User, Patient, Radiopharmacy, Quality Control, Reports,* and *Maintenance.* Each module encapsulates discrete functionalities while maintaining well-defined interfaces, fostering cohesive intermodule interactions, and facilitating efficient system scalability.

User Module: This module is defined to manage all data of the users registered in the system. We employed the Spring Boot security framework to establish a robust and comprehensive security infrastructure for fortifying the application. All user actions are saved on the server to maintain traceability.

Patient Module: The Patient Module within the application serves as a pivotal framework designed to manage patient-related aspects holistically. This module encapsulates a range of functionalities encompassing patient information acquisition, storage, retrieval, and facilitates seamless interactions with other modules. By leveraging the Patient Module, healthcare practitioners can proficiently register patients, record pertinent medical data, track their treatment journeys, and ensure accurate data integration throughout the application ecosystem, ultimately contributing to enhanced patient care and comprehensive healthcare management. In the new version new fields to register are added as

---

[3] https://react.dev/.

[4] https://spring.io/projects/spring-boot.

[5] https://www.mysql.com/.

the family member's contact information. In the explorations, we can easily filter by the main one and see the administrations. In the exploration tab we have new fields for dosimetry.

Quality Control Module: Within the realm of the quality control module, we have introduced the concept of "Group of Equipment", which serves to categorize equipment and stipulate the recurring tests conducted on them. This strategic linkage facilitates tracking pending tests within specified time intervals during assessments. Consequently, the efficacy of quality control searches has been notably enhanced.

Reports Module: This module serves to generate reports in.pdf format. We have improved the reports by adding new fields or improving the structure of the reports. We also have a new report (Dispense List) customized to print patient labels.

Radiopharmacy Module: In this module we have introduced new concepts that were not in the previous version. The current application used by the hospital (*Calysap*) has two types of products or raw materials ("hot" and "cold"). The *CalysapFénix* product entity now has a relationship with the isotope, and in this way, the hot products are distinguished, also in the new system we have new entities such as Radiopharmaceutical Ready, Generator, Precursor, and Reactive Equipment. Among these products, solely the Radiopharmaceutical Ready is eligible for dispensation after batch receipt. Conversely, the remaining products necessitate undergoing preparatory procedures encompassing Elutions, Compounds, and Autologous Samples. In addition, a new section has been introduced in CalysapFénix to dispense single doses. The data of each product is recorded at the reception and before proceeding to the preparation or dispensing the batch has to pass the quality control.

Maintenance Module: The Maintenance Module within the application is an integral component dedicated to the management and upkeep of the system's operational integrity. Through this module, the users can effectively configure different lists and configurations used in different modules.

The clinical validation tests of the *CalysapFénix* application have been carried out simultaneously with its development by the different professionals involved in each process. Thus, within the context of our in-person meeting, the advancement of the application was showcased, and the clinical team allocated the time leading up to the subsequent meeting to examine the specific segment of the application.

This has made it possible to monitor the progress and assess its applicability in daily practice while the application was being developed, and therefore, to detect failures or errors before continuing with each module, assess the possible improvements of each one and even, to go back and redo sections of the interface to make them more practical and applicable.

5   Conclusions and Future Works

This research paper seeks to bridge the gap between the intricate realms of nuclear medicine and radiopharmacy by presenting a pioneering CMNMR. By providing a unified framework for data integration and interpretation, this model aims to empower healthcare professionals to harness the full potential of nuclear medicine and radiopharmacy in the pursuit of precision medicine. Ultimately, we anticipate that this innovative approach will pave the way for enhanced clinical outcomes and an elevated patient care standard in the ever-evolving medical science landscape.

In the upcoming meeting with the doctors, we will complete and refine the CMNMR Radiopharmacy quality control section. This task encompasses augmenting the quality control segments associated with each radiopharmaceutical or radiopharmaceutical preparation. When the quality control of the radiopharmaceuticals and the preparation are done, we will conduct a thorough validation process encompassing the entirety of the radiopharmaceutical workflow, starting from batch discharge to the personalized allocation of doses to individual patients.

Subsequently, the patient information will be added to the new system following a specific order: first, the treatment details, then the PET CT scan images, and finally, other test results. Older patient records with no recent activity will be added last.

Acknowledgments. The author thanks *Saira Muhammad Iqbal, Héctor Santandreu Martínez, Carles Siscar Gelo* and the members of the *PROS Center* Genome group for fruitful discussions regarding the application of CM in the medical field. This work was supported by the Generalitat Valenciana through the **CoMoDiD** project (CIPROM/2021/023) and the Spanish State Research Agency through the **DELFOS** (PDC2021-121243-I00) and **SREC** (PID2021-123824OB-I00) projects, MICIN/AEI/https://doi.org/10.13039/501100011033, and co-financed with ERDF and the European Union Next Generation EU/PRTR.

References

1. Djekidel, M.: The changing landscape of nuclear medicine and a new era: the "NEW (Nu) CLEAR Medicine": a framework for the future. Front. Nucl. Med. **3**, 1213714 (2023). https://doi.org/10.3389/fnume.2023.1213714
2. Pastor, O., Molina, J.C.: Model-Driven Architecture in Practice: A Software Production Environment Based on Conceptual Modeling. Springer Science & Business Media, New York (2007). https://doi.org/10.1007/978-3-540-71868-0
3. Muhammad Iqbal, S.: Diseño y Desarrollo de una Aplicación Dirigida por Modelos para la Gestión de los Datos de un Servicio de Medicina Nuclear y Radiofarmacia. https://riunet.upv.es/handle/10251/188620. Bachelor degree Project (2023)
4. Villanova, P., et al.: Orion Clinic. Un sistema de información orientado a transformar el uso de la información en la práctica clínica, administrativa y asistencial de los hospitales. IS Inf. Salud **85**, 18–26 (2011)
5. Tecnicos   Radiologos.   http://www.tecnicosradiologia.com/p/medicina-nuclear.html. Accessed 09 Aug 2023
6. Burriel, V., et al.: Design and development of an information system to manage clinical data about usher syndrome based on conceptual modelling. In: BIOTECHNO (2013)
7. Burriel, V., Pastor, Ó.: Conceptual schema of breast cancer: the background to design an efficient information system to manage data from diagnosis and treatment of breast cancer patients. In: IEEE-EMBS International Conference on Biomedical and Health Informatics (BHI) (2014)
8. Arevshatyan, S., et al.: Integration and analysis of clinical and genomic data of neuroblastoma applying conceptual modelling. In: Workshop 38th International Conference on Conceptual Modeling (2019)

9. Reyes Román, J.F., Pastor, Ó.: Use of GeIS for early diagnosis of alcohol sensitivity. In: Proceedings of the International Joint Conference on Biomedical Engineering Systems and Technologies, pp. 284–289. SCITEPRESS-Science and Technology Publications, Lda (2016)
10. Reyes Román, J.F., Pastor, Ó., Casamayor, J.C., Valverde, F.: Applying conceptual modeling to better understand the human genome. In: Comyn-Wattiau, I., Tanaka, K., Song, I.Y., Yamamoto, S., Saeki, M. (eds.) ER 2016. LNCS, vol. 9974, pp. 404–412. Springer, Cham (2016). https://doi.org/10.1007/978-3-319-46397-1_31
11. Reyes Román, J.F.: Diseño y Desarrollo de un Sistema de Información Genómica Basado en un Modelo Conceptual Holístico del Genoma Humano. PhD Thesis, Universitat Politècnica de València. https://riunet.upv.es/handle/10251/99565 (2018)
12. Reyes Román, J.F., García, A., Rueda, U., Pastor, Ó.: GenesLove.Me 2.0: Improving the Prioritization of Genetic Variations. In: Damiani, E., Spanoudakis, G., Maciaszek, L.A. (eds.) ENASE 2018. CCIS, vol. 1023, pp. 314–333. Springer, Cham (2019). https://doi.org/10.1007/978-3-030-22559-9_14

Comprehensive Representation of Variation Interpretation Data via Conceptual Modeling

Mireia Costa⁽⊠⁾ , Alberto García S. , Ana León , and Oscar Pastor

PROS Group, Valencian Research Institute (VRAIN), Universitat Politècnica de
València, Valencia, Spain
micossan@vrain.upv.es, {algarsi3,aleon,opastor}@pros.upv.es

Abstract. The study of a DNA variation's impact on an individual's
health status is known as variation interpretation. An imprecise interpretation may result in incorrect clinical actions that endanger the patient's
health. Despite its obvious importance, variation interpretation remains
an unresolved challenge due to the wide dispersion and heterogeneity of
the data necessary for interpretation. Conceptual modeling has previously been demonstrated to be an effective solution to define complex
domains, achieving precise and consistent representations of dispersed
and heterogeneous data. This work presents the results of applying conceptual modeling to define a conceptual model that describes the required
data for conducting variation interpretation. This conceptual model represents the primary data dimensions required for the variation interpretation process and how they are related, resulting in a precise domain
description that will help make variation interpretation a systematic,
explainable, and reproducible procedure. To demonstrate how our conceptual model assists in achieving a more precise and consistent variation
interpretation process, examples of its instantiation to represent the data
required for evaluating the ACMG-AMP 2015 variation interpretation
guidelines criteria are presented.

Keywords: Conceptual Modeling · Variation interpretation ·
Precision medicine

1 Introduction

Precision medicine is a medical approach that promises to transform healthcare
by analyzing each patient's unique characteristics to provide personalized diagnosis and treatments [17]. Our DNA sequence is our most distinguishing characteristic as individuals. Therefore, determining how DNA variations can affect
each individual's health status is critical in precision medicine. This process is
known as variation interpretation [10].

Supported by ACIF/2021/117, CIPROM/2021/023, INNEST/2021/57, PID2021-
123824OB-I00 and PDC2021-121243-I00, MICIN/AEI/10.13039/501100011033 grants.

T. P. Sales et al. (Eds.): ER 2023 Workshops, LNCS 14319, pp. 25–34, 2023.
https://doi.org/10.1007/978-3-031-47112-4_3

A correct variation interpretation is critical for ensuring that the appropriate clinical actions are taken and that the patient's health is not jeopardized. Due to its importance, the scientific community has made numerous efforts to define guidelines for carrying out the variation interpretation process in a systematic, explainable, and reproducible manner. As a result, more than 20 variation interpretation guidelines have been developed, including the ACMG-AMP 2015 [14].

Although these guidelines are an important contribution to achieve a proper variation interpretation, this process still remains a challenge. One of the main reasons is the heterogeneity and dispersion that characterize the data required for variation interpretation [5]. Currently, 1.764 genomic data sources with potentially relevant information for variation interpretation are available [15]. Each data source has unique information, their particular terminology and potentially contradictory information. As a consequence, experts rely on data sources they know and trust instead of analyzing all the information available and, even when consulting the same data sources, different interpretations of the same piece of data occur. This makes it impossible to make the variation interpretation process systematic, explainable, and reproducible, which was the original intent of the variation interpretation guidelines. At the end, this leads to frequent disagreements when interpreting variations [6].

Conceptual modeling has previously been demonstrated to be an effective solution to define complex domains, and achieving precise and consistent representations of dispersed and heterogeneous data. Several works have used conceptual modeling to describe concrete domains. Focusing on recent works, Bernasconi et al. used conceptual models to describe meta-data of experimentally collected genomic data [2], and García et al. presented a conceptual model for representing omics data in precision medicine [16].

Despite the benefits that conceptual modelling offers, no conceptual model for variation interpretation data has been described before. This work aims to fill this knowledge gap by developing a conceptual model representing all of the data dimensions needed for variation interpretation. The proposed conceptual model is an important step toward making variation interpretation a systematic, explainable, and reproducible procedure.

The remaining of the paper is structured as follows. Section 2 provides the background for the conceptual model. Section 3 describes the proposed conceptual model. Section 4 presents examples of the instantiation of the model for ACMG-AMP guidelines criteria evaluation. Finally, Sect. 5 concludes the paper and discusses the future outlook.

2   Background

Several types of data are carefully evaluated by clinical professionals during the variation interpretation process. This data is typically classified into five categories: population frequency data, variation type and location data, case-level data, functional and computational data, and reputable source data [9].

The *population frequency* category determines the frequency at which a variation appears in a specific population. This is important because rare variations (i.e., those with a low frequency in the population under study) are more likely to cause disease than those that appear more frequently [8]. One well-known data source providing population frequency data is GnomAD [11].

The *variation type and location* category identifies the regions in which a variation is located and the effects it has on these regions. Variations that change specific regions of the protein (e.g., protein domains) or cause harmful effects (e.g., disrupt the protein sequence) are more likely to cause disease [14]. Data associated with this category is available, for instance, on Uniprot [4].

The *case-level* category has two dimensions: the patient's clinical context and genomic case-control studies. In the first dimension, the patient's clinical record is sought for a potential diagnosis and family history. This information can affect everything from the genetic analysis performed on the patient to the interpretation of the variation itself [7]. Due to privacy concerns, no public data sources contain this information. The second dimension focuses on genomic case-control studies, in which they compare the genome of people with a disease (i.e., cases) and healthy people (i.e., controls) and use a statistical analysis to discover links between DNA variations and diseases [13].

The *functional and computational* category includes data from functional studies and computational predictors. Functional studies empirically measure the impact of the variation on protein function, which is an important indicator of the variation's potential pathogenicity [3]. The data for functional studies can be obtained from the literature. Computational predictors are statistical or artificial intelligence models that attempt to predict the impact of a variation on the function of biological entities (e.g., a gene or a protein). An example of a computational predictor is Polyphen-2 [1].

The final category is *reputable sources*. It includes information about the role of a variation concerning the patient's health status based on the experience of various clinical experts. This allows for determining whether other experts have previously interpreted a variation. ClinVar is one data source that provides this information [12].

3   A Conceptual Model for Variation Interpretation

The conceptual model presented herein is divided into five views to reflect the five-category classification described above.

The five views are linked through the most relevant class of the model, namely, the VARIATION. A VARIATION represents the change in an individual's DNA that is being interpreted. The VARIATION is defined by a *name*, *type*, and the specific alteration that occurs at three levels: the genome (i.e., the *altGenome* attribute), the transcriptome (i.e., the *altTrans* attribute), and the proteome (i.e., the *aaAlt* attribute).

Since we cannot describe each view of the model here, in the subsections below we focus on the three most complex views of the model: i) the case-level, ii) the functional and computational, and iii) the reputable source.

3.1   The Case-Level View

As stated in Sect. 2, this view (see Fig. 1) focuses on two dimensions: the clinical context of the patient under study and existing case-control studies related to the variation under investigation.

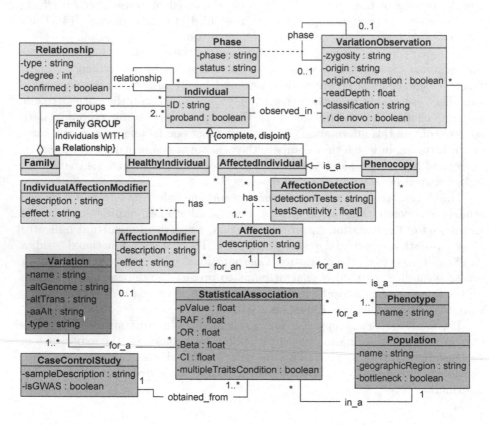

Fig. 1. The Case-level view. The Variation class is depicted in red. The patient's clinical context is depicted in light orange. Case-control studies are depicted in dark orange. (Color figure online)

The central class of the first dimension (i.e., the clinical context of the patient) is the INDIVIDUAL. An INDIVIDUAL can be either the patient if the *proband* attribute is set to true; otherwise, it is a family member or a non-related INDIVIDUAL. Two INDIVIDUALS can be members of the same family, represented in our model through the RELATIONSHIP and the FAMILY classes. The former specifies the *type* of relationship, whereas the latter aggregates INDIVIDUALS of the same family.

Evaluating the patient's family history to identify other family members affected with the same patient's disease and VARIATION strongly indicates that the VARIATION under study is potentially pathogenic [14]. To do this, the model must discriminate between healthy individuals (i.e., HEALTHYINDIVIDUAL) and those suffering from a specific Affection (i.e., AFFECTEDINDIVIDUAL). Affected individuals must be tested with a method (i.e., AFFECTIONDETECTION) comprising a set of *detectionTests*, each with a certain *testSensitivity*.

Affected family members may not have the VARIATION under study, possibly leading clinical experts to consider the VARIATION as benign. However, this is not an adequate assessment. Consider the case of Dilated Cardiomyopathy, which has an inherited origin in only up 20% to 50% of the cases and an environmental origin in the remaining cases. In this case, discarding a VARIATION as benign because an affected family member does not have it is an incorrect assessment. To avoid this, we use the well-known concept of PHENOCOPY, which represents an affected family member due to an environmental origin rather than a genetic one.

Another important consideration here is that no AFFECTION manifests the same way in two individuals. Factors such as the environment, the presence of other affections, and the genomic context of each individual can all influence how an AFFECTION manifests. In order to precisely characterize the effect that a Variation has on a particular INDIVIDUAL, it is important to represent such variability. Our model has introduced this variability with the AFFECTIONMODI-FIER and the INDIVIDUALAFFECTIONMODIFIER classes. AFFECTIONMODIFIERS are clinical features that alter how an AFFECTION manifests. For instance, high cholesterol levels aggravate heart diseases. The INDIVIDUALAFFECTIONMODI-FIER represents how an AFFECTIONMODIFIER manifests on a specific individual.

The last relevant class of this first dimension is the VARIATIONOBSERVATION class, which represents a VARIATION observed in a given INDIVIDUAL. This class defines properties of the VARIATION dependent on the INDIVIDUAL. For instance, if the VARIATION is inherited from the parents (i.e., germline) or acquired over time (i.e., somatic) is represented with the *origin* attribute.

Regarding the case-control dimension, its central class is the CASECONTROL-STUDY. This class is characterized by the *sample* of the study (i.e., a combination of cases and controls) and whether or not the study is a *GWAS* (i.e., a particular type of CASECONTROLSTUDY). A CASECONTROLSTUDY allows for establishing STATITISCALASSOCIATIONS between a VARIATION, a PHENOTYPE, and a specific POPULATION.

3.2   Functional and Computational Data View

The second view herein considered is the functional and computational one (see Fig. 2). Functional data is obtained from a FUNCTIONALSTUDY that computes a FUNCITONALASSESSMENT for a VARIATION regarding the *outcomeEvaluated* in the study. These studies were characterized following the recommendations of The Clinical Genome Resource Sequence variation Interpretation Working Group (ClinGen-SVI)[1], a relevant resource founded by the National Institutes

---

[1] https://clinicalgenome.org/working-groups/sequence-variation-interpretation/.

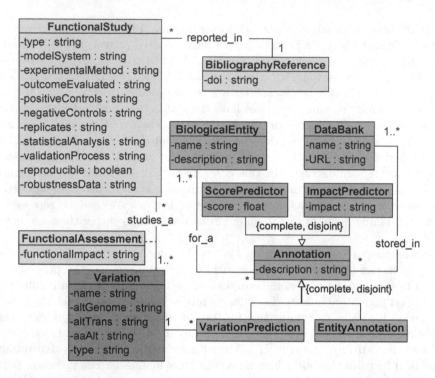

Fig. 2. Functional and computational data view. The Variation class is depicted in red. The functional data representation is depicted in lilac. Computational data is depicted in purple. (Color figure online)

of Health (NIH) to help in variation interpretation. These recommendations [3] allowed us to define the most relevant elements of a functional study in the context of variation interpretation, and they are incorporated into our model as attributes of the FUNCTIONALSTUDY class. A complete description of such attributes can be found in [3]. Each FUNCTIONALSTUDY is reported in the scientific literature, which we represented by means of the BIBLIOGRAPHYREFERENCE class.

In contrast to functional data, computational data is generated automatically by computational software. This kind of software generates ANNOTATIONS that in our model have been specialized twice. The first specialization is by the type of generated ANNOTATION. Here, a SCOREPREDICTOR provides a numerical prediction (i.e., a *score*), whereas an IMPACTPREDICTOR provides a categorical label describing the computational result (i.e., an *impact*). The second specialization identifies the annotated element. ANNOTATIONS are created for VARIATIONS (i.e., VARIATIONPREDICTION) or BIOLOGICALENTITY (i.e., ENTITYANNOTATION) such as genes and proteins.

3.3   Reputable Source Data View

This last view evaluates the existing information provided by other clinical experts. The main concept of this view is the SIGNIFICANCE class, which is used to assess the pathogenicity of a VARIATION for a PHENOTYPE. It is important to note that the *origin* of a VARIATION can condition its *clinicalSignificance*. Each SIGNIFICANCE requires at least one SUBMISSION, which is stored on external DATABANKS. A SUBMISSION describes the following information about a significance: *who* submitted it, the *method* used to assess it (e.g., literature review, clinical testing, etc.), and the *date* where it was performed (Fig. 3).

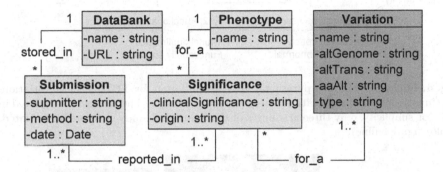

Fig. 3. Reputable source view. The variation is depicted in red. Interpretation data is depicted in pink. (Color figure online)

4   Use Case: ACMG-AMP 2015 Guidelines

To demonstrate how our conceptual model assists in achieving a more precise and consistent variation interpretation process, we show how it can be instantiated to represent data requirements associated with two criteria of the ACMG-AMP 2015 guidelines. The ACMG-AMP 2015 Guidelines are the gold standard for interpreting variations [14]. They define 28 criteria for variation interpretation based on the five data categories specified in Sect. 2. The examples presented here consider using the PS3 and PP5 criteria to interpret a particular variation named TPM1-D230N.

The PS3 criterion determines whether or not a *well-established in vitro or in vivo functional studies supportive of a damaging effect on the gene or gene product* exists. According to this definition, this criterion is based on evaluating functional data (see Sect. 3.2).

In this example (see Fig. 4), our instance reveals two in vitro studies assessing the effects of the TPM1-D230N variation, both of which conclude that the variation negatively affects the function of a protein (i.e., it is functionally abnormal). The PS3 criterion is met because all in vitro studies conclude that the variation has a negative effect.

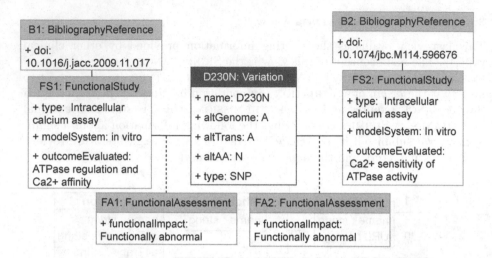

Fig. 4. Functional and computational view instantiation for TPM1-D230N variation. The functional data instantiation is depicted in light purple. The variation is depicted in red. For simplicity, only three attributes of the Functional study class are instantiated. (Color figure online)

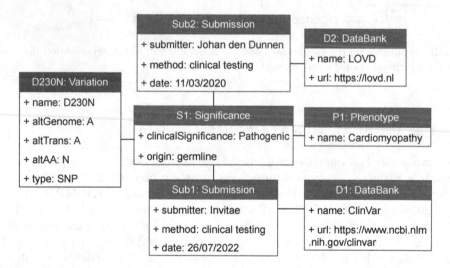

Fig. 5. Reputable source view instantiation for TPM1-D230N variation. The reputable data source data instantiation is depicted in pink. The variation is depicted in red. (Color figure online)

Regarding the second criterion (i.e., PP5), it evaluates if a *reputable source recently reported the variation as pathogenic, but the evidence is not available to the laboratory to perform an independent evaluation.*

Figure 5 depicts how two sources (described as databank in our model) have reported the variation under study as pathogenic. Both sources (i.e., ClinVar and LOVD) are highly recognized and reputable. As a result, the PP5 criterion is also met.

This use case shows how our model improves data representation, resulting in a clearer and easier evaluation of the complex criteria that are evaluated during the variation interpretation process. The division of the model in five views has allowed for an accurate delimitation of the data used for each criterion evaluation. This allows for precise traceability and reproducibility of variation interpretation results, which can be a first step towards more precise interpretation processes and conflict resolution among clinical experts. Our approach also makes the evaluation process of each criterion understandable and transparent. This level of transparency allows for a better understanding of variation interpretation outcomes and fosters trust and confidence in the process's validity.

5   Conclusions and Future Outlook

Variation interpretation is critical for providing diagnoses and treatments tailored to each patient's unique characteristics. It is a complicated process in which systematization, explainability, and reproducibility are critical factors. However, these factors are difficult to achieve due to the data's dispersion and heterogeneity. In this work, we created a conceptual model that clearly and explicitly represents all of the data dimensions required for a correct variation interpretation. The use case presented in the previous section demonstrated the advantages of using our conceptual model and how it can serve as the foundation for making variation interpretation systematic, explainable, and reproducible.

Future work will focus on performing a validation of the model and using the model to increase the systematization of the variation interpretation process. To validate the model, we plan to perform an empirical assessment with geneticists and clinical experts in the domain. Once the model is validated, we plan to connect the model with a previously defined meta-model representing the most relevant constructs underlying variation interpretation guidelines. This connection will allow us to automate and operationalize the variation interpretation process completely. We will also explore how to automate clinical guideline evaluation once we complete systematization.

References

1. Adzhubey, I., et al.: Predicting functional effect of human missense mutations using PolyPhen-2. Curr. Protoc. Hum. Genet. **76**(1), 7–20 (2013). https://doi.org/10.1002/0471142905.hg0720s76
2. Bernasconi, A., Ceri, S., Campi, A., Masseroli, M.: Conceptual modeling for genomics: building an integrated repository of open data. In: Mayr, H.C., Guizzardi, G., Ma, H., Pastor, O. (eds.) ER 2017. LNCS, vol. 10650, pp. 325–339. Springer, Cham (2017). https://doi.org/10.1007/978-3-319-69904-2_26

3. Brnich, S., et al.: Recommendations for application of the functional evidence PS3/BS3 criterion using the ACMG/AMP sequence variant interpretation framework. Genome Med. **12**, 3 (2019). https://doi.org/10.1186/s13073-019-0690-2
4. The UniProt Consortium: UniProt: the universal protein knowledgebase. Nucleic Acids Res. **45**(D1), D158–D169 (2016). https://doi.org/10.1093/nar/gkw1099
5. Costa, M., García S, A., Pastor, O., et al.: A comparative analysis of the completeness and concordance of data sources with cancer-associated information. In: Guizzardi, R., Neumayr, B. (eds.) ER 2022. LNCS, vol. 13650, pp. 35–44. Springer International Publishing, Cham (2022). https://doi.org/10.1007/978-3-031-22036-4_4
6. Furqan, A., et al.: Care in specialized centers and data sharing increase agreement in hypertrophic cardiomyopathy genetic test interpretation. Circ. Cardiovasc. Genet. **10**(5), e001700 (2017)
7. Garrett, A., et al.: Phenotype evaluation and clinical context: application of case-level data in genomic variant interpretation. In: Lázaro, C., Lerner-Ellis, J., Spurdle, A. (eds.) Clinical DNA Variant Interpretation. Translational and Applied Genomics, pp. 251–274. Academic Press (2021). https://doi.org/10.1016/B978-0-12-820519-8.00017-X
8. Gudmundsson, S., et al.: Variant interpretation using population databases: lessons from gnomAD. Hum. Mutat. **43**, 1012–1030 (2021). https://doi.org/10.1002/humu.24309
9. Harrison, S., et al.: Overview of specifications to the ACMG/AMP variant interpretation guidelines. Curr. Protoc. Hum. Genet. **103** (2019). https://doi.org/10.1002/cphg.93
10. Jackson, M., et al.: The genetic basis of disease. Essays Biochem. **62**, 643–723 (2018). https://doi.org/10.1042/EBC20170053
11. Karczewski, K.J., et al.: The mutational constraint spectrum quantified from variation in 141,456 humans. Nature **581**(7809), 434–443 (2020)
12. Landrum, M.J., et al.: ClinVar: improving access to variant interpretations and supporting evidence. Nucleic Acids Res. **46**(D1), D1062–D1067 (2017). https://doi.org/10.1093/nar/gkx1153
13. Lewallen, S., et al.: Epidemiology in practice: case-control studies. Community Eye Health **11**, 57–58 (1998)
14. Richards, S., et al.: Standards and guidelines for the interpretation of sequence variants: a joint consensus recommendation of the American College of Medical Genetics and Genomics and the Association for Molecular Pathology. Genet. Med. **17**(5), 405–423 (2015)
15. Rigden, D.J., et al.: The 2023 Nucleic Acids Research Database Issue and the online molecular biology database collection. Nucleic Acids Res. **51**(D1), D1–D8 (2023). https://doi.org/10.1093/nar/gkac1186
16. García S, A., et al.: A conceptual model-based approach to improve the representation and management of omics data in precision medicine. IEEE Access **9**, 154071–154085 (2021). https://doi.org/10.1109/ACCESS.2021.3128757
17. Zeggini, E., et al.: Translational genomics and precision medicine: moving from the lab to the clinic. Science **365**(6460), 1409–1413 (2019)

Enhancing Precision Medicine: An Automatic Pipeline Approach for Exploring Genetic Variant-Disease Literature

Lidia Contreras-Ochando[1]([✉])[iD], Pere Marco Garcia[1][iD], Ana León[1][iD], Lluís-F. Hurtado[1][iD], Ferran Pla[1][iD], and Encarna Segarra[1,2][iD]

[1] Valencian Research Institute for Artificial Intelligence (VRAIN), Universitat Politècnica de València, Camí de Vera s/n, 46020 Valencia, Spain
{liconoc,anleopa,lhurtado,fpla,esegarra}@upv.es
[2] ValgrAI: Valencian Graduate School and Research Network of Artificial Intelligence, Camí de Vera s/n, 46020 Valencia, Spain

Abstract. Advancements in genomics have generated vast amounts of data, requiring efficient methods for exploring the relationships between genetic variants and diseases. This paper presents a pipeline approach that automatically integrates diverse biomedical databases, including NCBI Gene, MeSH, LitVar2, PubTator, and SynVar, for retrieving comprehensive information about genes, variants, diseases, and associated literature. The pipeline consists of multiple stages: querying and searching across the different databases, extracting relevant data, and applying filters to refine the results. Its goal is to bridge the gap in information retrieval related to genetic variants and diseases by providing a systematic framework for discovering relevant literature. The pipeline uses open-access sources to uncover additional articles not referenced in expert reports that mention the genetic variants of interest. In this paper, we present the methodology of the pipeline, discuss its limitations and highlight its potential for advancing information systems, data management, and interoperability in the domains of genomics and precision medicine.

Keywords: Genomic variant-disease literature · Precision medicine · Knowledge integration

1 Introduction

The fields of genomics, precision medicine, and genetic variant-disease associations have advanced significantly, reshaping our understanding of human biology and healthcare [7,8,11]. Precision medicine, driven by genomics, aims to customize medical interventions based on individuals' unique genetic profiles, resulting in more accurate diagnoses and improved treatment outcomes [18]. Investigating the associations between genetic variants and diseases is crucial

T. P. Sales et al. (Eds.): ER 2023 Workshops, LNCS 14319, pp. 35–43, 2023.
https://doi.org/10.1007/978-3-031-47112-4_4

for uncovering disease mechanisms, identifying therapeutic targets, and developing targeted interventions [3].

Research in the field of genetic variant-disease associations has rapidly expanded with the availability of large-scale genomics datasets and advanced computational methods [19]. Understanding these associations holds immense implications for healthcare, including improved disease risk assessment, personalized treatment strategies, and advancements in drug development [9]. The exploration of genetic variant-disease associations requires the efficient extraction of relevant literature that describes the relationships studied between the variant and the disease. However, the large amount of existing literature and the different nomenclatures for genes, variants and diseases make it extremely difficult to find the relevant information in each case [2].

Our research focuses on developing a pipeline approach to facilitate the discovery of genetic variant-disease associations by finding the relevant literature in each case. By integrating diverse data sources and considering semantic annotations, our pipeline streamlines the identification of relevant literature and the extraction of pertinent information, allowing researchers and clinicians to gain deeper insights into the genetic foundations of diseases. In the era of big data, where the analysis and interpretation of vast genomics and biomedical information are crucial [10,13], our pipeline offers a valuable solution to navigate and extract meaningful insights from this wealth of data.

The main contributions of this paper are: (1) A novel and automatic system with a pipeline approach for the efficient retrieval of literature related to genetic variants; (2) The use of semantic annotations and disease synonym matching to filter and prioritize articles based on variant-disease co-occurrence; (3) By enabling efficient exploration of genetic variant-disease literature, the pipeline provides valuable insights for personalized treatment strategies, disease risk assessment, targeted interventions and population health management.

The following section contains a summary of relevant works in the field. Section 3 defines the problem we address in this paper. Section 4 gives details of our approach. Section 5 includes the experiments carried out to test the pipeline. Finally, Sect. 6 closes the paper with the conclusions and future work.

2   Related Work

The field of genomics and precision medicine has witnessed the emergence of several tools and resources aimed at exploring genetic variant-disease literature, supporting researchers and clinicians in uncovering meaningful insights from genomics data. However, many of these tools focus on specific aspects of the exploration process and may not offer a comprehensive framework.

Variomes [14] is a curation-support tool for personalized medicine that enables the triage of publications relevant to support an evidence-based decision. This tool's limitation remains in that it does not consider the supplementary material of the publications. Since 80% of the variants appear only in the supplementary material [2], Variomes is not able to find them.

On the other hand, LitVar2 [1] is a database specifically dedicated to exploring literature mentioning genetic variants. It provides publications related to the variants, even when they appear in the supplementary material. LitVar2 is a valuable resource for researchers interested in the literature aspect of variant-disease associations, but it lacks a disease filter and a more extensive range of variant syntactic variations in its search.

ClinVar [12] is a widely used database that aggregates and curates information about genetic variants and their clinical significance. It serves as a valuable resource for researchers and clinicians to assess the potential impact of variants on human health. While ClinVar provides extensive information on gene-variant-disease associations, its focus is primarily on clinical significance and may not provide comprehensive exploration capabilities. The cites provided in Clinvar are retrieved from Litvar2.

Mastermind [4], a search engine specifically designed for genetic variant exploration, allows users to access a vast collection of scientific literature and identify publications related to specific genetic variants. It provides a valuable resource for researchers to find relevant articles. Even though Mastermind is the most complete tool for the goal of this research, it requires a paid subscription.

Compared to these existing tools, our proposed system encompasses a broader scope by integrating multiple databases, using semantic annotations and facilitating disease-specific filtering in an open-access manner. In addition to the comprehensive integration of databases, semantic annotations and disease-specific filtering, our proposed pipeline stands out for its automation, minimizing the manual effort required from the user. The system takes care of the entire process, from data retrieval to the delivery of filtered articles.

3   Problem Definition

Our research addresses the problem of efficiently retrieving relevant literature where a genetic variant and a disease are mentioned. The system automatically integrates diverse data sources, utilizes semantic annotations, and incorporates disease synonym matching to provide a comprehensive and curated list of articles that mention the gene and variant and are filtered by the specified disease. In this section, we define the problem, outlining the inputs, outputs, and formulation of the problem our research aims to address.

1. The inputs are: A gene, a variant and a disease.
2. Given the inputs, the problem can be formulated as follows:
   (a) Identifying and collecting all the aliases associated with the gene [15].
   (b) Generating all possible syntactic variations of the variant [5,6,16,17].
   (c) Retrieving the synonyms associated with the disease.
   (d) Searching for variant-associated publications that match the gene aliases.
   (e) Obtaining annotations for each identified publication.
   (f) Filtering the articles based on the occurrence of the disease.
3. We produce as output a list of articles of interest.

4   Pipeline Methodology

The pipeline methodology presented in this study provides a systematic and automated approach for efficiently exploring literature related to genetic variants and diseases. The system integrates multiple data sources[1] and uses semantic annotations to automatically guide the retrieval of relevant information which can be further analyzed. The following sections outline the key components and steps involved in the pipeline methodology.

4.1   Data Sources

The system's pipeline utilizes diverse biomedical databases to gather comprehensive information about genes, variants, diseases and associated literature. The main data sources include:

- Gene: The Gene database[2], provided by the National Center for Biotechnology Information (NCBI), is a widely used resource for gene-related information.
- SynVar: The SynVar database[3] focuses on providing syntactic variations of genetic variants associated with specific genes.
- MeSH: The MeSH (Medical Subject Headings)[4] database is a controlled vocabulary resource developed by the National Library of Medicine (NLM).
- LitVar2: The LitVar2 database[5] is a resource that focuses on providing relevant literature for each genetic variant.
- PubTator: The PubTator database[6] provides bioconcepts annotated in biomedical literature.

4.2   Pipeline Workflow

Figure 1 shows the pipeline that works as follows: (1) The pipeline begins by querying the NCBI Gene database using the gene of interest to retrieve gene information; (2) Using the gene aliases obtained in the previous step, the pipeline queries the SynVar database to obtain a list of all unique syntactic variations associated with the specified variant; (3) The pipeline retrieves from the MeSH database synonyms of the specified disease; (4) The pipeline searches the LitVar2 database for variant-associated publications. It retrieves articles that mention the gene and the variant; (5) For all the publications found with LitVar2, the pipeline retrieves the semantic annotations of diseases from Pubtator; (6) Finally,

---

[1] It is worth noting that the pipeline methodology is flexible and can be adapted to accommodate additional data sources or specific requirements based on the research objectives and available resources.

[2] NCBI website: www.ncbi.nlm.nih.gov/gene.

[3] Synvar website: https://goldorak.hcsge.ch/synvar/.

[4] MeSH website: https://www.ncbi.nlm.nih.gov/mesh/.

[5] Litvar2 website: https://www.ncbi.nlm.nih.gov/research/litvar2/.

[6] Pubtator website: https://www.ncbi.nlm.nih.gov/research/pubtator/.

the pipeline filters the retrieved articles based on disease occurrence. Articles not mentioning the specified disease or its synonymous terms are excluded from further analysis.

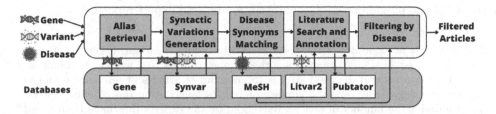

Fig. 1. Pipeline for retrieving filtered publications.

5 Experimental Work

In this section, we present the experiments conducted to evaluate the effectiveness and performance of the proposed system in retrieving variant-associated publications and filtering them by disease occurrence.

Due to the lack of space, Table 1 shows a summary of all the data and results that we will comment in the following sections. We refer the reader to our Github repository for further information, data and code[7] (in Python). All the pipeline connections to databases were performed using Entrez API[8], except for Litvar2 and Synvar, which have their own APIs.

5.1 Test Dataset

For the experiments, we constructed a dataset consisting of 16 examples, each comprising a variant, a gene, a disease, and a list of publications identified by their PubMed IDs (PMIDs). Each example is extracted from one report generated by expert companies in the field of medicine and genetics. Table 1 shows all the examples used.

5.2 Methodology

In the experiments, we applied the pipeline to each of the 16 examples in the dataset. Using the pipeline, we retrieve publications where the genetic variant is mentioned and then filtered the retrieved articles based on the disease occurrence. The results were compared against the expert-generated reports to evaluate the pipeline's accuracy and effectiveness in identifying relevant articles.

---

[7] Github repository: https://github.com/liconoc/variant_disease_pipeline/.

[8] Entrez website: https://www.ncbi.nlm.nih.gov/Web/Search/entrezfs.html.

5.3   Results

We assessed the pipeline's ability to identify articles referenced in the reports, as well as the identification of additional articles that were not mentioned in the reports. This comparison allowed us to evaluate the pipeline's performance and the ability to provide up-to-date information beyond expert reports.

Table 1. List of examples of the test dataset, including a variant, a gene, a disease and a list of publications. #names are the number of alias and synonyms found for the variants, genes and diseases; #ref is the number of articles of reference found in the expert's reports; #sys is the number of articles found using the pipeline system; #exp_found is the number of articles referenced by the experts that have been found with the pipeline.

Variant	#names	Gene	#names	Disease	#names	#ref	#sys	#ref_sys
Asn1271Lys	61	MYH7	9	Dilated Cardiomyopathy	60	1	3	0
Asp1272His	61	MYH7	9	Dilated Cardiomyopathy	60	1	0	0
Val411Met	61	SCN5A	15	Long QT Syndrome	48	25	34	7
Asp2Asn	62	PAX5	4	Hepatoblastoma	15	1	0	0
Arg1012Ter	1	STAG2	8	Sarcome Ewing	86	4	0	0
Ile258Val	62	PDGFRA	4	Sarcome Ewing	86	4	2	0
Pro1685Leu	18	PLXNB2	6	Hepatic Sarcoma	1	4	0	0
Arg2072Gln	62	ROS1	4	Hepatic Sarcoma	1	5	4	0
Ala670Val	60	HCN1	10	Ependynoma	1	1	0	0
Ser333Phe	18	ATM	9	Ependynoma	1	1	35	0
Arg365Gln	59	RAD50	4	Ependynoma	1	1	22	0
Ala670Val	60	HCN1	10	Neuroblastome	1	1	0	0
Tyr776Ter	1	TSC2	4	RHABDOMYOSARCOMA	51	1	0	0
Ser273Cys	59	MRE11	5	Nefroblastome	1	1	7	0
Lys1296Arg	61	DOCK3	4	Acute Myeloid leukemia	175	1	2	0
Thr1010Ile	18	MET	17	Malignant tumor of peripheral nerve sheath	1	9	118	1
						61	227	8

Aliases and Synonyms. To find all the articles that name the variants and to be able to filter by disease, we first need to know all the possible names that the genes, variants and diseases in the examples may have. Table 1 shows the number of names the pipeline could find for them. Note that the number of existing names for a variant depends on which gene it is related to. The same variant can have different names for two different genes.

Identification of Articles Referenced by the Experts. In our study, we evaluated the performance of the pipeline in retrieving variant-associated publications and compared the results with expert-generated reports. Out of the 61 publications mentioned in the expert reports, the pipeline successfully retrieved 8 publications (abstracts). However, it is important to note that, upon further investigation, we discovered that some of the articles were not available in the PubMed Central (PMC)[9] database. LitVar2 relies on Pubmed[10] and PMC as

---

[9] PMC website: https://www.ncbi.nlm.nih.gov/pmc.
[10] PubMed website: https://pubmed.ncbi.nlm.nih.gov/.

its primary sources of information[11]. However, the articles can be published in other repositories that may not be open access. As a result, it is possible that Litvar2 does not retrieve certain publications referenced in the expert reports. Our results indicate that a significant portion of the expert-generated references were not publicly accessible. In addition, a manual scan of the named articles in the expert reports revealed that only 14 mentioned the variant in the title, abstract, or PMC full-text. One of them would not even name the gene.

Identification of Additional Articles. Contrary to the previous results, the pipeline yielded a significantly higher number of articles than those referenced by the expert reports. In total, we identified 227 publication abstracts and 190 PMCs (full-text). Among these 227 publications, 202 articles mentioned the specific variant of interest accurately[12]. The 25 articles discarded shared the same rsID, but differed in the nucleotide of the mutation. When using the rsID nomenclature[13] on the articles, it may indicate a different mutation. Table 2 shows the different locations where the variants were found within the articles. This comprehensive coverage across different sections of the articles enhances the reliability and robustness of the pipeline.

Table 2. The number of articles where the variant was mentioned depending on the section of the article where it was found.

Title	Abstract	Text	Sup.Mat.	Total
6	12	77	96	**202**

Filtering by the Disease. Upon applying the disease filtering step, we observed that only one disease maintained a substantial number of articles out of the total found. Specifically, this disease retained 28 out of the 34 articles identified. The effectiveness of the disease filtering process highlights the importance of considering the specific disease context when exploring gene-variant literature. By filtering the articles based on disease relevance, we ensure that the final set of articles specifically focuses on the intersection of the gene, variant and disease of interest.

---

[11] PubMed is a repository of publication abstracts in the field of biomedicine and life sciences, while PMC is a repository of full-text articles.

[12] Variant names indicate the position in which the mutation occurs in the DNA, RNA or protein sequence, as well as the original nucleotide or amino acid (wild type) and the mutated one.

[13] rsID nomenclature indicates the position and wild type but not the specific mutation. Therefore, one rsID can identify all mutations with the same wild type and the same position.

5.4   Discussion

Despite the dependency on PMC, LitVar2 remains a valuable resource for capturing a substantial number of gene-variant-disease open-access literature. Its integration with other databases and resources, along with continuous updates and advancements in literature curation, ensures a comprehensive and up-to-date collection of relevant publications. The pipeline's ability to identify and retrieve a portion of the expert-referenced articles and many up-to-date new ones demonstrates its utility in complementing expert reports and providing an efficient tool for exploring gene-variant-disease literature.

The disparity in the number of articles found and retained for each variant and disease underscores the variability in the availability of literature and its explicit mention of diseases. It is important to note that while some variants and diseases may have fewer associated articles, these articles are expected to provide valuable insights and contribute to understanding the specific gene-variant-disease connections.

6   Conclusions and Future Work

In this paper, we have presented a pipeline to enhance the exploration of gene-variant-disease literature by leveraging advanced search techniques and integrating multiple databases. We identified a significantly large number of articles that contribute valuable scientific evidence to the field of precision medicine and genomics. Our pipeline provides researchers with a comprehensive and efficient tool to investigate gene-variant-disease co-occurrences. However, it is important to consider the limitations imposed by the availability of full-text content. The pipeline's performance relies on the openness and accessibility of literature, which can impact the comprehensive retrieval of all relevant publications.

As future work, ongoing enhancements to the pipeline's algorithms and data integration techniques will contribute to improving its accuracy and efficiency. Furthermore, including additional open data sources and utilizing emerging technologies, such as natural language processing, can further enhance the pipeline's capabilities to extract relevant information from publications and uncover relationships among biomedical entities. A key aspect to address is retrieving information and relationships from supplementary materials, as they often contain valuable data regarding the variants of interest.

Acknowledgment. This work is partially supported by MCIN/AEI/10.13039/501100011033, by the "European Union" and "NextGenerationEU/MRR", and by "ERDF A way of making Europe" under grants PDC2021-120846-C44 and PID2021-126061OB-C41. It is also partially supported by the Generalitat Valenciana under project CIPROM/2021/023.
We would like to thank the authors of LitVar2 for their valuable assistance.

References

1. Allot, A., Peng, Y., Wei, C.H., Lee, K., Phan, L., Lu, Z.: LitVar: a semantic search engine for linking genomic variant data in PubMed and PMC. Nucleic Acids Res. **46**(W1), W530–W536 (2018)
2. Allot, A., et al.: Tracking genetic variants in the biomedical literature using LitVar 2.0. Nat. Genet. **55**, 901–903 (2023)
3. Cano-Gamez, E., Trynka, G.: From GWAS to function: using functional genomics to identify the mechanisms underlying complex diseases. Front. Genet. **11**, 424 (2020)
4. Chunn, L.M., et al.: Mastermind: a comprehensive genomic association search engine for empirical evidence curation and genetic variant interpretation. Front. Genet. **11**, 577152 (2020)
5. Den Dunnen, J.T., et al.: HGVS recommendations for the description of sequence variants: 2016 update. Hum. Mutat. **37**(6), 564–569 (2016)
6. den Dunnen, J.T.: Sequence variant descriptions: HGVS nomenclature and mutalyzer. Curr. Protoc. Hum. Genet. **90**(1), 7–13 (2016)
7. NHS England: Accelerating genomic medicine in the NHS. NHS England Website (2022). www.england.nhs.uk/long-read/accelerating-genomic-medicine-in-the-nhs. Accessed 23 Nov 2022
8. Ginsburg, G.S., Phillips, K.A.: Precision medicine: from science to value. Health Aff. **37**(5), 694–701 (2018)
9. Goetz, L.H., Schork, N.J.: Personalized medicine: motivation, challenges, and progress. Fertil. Steril. **109**(6), 952–963 (2018)
10. Hassan, M., et al.: Innovations in genomics and big data analytics for personalized medicine and health care: a review. Int. J. Mol. Sci. **23**(9), 4645 (2022)
11. Krainc, T., Fuentes, A.: Genetic ancestry in precision medicine is reshaping the race debate. Proc. Natl. Acad. Sci. **119**(12), e2203033119 (2022)
12. Landrum, M.J., et al.: ClinVar: public archive of relationships among sequence variation and human phenotype. Nucleic Acids Res. **42**(D1), D980–D985 (2014)
13. Luo, J., Wu, M., Gopukumar, D., Zhao, Y.: Big data application in biomedical research and health care: a literature review. Biomed. Inform. Insights **8**, BII-S31559 (2016)
14. Pasche, E., Mottaz, A., Caucheteur, D., Gobeill, J., Michel, P.A., Ruch, P.: Variomes: a high recall search engine to support the curation of genomic variants. Bioinformatics **38**(9), 2595–2601 (2022)
15. Povey, S., Lovering, R., Bruford, E., Wright, M., Lush, M., Wain, H.: The HUGO gene nomenclature committee (HGNC). Hum. Genet. **109**, 678–680 (2001)
16. Saberian, N.: Text Mining of Variant-Genotype-Phenotype Associations from Biomedical Literature. Wayne State University (2020)
17. Smigielski, E.M., Sirotkin, K., Ward, M., Sherry, S.T.: dbSNP: a database of single nucleotide polymorphisms. Nucleic Acids Res. **28**(1), 352–355 (2000)
18. Striancse, O., et al.: Precision and personalized medicine: how genomic approach improves the management of cardiovascular and neurodegenerative disease. Genes **11**(7), 747 (2020)
19. Uffelmann, E., et al.: Genome-wide association studies. Nat. Rev. Methods Primers **1**(1), 59 (2021)

CMOMM4FAIR

CMOMM4FAIR – 3rd Workshop on Conceptual Modeling, Ontologies and (Meta)data Management for Findable, Accessible, Interoperable, and Reusable (FAIR) Data

João Moreira, Luiz Olavo Bonino, and Pedro Paulo F. Barcelos

University of Twente, The Netherlands
{j.luizrebelomoreira, l.o.boninodasilvasantos,
p.p.favatobarcelos}@utwente.nl

The Findable, Accessible, Interoperable, and Reusable (FAIR) data principles condense decades of work and challenges on data usage into four main aspects. The guidance provided by the principles on how to make data or any other digital object findable, accessible, interoperable, and reusable is being seen as the basis for how organizations are tackling their modern informational challenges. The goal of FAIR is to promote optimal reusability of digital assets, mainly through the use of machines. A common expression is that "in FAIR, the machine knows what I mean". In other words, we would like to have computation systems that can properly interpret information found in their self-guided exploration of the data ecosystem.

Before machines are able to "know what we mean," the meaning in our informational artefacts should also be clear for us, thus, addressing semantic interoperability. Here enters conceptual modeling as a means to improve semantic interoperability by capturing knowledge about a particular universe of discourse in terms of various semantic artefacts such as ontologies, semantic data models, and semantic metadata models. The workshop on Conceptual Modeling, Ontologies and (Meta)Data Management for Findable, Accessible, Interoperable, and Reusable Data (CMOMM4FAIR) aims at investigating, discussing, and improving conceptual modeling practices towards improved FAIRness.

In this 3rd edition of the CMOMM4FAIR workshop, co-located with the 42nd International Conference on Conceptual Modeling (ER 2023), we accepted one paper, entitled "euFAIR: a digital tool for assessing the FAIR principles". This paper discusses a tool (euFAIR) for the automatic assessment of the FAIRness level of datasets in the Italian Public Administration.

Acknowledgments. We would like to express our deepest appreciation to the authors of the submitted papers and to all Program Committee members for their diligence in the paper review and selection process. We would also like to thank the Organizing Committee of ER 2023, in particular the workshop chairs for all their support.

CMOMM4FAIR Organization

Workshop Chairs

João Moreira   University of Twente, The Netherlands
Luiz Olavo Bonino   University of Twente, LUMC, The Netherlands
Pedro Paulo F. Barcelos   University of Twente, The Netherlands

Website and Publicity Chair

Asen Pantov   University of Twente, The Netherlands

Program Committee

João Moreira   University of Twente, The Netherlands
Maya Daneva   University of Twente, The Netherlands
Nathalie Aussenac-Gilles   IRIT CNRS, France
Patricio de Alencar Sliva   Universidade Federal Rural do Semi-Árido,
                                       Brazil
Peter Mutschke   Leibniz Institute for the Social Sciences,
                                       Germany
Robert Pergl   Technical University of Prague,
                                       Czech Republic
Veruska Zamborlini   Federal University of Espirito Santo, Brazil
Wanderley Lopes de Souza   Federal University of São Carlos, Brazil

euFAIR: A Digital Tool for Assessing the FAIR Principles

Matteo Lia[1,2(✉)], Davide Damiano Colella[1], Antonella Longo[1,3] (iD),
and Marco Zappatore[1,3] (iD)

[1] Department of Innovation Engineering, University of Salento, via Monteroni sn, 73100 Lecce,
Italy
{matteo.lia,antonella.longo,
marcosalvatore.zappatore}@unisalento.it,
matteo.lia@parsec326.it,
davidedamiano.colella@studenti.unisalento.it
[2] Parsec 3.26 Srl, via del Platano 7, 73020 Cavallino – Castromediano, Lecce, Italy
[3] Italian Research Center on High Performance Computing, Big Data and Quantum Computing
(ICSC), Bologna, Italy

Abstract. Over the last decade, the importance of FAIR principles as a reference for data reusability and openness has increased constantly. Various tools for FAIR data assessment exist, including manual approaches like the FAIR Data Self-Assessment Tool (SAT) and automated tools as the FAIR Evaluation Services and Generic Automatic Tool (GAT). However, subjectivity in manual assessment and limited guidance in automated tools represent significant limitations. In such a context, the Italian "*Piano Triennale per l'Informatica nella Pubblica Amministrazione*" has laid the foundations for open data practices in the Italian Public Administration (PA) since 2017. In this work, we propose a tool called euFAIR for the automatic assessment of dataset FAIRness in the Italian PA. The tool incorporates European Data Quality Guidelines for a more comprehensive dataset evaluation and implements a set of specific ad-hoc designed metrics. With this tool, we aim at improving data quality and sharing, as well as the adherence to community standards in the Italian PA. To validate our results, we compared euFAIR with SAT and GAT, highlighting significant differences.

Keywords: FAIR · Tool · Assessment · Public Administration

1 Introduction

The Italian "*Piano Triennale per l'Informatica nella Pubblica Amministrazione*" (Three-Year Plan for Information Technology in Public Administration) [1] aligns with European and Italian regulatory developments and sets the groundwork for nationwide management of open data. Since its 2017 inception and subsequent updates, the plan enhances transparency of the Italian Public Administration (PA), facilitates citizens' access to information, and fosters new products and services by businesses. All PAs' open data is

T. P. Sales et al. (Eds.): ER 2023 Workshops, LNCS 14319, pp. 49–58, 2023.
https://doi.org/10.1007/978-3-031-47112-4_5

consolidated in a dedicated Web portal [2], connected to its European counterpart [3]. However, for effective and reliable data sharing and reuse, the datasets in the portal must adhere to the FAIR (Findable, Accessible, Interoperable and Reusable) principles [4] and be enriched accordingly. The FAIR principles are intended to be a guideline to enhance data reusability with specific emphasis on automation (e.g., machine readable formats). One way to support the provision of "more FAIR" data, is the development of tools to assess or even score the degree to which a dataset complies with the FAIR principles. In recent times, the scientific community has made significant strides in applying and evaluating the FAIR principles on Digital Objects (DO), encompassing publications, datasets, and research software. As a result, there are several openly available automated and manual FAIR assessment services, including FAIR Evaluation Services and FAIR Data Self-Assessment Tool (SAT). While the versatile manual FAIR assessment tools totally depend on the user, the automated FAIR assessment tools search for known structures that could be used in the metadata. This makes the automatic tools compatible with most resources, but its precision can vary depending on the resources. Automatic tools allow higher precision at the expense of less coverage, by limiting the tool's context of use. Nevertheless, this does not represent a deficiency of the tools of this category. It is a feature deriving from the real nature of FAIRness, since «the final FAIR Principle speaks directly to the fact that "FAIR" will have different requirements for different communities» [5].

In this scenario, we developed euFAIR, an automatic tool for assessing the FAIRness level of the Italian PA datasets, thanks to ad-hoc metrics. The tool goes beyond the assessment based solely on the FAIR principles and also incorporates the European Data Quality Guidelines (DQG) [6]. By embracing this all-encompassing approach, our primary objective is to enhance overall data quality, facilitate data sharing, and encourage adherence to community standards within the Italian PAs. In order to evaluate euFAIR efficacy, we conducted a thorough comparison between euFAIR, and SAT and GAT methodologies. The outcome of this comparison exposed noteworthy disparities in the evaluation results, signifying the potential superiority and effectiveness of our proposed euFAIR framework. The paper is organized as follows: the second section will focus on the related work data FAIRness evaluation tools. The proposed euFAIR tool will be described in the third section. The comparison evaluation is described in the fourth section, while conclusions and future work will be drawn in the fifth section.

2   Related Work

Over the years, a number of tools have been developed to evaluate data FAIRness, mainly based on manual assessment, involving methods such as case studies, checklists, and templates. Other approaches leverage Web scraping and a series of tests to automatically do the assessment. In this section, a subset of those tools will be examined, as they will be compared with the proposed euFAIR tool.

The Australian Research Data Commons (ARDC) actively promotes the adoption of the FAIR principles, providing valuable resources to the global research community. Among these resources is a FAIR Data Self-Assessment Tool (SAT) [7], available on the ARDC website. The tool aims to help researchers gauge the "FAIRness" of their data and encourage discussions about improving data practices. The assessment tool is a survey with 12 questions, aligned with the four FAIR principles, to quantify compliance with each category. Some questions require single-choice responses, while others use a Yes/No format. The tool evaluates answers for consistency, linking each question to a specific FAIR principle. As researchers respond, the corresponding section's bar fills up, showing the strength of their answers in relation to the FAIR principle. The "Total across F.A.I.R" bar aggregates scores from all sections, providing an overall "FAIRness" score for the data.

FAIR Evaluation Services is an open-source framework for conducting and developing maturity indicator tests [8], but its documentation focuses primarily on creating new indicators rather than providing comprehensive guidance on metrics and results. To use the web interface for automated assessment, users must provide a GUID, title, and ORCID. The output offers an overview of passed and failed tests for each indicator, but concerns arise regarding data retention policies as all tests are stored, leading to a substantial database of evaluations [9]. The tool lacks clear guidance on dataset FAIRness improvement, and understanding the detailed output requires technical expertise. Ongoing development efforts are being made to implement further metrics, but advice mechanisms for enhancing dataset FAIRness are currently absent [10, 11].

3   The euFAIR Tool

To cope with the limitations briefly sketched in Sect. 2, the euFAIR[1] tool is proposed: it has two complementary purposes. First, to represent a suitable component for the Italian open data portal supporting the assessment *after* the datasets are harvested, to identify their weaknesses and limitations. Second, to allow PAs assessing their datasets *before* publication, so to provide stronger and more community-compliant data.

3.1   Design Methodology

The methodology followed for the definition of the requirements and the whole evaluation is the Design Science in Information Systems Research [12]. The design of the tool was inspired by the gaps found in SAT and GAT, having in mind the main user which is the open data manager of a PA. The seven research steps, according to the referenced paradigm [12] have been followed; for the lack of space, they are not described here. Here we focus only on the euFAIR functional and technical description.

3.2   Data Model

The Italian open data portal is a key component to understand the data lifecycle of Italian PA data and to analyze their FAIRness, as it provides a single point of access to datasets that can be further shared in multiple formats.

The most comprehensive path that data can follow is when: 1) they are loaded onto the proprietary portal of the Italian PA that produces them; 2) (if agreements have been

---

[1] The tool is available on this repository: https://github.com/datalab-unisalento/euFAIR-tool/.

made with the Italian portal) they are uploaded onto it; 3) they are (generally) loaded onto the European portal. However, not all the datasets follow these procedures, not all the PAs are yet to be harvested from the Italian portal and not all the Italian PA datasets are available in the European portal.

Therefore, we modeled a dedicated database where every dataset as an ID associated with the Italian portal and some datasets have an ID for the European portal. The proposed database behind the euFAIR tool (see Fig. 1) is designed to store the (meta) data evaluations of the datasets. To do so the database is structured as follows. The *Dataset* table holds all the IDs of the dataset that we have tested or intend to test; it also keeps track of their Italian and EU IDs, as well as the dataset name and description, the timestamp of last update/check, and the accrual periodicity. The *Holder* table currently holds only the ID of the dataset holder, but it is intended to be further extended. The *Data Evaluation* and *Metadata Evaluation* tables, deriving from weak entities (see Fig. 1), strictly depend on the Dataset they refer to. Other attributes are: the metadata profile used to make the evaluation, the date, and the version of the tool. For the metadata we also keep track of the portals (origin, European, or) that produce them.

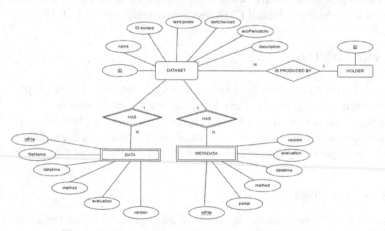

Fig. 1. Entity Relationship model of euFAIR Database

Since euFAIR uses the database to obtain the list of datasets that need to be evaluated or whose evaluation needs an update, we populated the database via the CKAN API method `current_package_list_with_resources`[2], thus retrieving all the datasets metadata from the portal, which we extract some of the dataset attributes from: ID, name, holder, last_update and accrual periodicity. To complete the data with the ID of the dataset in the European portal we need to take into consideration some details. The European ID is created based on the agreements made with the portal that offers the datasets when the harvesting has been programmed. There is no unique way to calculate the ID based on its original one. However, in our case the agreements made seem to suggest that the European ID is the name of the dataset.

---

[2] `CKAN: API Guide`, https://docs.ckan.org/en/latest/api/index.html, `last accessed 2023/04/22`.

In case of a double entry with the same ID a suffix is given to the dataset followed by an incremental number starting with 1. For example, if a dataset with the name *'Global_Temperature_Trends'* is being harvested from the Italian portal (Ip) but on the European portal (EUp) a dataset with that ID is already present, then the dataset is uploaded with the ID *'Global_Temperature_Trends 1'*.

To get this information we obtain a list of all the IDs in the Ip catalog present in the EU portal through the hub-search API using the list dataset method. Every ID in the list is then searched on our database and associated with the dataset with the given name through query. If an ID has the aforementioned suffix, it is removed before searching.

3.3 Evaluation Metrics

The core of the euFAIR tool is the series of metrics designed to specifically evaluate the FAIRness of the data in the context of the Italian PA and take into account the current legislation (that encourages the FAIRness of the data and mandates its openness). Every FAIR metric is evaluated through one or more tests meant to give a percentage of accomplishment. For every FAIR principle, a series of tests is then performed to check the dataset compliance with the EU Data Quality Guidelines.

In this section, we describe the test for the FAIR F2 principle: *"data are described with rich metadata"*. The best way to determine how rich our metadata are is to take into account the guidelines provided by the Ip and EUp, to refer to the metadata profiling system provided both by the European Commission (DCAT-AP [13]) and the AGID (DCAT-AP_IT [14]). These profiles give a set of attributes to be implemented in the metadata, and each of them is expressed as either mandatory, recommended, or optional. The euFAIR tool support multiple profiling systems. Similarly, the tool can be configured according to different and country-specific profiling systems. At present, both the previously mentioned profiles are implemented and a "merged" profile is automatically created (a profile that for each field takes the stricter requirement).

Implementation. In order to achieve a valid evaluation, we have devised a scoring mechanism that assigns a maximum score of $1, 0.5$, or 0.2 points to each field, depending on whether it is mandatory, recommended, or optional, respectively. The total achievable score in the test is the sum of the max scores for all the required fields. Additionally, an actual score indicator is considered, starting at 0 and increasing with each well-evaluated field. When a field is successfully implemented, the actual score grows by the value of that particular field; otherwise, no points are added. If a field contains sub-fields, the evaluation begins with them: 1) we first calculate a maximum score by summing up the max points obtainable for each sub-field while considering whether they are mandatory; 2) we then evaluate the actual score for the field as previously described; 3) dividing the actual score by the max score yields the 'percentage of completeness' for that field.

Taking into consideration both the max score of the field and its "percentage of completeness", we can determine the field point by multiplying the two values. For example: a mandatory field (max point 1.0) is expected to implement 2 mandatory fields, 1 recommended, 3 optional. The max score obtainable considering the sub fields is $maxScore = *1 + 1 * 0.5 + 3 * 0.2 = 3.1$. It follows that only one mandatory field is however implemented, 1 recommended and 2 optional. The actual score of the field will therefore be: $actualScore = 1 * 10 + 1 * 0.5 + 2 * 0.2 = 1.9$. The percentage of completeness will be: $percCompl = 61\%$. To the overall actual score will then be added 1*0.61 (i.e., max point of the field times the percentage of completeness of the field considering the subfields implemented). This procedure is repeated in case of sub-sub-fields, starting with the evaluation of the innermost section, and going up. Once all the fields are evaluated the test gives the $finalScore = actualScore/maxScore$.

Now, let us shift our attention to the European Data Quality Guidelines Test, which aligns strongly with the project's European context. Several tests are utilized to evaluate dataset FAIRness, following the EU Data Quality Guidelines (DQG). These guidelines encompass relevant metrics for assessing the FAIRness of (meta)data and offer specific formats for distributions. In this work, we focus on describing the DQG findability test as follows. Regarding findability, the DQG recommends minimizing NULL values in the metadata and emphasizes the significance of specific fields (e.g., *title*, *description*, *keywords*, *categories*, *spatial*, and t*emporal*) to ensure easy discoverability of the dataset. To implement this, the tool calculates the number of fields and empty fields, deriving a percentage of completeness. It then checks the specified fields for their implementation, assigning a maximum score of 6, with 1 point for each checked aspect. For each specific field, a point is awarded for filled-in fields, and a score between 0 and 1 (based on the completeness percentage) for the number of NULL values.

3.4 Tool Architecture

The tool is divided into two components, corresponding to the dataset update and to its assessment. The *Update* component allows to check for updates in the dataset and to proceed with new evaluations for those requiring it. The user is presented with 3 functions. First, searching for new datasets by consulting the Ip to check if new datasets have been produced; and then adding the new datasets to the database. Second, checking for new properties in the dataset. Third, updating the FAIRness evaluations of the dataset. The tool also shows how many datasets need update, the current evaluation stage (i.e., the portal in which the evaluation is being made and the method). More specifically, the tool starts by retrieving the list of the datasets that need to be updated and exploits the dataset attributes in the database by looking for all the datasets: 1) for which the accrual periodicity is passed, 2) that are not yet evaluated, 3) that have no declared accrual periodicity. Ordering by accrual periodicity allows us not to give priority to those datasets that might not require updating and/or have been updated already. Once the list is retrieved the tool starts the assessment and scoring, for every available portal (native, Ip, EUp) and for every available method (DCAT_AP, DCAT_AP-IT, merged). For every dataset a call is made through CKAN API to retrieve the metadata from the Ip and metadata evaluation tests are performed. The same thing is done through hub-search API for the EUp. Finally, the webpage of the PA that shared the dataset is retrieved and the assessment is made on the native portal. The scenario is highly unpredictable due to the numerous data providers contributing to the portal and their varied data sharing methods. Initially, the tool attempts to retrieve metadata using the common CKAN API, subjecting it to the same tests as for Ip. Otherwise, Web scraping is performed, and generic tests are conducted on the scraped metadata to identify known fields and properties. However, the reliability of these tests is reduced due to the lack of knowledge about the website's structure. If data distribution is found, available files are listed, grouped to identify similar files with different extensions, and the best format is evaluated. FAIR principle tests are format-independent, while DQG-based tests are format-specific (currently supporting only CSV but with plans to expand). Lastly, the dataset or distribution's accrual periodicity is checked for the EUp, Ip, and native portal metadata.

The second component allows to interact with the database. The euFAIR UI offers the user three options. First, the *database query module* allows the user to choose between a single dataset evaluation or a Holder evaluation. When a selection is confirmed the frame with the evaluations graph is shown. The database status is also indicated (i.e., red for preliminary checks, yellow for dataset update, and green for database updated). When the user searches for a dataset ID or name, all its features are shown and the retrieved results are then used to calculate its avg, min and max scores (Fig. 2).

Fig. 2. Graphical user interface of the database query module

For a single dataset, the user inputs the Holder ID and selects from the results list. The assessment page displays when the selection is complete (Fig. 3). The user can choose the vocabulary profiling system and the portal to be evaluated. The assessment page summarizes results for every FAIR criterion and metric, including max, min, and avg scores. If files are linked to the dataset and have been evaluated, a clickable label provides access to those evaluations. Users can select between the evaluation results for the dataset and the holder. Additionally, users have the option to retrieve a dataset as via CKAN API by specifying the source URL or import a dataset as a CSV file.

3.5 Tool Settings

The euFAIR tool settings are provided via a dedicated JSON file. The file allows the user to specify: 1) *properties_settings*, to store the metadata properties as for F2 ad their weight, as well as new metadata profiles; 2) *Vocabularies*, to indicate, as for I2, the vocabulary format and properties; 3) *"format"*, to specify the best dataset format when multiple distribution formats are expected (every format is given a score based on its machine-readability and whether it is proprietary or not, according to the data quality guidelines of the European Commission); 4) *"permanent_link"*, which provides the structure of the tool's permalink.

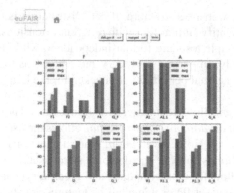

Fig. 3. Vocabulary profiling system interface.

4 Comparative Evaluation

To better understand euFAIR's potential, we assessed the same dataset with different tools, through the EUp, under a DCAT_AP profile. To carry out the comparison of tools we're taking into account the metadata of one of the datasets available in the Ip. The sample dataset contains (in JSON and CSV formats) the railway lines and stations of the city of Porto Recanati[3], located in the Italian administrative region of Marche.

The comparison between euFAIR and the other tools (SAT and GAT), highlighted several differences. The proposed euFAIR tool present additional advantages if compared to SAT, and GAT as it euFAIR offers additional features such as recognizing the validity of the EUp permalink, quantifying data richness, considering metadata persistence through data nature and portal, assessing the quantity of FAIR vocabulary usage, and evaluating license, provenance, and community standard adherence.

The Self-Assessment Tool (SAT) shows immediately its weakness as a deep knowledge of the dataset is required to the user in order to give the information required. Furthermore, answering these questions is rather difficult since questions such as "Did you provide sufficient metadata (information) about your data for others to find, understand and reuse your data?" are liable to non-objective answers.

The generic automatic tool (GAT) shows great potential, correctly scraping much of the data but analyzing it through generic metrics.

Table 1 shows the differences in the assessment score, which are mainly due to the different metrics and methods used by the tools, while Table 2 and Table 3 examine the differences between SAT and euFAIR and between GAT and euFAIR, respectively.

---

[3] The sample dataset is avaiable at the following link: https://dati.gov.it/view-dataset/dataset?id=1653a9ef-529f-4b65-91d7-894d19dea2e5, last accessed 2023/04/22.

Table 1. Assessment of the same dataset with different tools

	Self-Assessment Tool	Generic Automatic Tool	euFAIR
F	70	50	57
A	80	80	80
I	100	71	29
R	100	0	29

Table 2. Main differences between SAT and euFAIR

	Self-Assessment Tool (SAT) vs euFAIR
F	SAT awards full points for a DOI permalink, and euFAIR additionally includes the EUp portal permalink, with more options for datasets. SAT assesses comprehensive metadata descriptions and structured language usage, whereas euFAIR quantifies the extent of metadata descriptions
A	No relevant differences as both the tools have similar behaviors
I	SAT only takes into consideration the data format and the vocabulary type used in the metadata; euFAIR also focuses on how much the vocabularies are used
R	SAT only considers whether a license and a provenance are given; euFAIR also focus on whether the license uses its vocabulary and how the provenance is made explicit

Table 3. Main differences between GAT and euFAIR

	Generic Automatic Tool (GAT) vs euFAIR
F	GAT does not consider the EUp permalink as valid and does not quantify data richness
A	GAT does not consider the metadata persistent because it cannot find a persistence Policy key; euFAIR considers the nature of the data and the portal itself
I	GAT evaluates if the resource uses FAIR vocabs.; euFAIR also assesses how much
R	GAT only looks for license; euFAIR looks also for provenance and community standard

5   Discussion and Conclusions

In the open science arena, a big effort is required to provide and assess FAIR data. Existing SAT and GAT tools show some drawbacks which have been analyzed and overcome in this paper by proposing euFAIR, a novel tool to assess data FAIRness.

As of now, euFAIR is still in its alpha version, more testing being needed in order to validate its usability and efficacy and we hope a discussion will open to further enrich the tool's metrics and possibilities. The tool can target different recipients and different versions could be created depending on the final user and more extensive validation procedures will also be performed accordingly. Forthcoming improvements will target better data analysis and visualization to support Ip's assessment and a reporting system to give feedback to the PA so to improve data quality prior to dataset publication. In addition, we will evaluate whether decoupling the dataset evaluation update feature from other tool versions, such as one designed for the PA, as background evaluations could impact performances and a PA may not be interested in the update of all its datasets. By adopting euFAIR as a tool for PAs to evaluate the FAIRness of their datasets, it becomes feasible to use euFAIR metrics tests and to develop a collection to share on an already existent FAIR evaluator tool. Finally, we plan to widen the number of accepted formats to provide a more complete assessment on the data side and to consider more Holders' attributes such as geographical data to provide a deeper data analysis.

Acknowledgements. This work has been partially developed with the financial support of the Puglia region under the RIPARTI projects - POC PUGLIA FESRTFSE 2014/2020 – CUP: F87G22000270002, with the support of CETMA DISHME project (EU Grant Agreement n. 101083699), and with the support from Italian Research Center on High Performance Computing, Big Data and Quantum Computing (ICSC) funded by EU – NextGenerationEU (PNRR-HPC, CUP: C83C22000560007).

References

1. AGID, Dipartimento per la Trasformazione Digitale: Piano Triennale per l'Informatica (2022)
2. AGID: I dati aperti della PA. https://www.dati.gov.it/. Accessed 02 May 2023
3. European Commission: The official portal for European data. https://data.europa.eu/en. Accessed 01 May 2023
4. Wilkinson, M.D., et al.: The FAIR guiding principles for scientific data management and stewardship. Sci. Data. **3**, 160018 (2016). https://doi.org/10.1038/sdata.2016.18
5. Fairsharing: Philosophy of FAIR testing. https://fairsharing.github.io/FAIR-Evaluator-FrontEnd. Accessed 27 Apr 2023
6. Publications Office of the European Union.: Data.europa.eu data quality guidelines (2021)
7. Australian Research Data Commons (ARDC): Fair Self-Assessment Tool. https://ardc.edu.au/resource/fair-data-self-assessment-tool/. Accessed 25 Apr 2023
8. FAIRmetrics, FAIRsharing: The FAIR Maturity Evaluation Service. https://fairsharing.github.io/FAIR-Evaluator-FrontEnd/#!/. Accessed 26 Apr 2023
9. FAIRMetrics: FAIR Maturity Indicators and Tools. https://github.com/FAIRMetrics/Metrics. Accessed 26 Mar 2023
10. Krans, N.A., Ammar, A., Nymark, P., Willighagen, E.L., Bakker, M.I., Quik, J.T.K.: FAIR assessment tools: evaluating use and performance. NanoImpact. **27**, 100402 (2022). https://doi.org/10.1016/j.impact.2022.100402
11. Gehlen, K.P., Höck, H., Fast, A., Heydebreck, D., Lammert, A., Thiemann, H.: Recommendations for discipline-specific fairness evaluation derived from applying an ensemble of evaluation tools. Data Sci J. **21** (2022). https://doi.org/10.5334/dsj-2022-007
12. Hevner, A.R., March, S.T., Park, J., Ram, S.: Design science in information systems research (2004)
13. Pavlina Fragkou (SEMIC EU): DCAT-AP 3.0. https://semiceu.github.io/DCAT-AP/releases/3.0.0/. Accessed 20 Mar 2023
14. AGID, Team Digitale: Linee guida per i cataloghi dati (2020)

EmpER

EmpER – 6th International Workshop on Empirical Methods in Conceptual Modeling

Dominik Bork[1], Miguel Goulão[2], and Sotirios Liaskos[3]

[1] TU Wien, Austria
dominik.bork@tuwien.ac.at
[2] Universidade Nova de Lisboa, Portugal
mgoul@fct.unl.pt
[3] York University, Canada
liaskos@yorku.ca

Conceptual modeling has enjoyed substantial growth over the past decades in fields ranging from Information Systems Analysis and Business Process Engineering to Biology and Law. A plethora of conceptual modeling practices (languages, frameworks, methods, etc.) have been proposed, promising to facilitate activities such as communication, design, or decision-making. Success in adopting a conceptual modeling practice is, however, predicated on convincingly demonstrating that it indeed successfully supports these activities. At the same time, the way individuals and groups produce and consume models gives rise to cognitive, behavioral, organizational, or other phenomena, whose systematic observation may help us better understand how models are used in practice and how we can make them more effective.

The act of building conceptual models is ideally informed by empirical evidence that is nowadays abundant in the form of digital data. This overabundance of data, combined with the advent of advanced data analysis and artificial intelligence (AI) techniques, introduces major opportunities and challenges in an empirically informed conceptual modeling practice.

The 6th International Workshop on Empirical Methods in Conceptual Modeling (EmpER 2023), co-located with the 42nd International Conference on Conceptual Modeling (ER 2023), aimed at bringing together researchers with an interest in the empirical investigation of conceptual modeling. Like its five predecessors, the 6th installment of the workshop invited three kinds of papers: full study papers describing a completed study, work-in-progress papers describing a planned study or study in progress, as well as position, vision, and lessons-learned papers about the use of empirical methods for conceptual modeling. The workshop particularly welcomed negative results as well as proposed empirical studies that are in their design stage so that authors could benefit from early feedback and adjust their designs prior to a real data collection.

In addition, starting from this year, the workshop broadened its scope to include technical contributions or evaluation studies that report on the use of data and empirical evidence to inform conceptual modeling practices, such as, for example, the use of learning-based Artificial Intelligence (AI) and advanced data analysis techniques for constructing, reasoning with, validating, comprehending, and otherwise utilizing conceptual models.

A total of twenty-three (23) reviewers were invited to serve on the program committee of the workshop based on their record of past contributions in the area of empirical conceptual modeling.

In total, eight (8) papers were submitted to the workshop, each of which underwent single review by three PC members. Following reviewer feedback, five (5) out of the eight (8) papers were selected for inclusion in the program. The accepted papers include an empirical exploration of open-source issues for predicting privacy compliance (Guber et al.), an application of human-centered conceptual modelling for re-designing urban planning (Lenzi et al.), a tool for generating SQL queries from NL specifications using knowledge graphs (Campêlo et al.), a systematic ontology-based approach for generating enterprise architecture artifacts from multiple sources (Guerreiro and Sousa), and a system that utilizes reinforcement learning for modeling and analysing the structure of web portals (Shukla et al). The workshop involved presentations of the papers followed by discussion and audience feedback to the authors.

EmpER Organization

Workshop Chairs

Dominik Bork   TU Wien, Austria
Miguel Goulão   Universidade Nova de Lisboa, Portugal
Sotirios Liaskos   York University, Canada

Program Committee

Anna Bernasconi   Politecnico di Milano, Italy
Robert Andrei Buchmann   Babeş-Bolyai University of Cluj Napoca,
                                Romania
Michel Chaudron   Eindhoven University of Technology,
                                The Netherlands
Marian Daun   University of Applied Sciences
                                Würzburg-Schweinfurt, Germany
Istvan David   McMaster University, Canada
Robson Fidalgo   Universidade Federal de Pernambuco,
                                Brazil
Antonio Garmendia   Universidad Autónoma de Madrid, Spain
Vincenzo Gervasi   University of Pisa, Italy
Sepideh Ghanavati   University of Maine, USA
Andrea Herrmann   Herrmann & Ehrlich, Germany
Jennifer Horkoff   Chalmers University of Technology,
                                Sweden
Katsiaryna Labunets   Utrecht University, The Netherlands
Tong Li   Beijing University of Technology, China
Grischa Liebel   Reykjavik University, Iceland
Lin Liu   Tsinghua University, China
Raimundas Matulevičius   University of Tartu, Estonia
Jeffrey Parsons   Memorial University of Newfoundland,
                                Canada
Geert Poels   Ghent University, Belgium
Jan Recker   University of Hamburg, Germany
Ben Roelens   Open Universiteit, Ghent University,
                                Belgium
Carla Silva   Universidade Federal de Pernambuco,
                                Brazil
Irene Vanderfeesten   KU Leuven, Belgium
Manuel Wimmer   Johannes Kepler University Linz, Austria

Empirical Exploration of Open-Source Issues for Predicting Privacy Compliance

Jenny Guber[1]([✉]) [iD], Iris Reinhartz-Berger[1] [iD], and Marina Litvak[2] [iD]

[1] Department of Information Systems, University of Haifa, Haifa, Israel
jguber@campus.haifa.ac.il, iris@is.haifa.ac.il
[2] Shamoon College of Engineering, Beer Sheva, Israel
marinal@ac.sce.ac.il

Abstract. In the last decade, privacy has gained a significant interest in software and information systems engineering mainly due to the emergence of privacy regulations, including the General Data Protection Regulation (GDPR). However, checking privacy compliance is challenging and depends on many factors, such as the programming language and the software architecture, as well as the underlying regulation. In this exploratory research, we aim to study whether positive discussions on privacy-related issues in Open-Source Software (OSS) environments can predict privacy compliance of the software. Such predictions are beneficial in different scenarios, including in software reuse. Our main contribution will lie in conceptually modeling and understanding the relations between privacy compliance and positive discussions of privacy-related OSS issues. The research comprises three parts: (1) identifying privacy-related issues using supervised machine learning techniques; (2) improving the identification of privacy-related issues utilizing ontologies; and (3) identifying the sentiment of privacy-related issues and analyzing relations to privacy compliance. This paper describes the design and results of part 1, as well as the design of parts 2 and 3.

Keywords: Privacy · Software Development · Software Reuse · Open Source

1 Introduction

Privacy gained a significant interest in the last decade due to the emergence of privacy regulations including the well-known General Data Protection Regulation (GDPR)[1] which became enforceable in EU countries from 2018. According to UNCTAD, an intergovernmental organization within the United Nations Secretariat, 137 out of 194 countries (~71%) had put in place legislation to secure the protection of data[2].

In parallel, the discussion on privacy-related issues has increased in Open-Source Software (OSS) repositories [1]. Today's software development follows an agile, issue-driven style in which feature requests, bug reports, functionality implementations and all other development tasks are usually handled as issues (e.g., in GitHub or JIRA).

---

[1] https://gdpr.eu.
[2] https://unctad.org/page/data-protection-and-privacy-legislation-worldwide.

T. P. Sales et al. (Eds.): ER 2023 Workshops, LNCS 14319, pp. 63–73, 2023.
https://doi.org/10.1007/978-3-031-47112-4_6

Manually analyzing 1,244 issue comments from six diverse projects, Khalajzadeh et al. [2] identified eight categories of human-centric issues discussed by developers in OSS environments. The most discussed category (with 4.98% of the issues) was Privacy & Security, covering discussions related to privacy, security, data protection, reliability, and trust.

Despite these progresses in privacy regulations and discussions, checking privacy compliance on code artifacts is challenging; it depends on many factors, including the programming language, the software architecture, and the underlying privacy regulation. Our goal is to explore to what extent privacy compliant projects can be identified based on the meta-data available in OSS environments. To achieve our goal, we phrased the following research questions:

RQ1: To what extent can privacy-related issues be automatically identified using Machine Learning (ML) methods?

While in some OSS environments, issues can be labeled, many developers do not use this feature or use it to a limited extent. Hence, we aim to explore whether ML methods, in particular, supervised ones, can automatically identify privacy-related issues.

RQ2: To what extent can the identification of privacy-related issues be improved utilizing privacy ontologies?

Several privacy ontologies and domain models have been proposed over the years. These include Gharib et al.'s [3] privacy requirements ontology, Tom et al. [4] domain model of the GDPR, and Torre et al.'s [5] UML model for automatically checking GDPR compliance. We hypothesize that building an issue representation based on content related to such ontologies or models may improve the accuracy of privacy-related issues identification.

RQ3: To what extent do automatically analyzed sentiments of privacy-related issues correlate with privacy compliance of those projects?

To this end, we intend to use annotations of OSS projects into [privacy-]compliant and non-compliant. The annotations will be collected from the projects' owners or privacy experts. Our hypothesis is that projects with positive discussion on privacy-related issues may have a higher potential to be considered (labeled) compliant.

Overall, our research will contribute to the conceptual modeling community by exploring novel ways to automate privacy issue identification, leveraging privacy ontologies to improve identification, and above all conceptually modeling and understanding the relations between positive sentiment and privacy compliance of OSS projects. The results of this study can further inform the development of new tools and methods to enhance privacy-compliant software development in general and reuse in particular.

The structure of the rest of the paper is as follows. Section 2 describes the study design. Section 3 presents preliminary results and discusses threats to validity. Section 4 briefly reviews related work, and Sect. 5 concludes and describes further activities.

2   Study Design

This section elaborates on the dataset and methods used for addressing the above three research questions. An overview of the study is presented in Fig. 1. In the following we elaborate on the current dataset, as well as on the inputs, techniques and outputs of each one of the high-level activities: identifying privacy-related issues, improving the identification with ontologies and sentiment analysis & relations to [privacy] compliance.

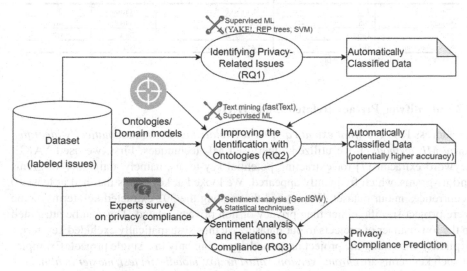

Fig. 1. The Study Overview

2.1 The Dataset and its Preprocessing

Our current dataset is composed of 2,556 issues from OSS projects. 1,374 of them (about 54%) are issue reports from Jira associated with two large-scale, popular and well-maintained software projects, Chrome and Moodle. These issue reports were explicitly labeled as privacy-related by their contributors; their status was "assigned", "fixed" or "verified"; and they were verified by the authors of [6] to have sufficient ("not limited") information. The other 1,182 are issues from six GitHub projects annotated by Khalajzadeh et al. [2] as non privacy-related.

Table 1 presents the relevant characteristics of our dataset. Although our dataset involves two different environments (Jira and GitHub), the basic fields for documenting each issue in those environments are quite similar and include issue ID, issue title/subject and issue body/description. For our analyses, we concatenated the issue title/subject and body/description into a single text document for each issue[3].

The dataset was preprocessed using common techniques. These include cleaning and removing irrelevant parts (e.g., html tags, numbers, punctuation marks and stop words) and performing tokenization and lemmatization. We extracted words from URLs that appear in the issues and left them for the analyses, assuming that some may be meaningful and relevant for classification.

---

[3] The dataset is available at https://zenodo.org/record/8351237.

Table 1. Dataset Characteristics

Project Name	Source	# Issues	Project Name	Source	# Issues
Chrome	Jira	896	Healthchecks	GitHub	288
Moodle	Jira	478	Firefox-ios	GitHub	361
Mobile-ios	GitHub	30	CWA[a]-app-ios	GitHub	312
Mobile-android	GitHub	31	CWA[a]-app-android	GitHub	160

[a]CWA stands for Corona Warn App.

2.2 Identifying Privacy-Related Issues (RQ1)

To address RQ1 (*To what extent can privacy-related issues be automatically identified using ML methods?*), we utilized supervised ML techniques. First, we used YAKE! keyword extractor [7] for extracting potential key-terms, namely, unigrams, bi-grams and tri-grams, which frequently appeared. We looked at key-terms that had at least 10 occurrences in our dataset. We noticed that among the 302 extracted key-terms, some were project-specific rather than privacy or non privacy-related. This can be attributed to the low number of projects in our dataset. We thus systematically excluded key-terms explicitly containing the project name and appearing only in a single project. Examples of such key-terms are *chrome version, safari firefox, moodle net test, master healthcheck* and *moodle privacy class*. As can be seen, some of these terms may refer to privacy, but we have no evidence whether they are generally privacy-related or particular concerns of the specific project. Thus, we prefer to omit them to avoid bias. We did not exclude sub-terms of excluded terms (e.g., *privacy class*). We further examined the key-terms to exclude terms originating from templates (e.g., file XXX line ## or version ## stable operating). This was done by systematically analyzing the issues related to frequently appearing key-terms. Such terms mainly characterize the development process or documentation practices.

Overall (after exclusions), we got 150 potentially relevant key-terms. We further associated these terms with frequency scores inversely weighted with YAKE! scores and normalized using Min-Max normalization [8]; the lower the YAKE! score is, the more relevant the key-term is and, consequently, the higher the overall score is. Formally expressed:

$$score_i = \frac{X_i - X_{min}}{X_{max} - X_{min}}, \tag{1}$$

where:

$$X_i = \frac{term_i \ frequency}{term_i \ YAKE! \ score}; X_{min} = min_i X_i; X_{max} = max_i X_i$$

We then applied two supervised ML classifiers: Reduced Error Pruning (REP) tree [9] and Support Vector Machine (SVM) [10]. We chose these classifiers because they differ in their underlying principles, learning approach, and decision-making processes: REP trees use a top-down, greedy approach to build decision trees with axis-aligned decision boundaries, while SVMs find an optimal hyperplane to separate classes in the feature space, allowing for more complex decision boundaries using kernel functions. For both classifiers, we used 70% of the dataset for training and 30% for testing; we kept the class distribution similar in both sets. We further compared the outcomes of the two classifiers to validate that the results are similar in terms of correctly classified items, precision, recall and F-measure, indicating that similar patterns or relationships in the data have been learned.

2.3 Improving the Identification with Privacy Ontologies (RQ2)

To address RQ2 (*To what extent can the identification of privacy-related issues be improved utilizing privacy ontologies?*), we plan to explore relevant ontologies and domain models, such as Gharib et al. [3], which contextualizes privacy requirements into four element groups (organizational concepts, risk, treatment, and privacy [requirements]), based on a systematic literature review (see a summary of the resultant dimensions and key concepts in Table 2). Another example is Torre et al. [5], which particularly targets the GDPR and identifies nine packages of classes (Enumerations, Data Processing, Data Subjects, Compliance, Data Subject Rights, Data Transfer, Administration, Principles and Main Actors).

Table 2. Dimensions and Key Concepts in Gharib et al.'s ontology

Dimension	Key Concepts
Organizational	Agentive entities: Actor, role, agent
	Intentional entities: Goal, decomposition
	Informational entities: Information, personal information, public information
	Information types of use: Produce, read, modify
	Information ownership and permissions: Own, permission
	Entity interactions: Information provision, goal delegation, permissions delegation
	Entity social trust: Trust/distrust
	Monitoring
Risk	Threat, impact, casual threat, intentional threat, attack method, vulnerability, threat actor
Treatment	Privacy goal, privacy constraint, privacy policy, privacy mechanism
Privacy	Privacy requirement, confidentiality, notice, anonymity, transparency, accountability, non-repudiation

Understanding that the key-terms extracted from the OSS issues and the ontological concepts may reside at different abstraction levels, the proposed contribution of this part of the research is to explore the gaps and suggest some conceptual bridges. To this end, we plan to identify key-terms related to the privacy concepts by examining the semantic similarity between them and excluding unrelated terms from the issues representation prior to classification. In particular, we will apply fastText [11] library for obtaining semantic representation of key-terms and privacy concepts. We will further use cosine similarity to detect concept-related terms. A text, composed of a sequence of words (such as a phrase or a concept description), will be represented by an average vector calculated from the vectors representing its words. Following this step, we will reapply the classifiers used in addressing RQ1 and compare the accuracies before and after utilizing the ontologies.

2.4 Sentiment Analysis and Relations to Privacy Compliance (RQ3)

Addressing RQ3 (*To what extent do automatically analyzed sentiments of privacy-related issues correlate with privacy compliance of those projects?*) encompasses two steps: performing sentiment analysis and conducting privacy compliance assessment surveys. The outcomes of these steps will be analyzed with common statistical techniques, such as regression or correlation.

In order to perform sentiment analysis, we intend to apply SentiSW [12], which was specifically developed for and validated on GitHub issues. SentiSW is a supervised entity level sentiment analysis tool consisting of both sentiment classification and entity recognition, where the entity is either 'Person' or 'Project'. We also plan to experiment with different representations, classifiers (including pre-trained language models) and entities to accurately identify the sentiments of privacy-related issues. We plan to use an annotated dataset provided in [12], for training our models.

Independently of the sentiment analysis, we plan to conduct a survey with OSS project owners or privacy experts to annotate the projects in our dataset to [privacy-]compliant and non-compliant. To this end, we plan to use the eight privacy design strategies that have been presented in [13]: (1) MINIMISE – the amount of personal data that is processed should be restricted to the minimal amount possible; (2) HIDE – any personal data, and their interrelationships, should be hidden from plain view; (3) SEPARATE – personal data should be processed in a distributed fashion, in separate compartments whenever possible; (4) AGGREGATE – personal data should be processed at the highest level of aggregation and with the least possible detail in which it is (still) useful; (5) INFORM – data subjects should be adequately informed whenever personal data is processed (transparency); (6) CONTROL – data subjects should be provided agency over the processing of their personal data ("intervenability"); (7) ENFORCE – a privacy policy compatible with legal requirements should be in place and should be enforced (accountability); and (8) DEMONSTRATE – a data controller should be able to demonstrate compliance with the privacy policy and any applicable legal requirements (accountability). Examples of relevant questions for our survey can be found in Table 3.

3 Preliminary Results and Threats to Validity

As noted, we have currently conducted only the first part of the research (addressing RQ1). In the following we present and discuss RQ1 outcomes and threats to validity.

3.1   RQ1 Outcomes

Table 4 presents the outcomes of RQ1. As indicated by the relatively high values of accuracy (85.9% and 86.7%) and F-measure (86.6% and 87.1%) for both classifiers, we can conclude that the classifiers have the potential for distinguishing between privacy- and non privacy-related issues.

Due to the simple, intuitive and visual format of REP trees, Fig. Depicts the obtained tree. The root term, *privacy*, appeared in only 20% of privacy-related issues, so the relatively high accuracy of the classifier cannot be attributed only to this term. We further observed that the resultant tree includes organizational terms, such as *user, datum* and *setting*; terms related to policies or guidelines for data privacy handling, e.g., *tool policy* and *tool data privacy*; and terms related to privacy risks, including *incognito window, incognito mode* and [third] *party cookie*.

Table 3. Potential questions for a privacy compliance assessment survey

#	Question	Strategy
1	Are there clear guidelines and policies in place to ensure that data collection is limited to what is strictly required?	Minimize
2	Does the software utilize techniques to anonymize or pseudonymize personal data to protect user identities?	Hide
3	Are sensitive user data concealed from unauthorized access or disclosure within the system?	Hide
4	Are there access controls in place to restrict access to personal data based on roles and permissions?	Separate
5	Are there mechanisms in place to prevent re-identification of individuals from aggregated data?	Aggregate
6	Are users provided with a comprehensive privacy policy and terms of service that explain data processing practices in a user-friendly manner?	Inform
7	Are there options for users to customize their privacy settings according to their preferences?	Control
8	Are there automated processes in place to ensure compliance with privacy policies and data protection regulations?	Enforce
9	Are there regular reviews or reports on the software's privacy practices available to stakeholders and users?	Demonstrate
10	How would you rate the software's privacy compliance on a scale from 1 to 5, with 1 being not compliant at all and 5 being fully compliant?	---

Our results show a high potential for using ML as a tool for identification of privacy-related issues within OSS repositories. These outcomes are not trivial as the development issues commonly refer to concrete technical problems and solutions and are phrased in a natural language by different community contributors, without having a single under-lying terminology. Moreover, diving into these results, we spotted that there were more

Table 4. Classifiers' evaluation metrics for RQ1

	REP Tree	SVM
Accuracy	85.9%	86.7%
Precision	90.2%	92.7%
Recall	83.3%	82.1%
F-measure	86.6%	87.1%

privacy-related issues classified as non (70) than non privacy-related issues classified as related (38). To improve these results, we will explore in the second part of the research how to explicitly utilize privacy ontologies and domain model. In the third part we will examine the sentiment of these issues and whether it can predict privacy compliance. These will enable us to conceptually model and understand the relations between privacy compliance and positive discussions of privacy-related issues in OSS projects (Fig. 2).

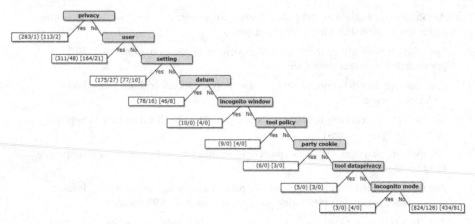

Fig. 2. The REP tree outcome received for RQ1

3.2 Threats to Validity

Our research is exploratory and still in progress. Currently, we have relatively small numbers of projects and issues, but the entire dataset was manually verified by researchers and experts. We further noticed that there may be a confounding variable – the OSS platform: the privacy-related issues were taken from Jira, whereas the non privacy-related issues – from GitHub. As previously noted, the basic fields of documenting each issue are quite similar in the two environments. Therefore, we put the emphasis on the data managed for each issue rather than on the platform managing it. However, to better establish our method and results, we intend to extend our dataset to include more projects and platforms and sequentially to increase the number of issues. It should also be noted that our privacy-related issues are taken from two specific projects (Chrome and Moodle) that mostly involve browsing and management of e-learning systems. Their privacy concerns may be relevant to diverse domains, but further evaluation in domains, such as healthcare and finance, is desirable.

Another threat refers to the specific methods (REP tree and SVM classifiers, YAKE! keyword extractor, fastText library for semantic representation, and SentiSW analyzer) we used or plan to use. We justify the selection of each method. Yet, future research with additional methods is planned, including unsupervised ML methods.

4 Related Work

Our main goal is to support privacy-compliance identification. However, checking privacy compliance of the software itself is challenging, due to the gap between the formulation of regulations and their realization in software artifacts. Thus, it is unclear whether traditional static code analysis-based vulnerability discovered can assist in compliance checking of regulatory requirements [14]. To this end, the authors of [14] suggest how to identify security and privacy requirements for Health Insurance Portability and Accountability Act (HIPAA) technical requirements. A model for automated privacy compliance checking of applications in the cloud is introduced in [15]. The authors claim that privacy compliance of applications hosted in cloud or fog computing platforms can and should be automatically carried by the platform itself; the platform may monitor privacy-oriented behavior through signals such as its network traffic characteristics; these signals can be analyzed and compared with the principles found in the application's privacy policy. The methods suggested in these studies are specific to the application domain, the software architecture/platform and/or the underlying privacy policies/regulations. In our work, we aim to explore whether meta-data of OSS projects (in particular, issues) can predict privacy compliance.

Enhancing the precision and correctness of categorization often depends on the size and quality of the available training dataset, as well as on which data the training has been performed (e.g., general vs. domain-specific). The researchers in [16] conducted experiments with various classifiers, both prior to and subsequent to utilizing a disease ontology. Notably, this integration led to enhancement in the outcomes. Moreover, the adoption of a domain-specific ontology has demonstrated improvement in the accuracy of text classification in situations where a sufficiently large and well-labeled training corpus is not at hand [17]. Another approach, outlined in [18] and employing Wikipedia as an ontology, suggests an automatic classification of text documents into a dynamically defined set of topics of interest. This method does not require a previously annotated training set of documents. It relies solely on ontological knowledge, thus avoiding the necessity for classifier retraining. Our research seeks to extrapolate these approaches to the realm of privacy, employing relevant ontologies and domain models, e.g., [3, 5].

Finally, considering the textual content of issues in the OSS community, studies have explored emotional expressions and sentiments. The work in [19], for example, found that applying emotion mining to developer issue reports can be useful to identify and monitor the mood of the development team, and thus predict and resolve potential threats to the team well-being, as well as discover factors that enhance team productivity. The work in

[12] created a domain-specific tool for sentiment analysis of the issues documented in software development, improving the accuracy of the analysis by enriching the classifier with domain-specific lexicon. The work in [20] reports on a positive correlation between favorable sentiments and improved practices in the context of software engineering and development. We intend to advance this body of research by analyzing the relations of the analyzed sentiments to privacy compliance.

5   Conclusions and Future Work

In this research, we attempt to explore whether positive discussions on privacy-related issues in Open-Source Software (OSS) environments can predict privacy compliance of the software. This will provide a rapid and easy way to identify privacy compliant software artifacts and assess their compliance. These assessments may be considered in reuse scenarios and can help rank the projects retrieved to a user query. The paper describes the method and preliminary results from its first step. The preliminary results (for addressing RQ1) show around 86% accuracy (correct classification) with respect to automatic identification of privacy-related issues utilizing both REP tree and SVM classifiers. The two additional planned steps, whose only design is presented in this paper, include enhancement of classification accuracy by utilizing privacy ontologies or domain models (for addressing RQ2) and checking the relations between analyzed sentiments and privacy compliance labels (compliant vs. non-compliant), annotated using responses from a privacy compliance assessment survey (for addressing RQ3).

In the future, we intend to implement and refine the entire method and evaluate it on larger and more diverse datasets. We intend to experiment with different methods for semantically-enriched representations (word vectors, topics, etc.) and Large Language Models. We also plan to explore unsupervised machine leaning methods, to avoid the need for large, reliable annotated data. The results of steps 2 and 3 will support the creation of a method for predicting privacy compliance of software projects and guiding reuse decisions in light of non-functional requirements. Additional further directions include extending of the suggested method to other types of non-functional requirements, such as security and maintainability, for which OSS discussions and sentiments may be valuable.

References

1. Hennig, A., Schulte, L., Mayer, P.: Understanding issues related to personal data and data protection in open source projects on GitHub. In: Proceedings of International Conference on Mining Software Repositories (MSR 2023) (2023)
2. Khalajzadeh, H., Shahin, M., Obie, H.O., Grundy, J.: How are diverse end-user human-centric issues discussed on GitHub? In: Association for Computing Machinery (2022)
3. Gharib, M., Giorgini, P., Mylopoulos, J.: Towards an ontology for privacy requirements via a systematic literature review. In: Mayr, H., Guizzardi, G., Ma, H., Pastor, O. (eds.) Conceptual Modeling, ER 2017, vol. 10650, pp. 193–208. Springer, Cham (2017). https://doi.org/10. 1007/978-3-319-69904-2_16

4. Tom, J., Sing, E., Matulevičius, R.: Conceptual representation of the GDPR: model and application directions. In: Zdravkovic, J., Grabis, J., Nurcan, S., Stirna, J. (eds.) Perspectives in Business Informatics Research, BIR 2018, vol. 330, pp. 18–28. Springer, Cham (2018). https://doi.org/10.1007/978-3-319-99951-7_2

5. Torre, D., Alferez, M., Soltana, G., Sabetzadeh, M., Briand, L.: Modeling data protection and privacy: application and experience with GDPR. Softw. Syst. Model. **20**(6), 2071–2087 (2021). https://doi.org/10.1007/s10270-021-00935-5

6. Sangaroonsilp, P., Dam, H.K., Choetkiertikul, M., Ragkhitwetsagul, C., Ghose, A.: A taxonomy for mining and classifying privacy requirements in issue reports. Inf. Softw. Technol. **157**, 107162 (2023). https://doi.org/10.1016/j.infsof.2023.107162

7. Campos, R., Mangaravite, V., Pasquali, A., Jorge, A., Nunes, C., Jatowt, A.: YAKE! keyword extraction from single documents using multiple local features. Inf. Sci. **509**, 257–289 (2020). https://doi.org/10.1016/j.ins.2019.09.013

8. Jayalakshmi, T., Santhakumaran, A.: Statistical normalization and back propagation for classification. Int. J. Comput. Theory Eng. **3**(1), 1–6 (2011). https://doi.org/10.7763/IJCTE.2011.V3.288

9. Quinlan, J.R.: Simplifying decision trees. Int. J. Hum. Comput. Stud. **27**, 221–234 (1987). https://doi.org/10.1006/ijhc.1987.0321

10. Cortes, C., Vapnik, V.: Support-vector networks. Mach. Learn. **20**, 273–297 (1995)

11. Joulin, A., Grave, E., Bojanowski, P., Douze, M., Jegou, H., Mikolov, T.: FASTTEXT.ZIP: compressing text classification models. In: ICLR 2017, pp. 1–13 (2017)

12. Ding, J., Sun, H., Wang, X., Liu, X.: Entity-level sentiment analysis of issue comments. In: IEEE/ACM 3rd International Workshop on Emotion Awareness in Software Engineering, SEmotion 2018, pp. 7–13 (2018). https://doi.org/10.1145/3194932.3194935

13. Hoepman, J.-H.: Privacy design strategies. In: Cuppens-Boulahia, N., Cuppens, F., Jajodia, S., Abou El Kalam, A., Sans, T. (eds.) SEC 2014. IAICT, vol. 428, pp. 446–459. Springer, Heidelberg (2014). https://doi.org/10.1007/978-3-642-55415-5_38

14. Farhadi, M., Haddad, H., Shahriar, H.: Compliance checking of open source EHR applications for HIPAA and ONC security and privacy requirements. In: 2019 IEEE 43rd Annual Computer Software and Applications Conference (COMPSAC), pp. 704–713 (2019). https://doi.org/10.1109/COMPSAC.2019.00106

15. Farhadi, M., Pierre, G., Miorandi, D.: Towards automated privacy compliance checking of applications in cloud and fog environments. In: 2021 8th International Conference on Future Internet of Things and Cloud, pp. 11–18 (2021). https://doi.org/10.1109/FiCloud49777.2021.00010

16. Malik, S, Jain, S.: Semantic ontology-based approach to enhance text classification. In: ISIC 2021 (2021)

17. Sanchez-pi, N., Martí, L., Cristina, A., Garcia, B.: Improving ontology-based text classification : an occupational health and security application. J. Appl. Log. **17**, 48–58 (2016). https://doi.org/10.1016/j.jal.2015.09.008

18. Allahyari, M., Kochut, K.J., Janik, M.: Ontology-based text classification into dynamically defined topics. In: 2014 IEEE International Conference on Semantic Computing (2014)

19. Murgia, A., Adams, B.: Do developers feel emotions ? an exploratory analysis of emotions in software artifacts. In: MSR 2014, pp. 262–271 (2014). https://doi.org/10.1145/2597073.2597086

20. Junior, R.S.C., Carneiro, G.D.F.: Impact of developers sentiments on practices and artifacts in open source software projects : a systematic literature review. In: Proceedings of the 22nd International Conference on Enterprise Information Systems (ICEIS 2020), vol. 2, pp. 978–989. (2020). https://doi.org/10.5220/0009313200310042

Human-Centred Conceptual Modelling for Re-Designing Urban Planning

Emilia Lenzi[1]([📧])[ID], Emanuele Pucci[1][ID], Federico Cerutti[2][ID],
Maristella Matera[1][ID], and Letizia Tanca[1][ID]

[1] Department of Electronics, Information, and Bioengineering, Politecnico di Milano,
Milan, Italy
{emilia.lenzi,emanuele.pucci,maristella.matera,letizia.tanca}@polimi.it
[2] University of Brescia, Brescia, Italy
federico.cerutti@unibs.it

Abstract. In traditional Urban Analysis frameworks, data collection is
a rather informal process, and data analysis is based on numerous redun-
dant and context-dependent measures. In addition, when automating the
analysis process, the knowledge of the domain experts is often scarcely
considered. This paper describes a successful example of *human-centred
conceptual design*, which allowed us to represent and organise the explicit
and implicit knowledge of a team of architects and urban planners, using
the ER model as a shared language between computer scientists and
domain experts.

Keywords: Conceptual Modelling · Human-Centred Design ·
Experience Report

1 Introduction

A data-driven culture is becoming increasingly important nowadays, trespass-
ing the traditional borders of technical disciplines and encompassing a variety
of subjects. The analysis of urban performance is no exception [4]. However, in
contrast to the spread of modern data analysis techniques, the existing data
analysis practices in this field still focus on standard techniques, often of a
statistical nature, seldom integrated within more sophisticated intelligent systems. Although adopting intelligent systems is beneficial from an evaluation
and formalisation point of view, it may entail risks concerning the responsibility of human beings in the decision-making process. When designing intelligent systems, including their final users in the design process and applying
Human-Centred Design (HCD) principles in the conceptual modeling phase is
of paramount importance to ensure that the resulting systems are tailored to
the needs of users and support them reliably, safely, and trustworthy [15].

In this paper, we show how the Entity-Relationship (ER) model, although
technical, was successfully used as *lingua franca* among HCD experts, database
designers, and application domain professionals, whose implicit and explicit

T. P. Sales et al. (Eds.): ER 2023 Workshops, LNCS 14319, pp. 74–84, 2023.
https://doi.org/10.1007/978-3-031-47112-4_7

knowledge was identified through focus groups and co-desing activities. We describe the many considerations that resulted from this choice regarding the model's potential and the impact that formative studies typical of HCD approaches had on the quality of the final schema. We also show how the resulting schema guided the domain experts through an iterative systematisation of the different methodological concepts at the basis of their analysis methodology and their relationships, which was unexpected before starting the process. Indeed, more than on ER's (well-known) ability to represent the expert's knowledge and requirements, the work illustrated in this paper focuses on its ability to stimulate discussion and elicit those requirements when the design is performed following a human-centred approach. The discussion revolves around a case study that applies our proposal, from now on denoted as Human-Centred Conceptual Design (HCCD), to the creation of a knowledge-based system for the Integrated Modification Methodology (IMM), a procedure developed at the Department of Architecture, Built Environment and Construction Engineering (DABC) of Politecnico di Milano, aligned to the 17 Sustainable Development Goalss (SDGs) promoted by the United Nations [10].

This paper is organised as follows: In Sect. 2, we report works comparing different frameworks for assessing sustainability in the urban environment and works describing the collaboration between human-centred design and data engineering. In Sect. 2.1, we describe the IMM methodology. Section 3 presents the results of the data inspection performed on the IMM datasets and shows the main steps of our work. In Sect. 4 and Sect. 5, we elaborate on human-centred conceptual design, and in Sect. 6, we discuss the results obtained, also summarised in the conclusions (Sect. 7).

2   Rationale and Background

Assessing sustainability in the urban environment is a complex task, mainly due to the ambiguous definition of the concept itself, the heterogeneity and the enormous amount of data and variables, the presence of context-dependent measures, and the absence of recognised techniques for assessing the adequacy of the data and the proposed interventions. These aspects highlight the need for comprehensive frameworks addressing all those aspects that are fundamental to the analysis of the city and the built environment in a more general, but at the same time efficient way [1].

Our work aims to address this lack by placing domain experts at the centre of the design process, through a human-centred re-design approach. A few recent streams of research highlighting the benefits of applying human-centred design to conceptual modelling. In the knowledge representation realm, Richards [14] focuses on supporting domain experts in many different activities, including knowledge acquisition, inferencing, maintenance, tutoring, critiquing, "what-if" analysis, explanation, and modelling. He uses knowledge acquisition and representation techniques known as Ripple-Down Rules (RDR). To support the exploration activities, he combines RDR with formal concept analysis, automatically generating an abstraction hierarchy from the low-level RDR assertions. From the knowledge management viewpoint, some methodologies use ontologies, [5, 8, 12]. Kotis *et al.* [6], for instance, developed a method to actively involve knowledge workers in the ontology life cycle and empower them to continuously manage their formal conceptualisations in their day-to-day activities and shape their information space.

These studies show that incorporating HCD methods can lead to more usable and practical conceptualisations that better support users' tasks. Our work extends this stream of research by proposing a methodology and applying it to the IMM analysis framework to represent the implicit knowledge of domain experts thanks to the ER model.

2.1 IMM Overview

The IMM methodology aims to enhance the city's environmental performance by modifying its structural characteristics [9]. IMM considers the built environment as a Complex Adaptive System (CAS), *i.e.*, a system with a significant number of agents that produce a non-linear dynamic behaviour which may not be predictable according to the agents' behaviour [3].

A city can be divided into four main structural subsystems representing elementary elements of the built environment: *Volume* (the built part), *Void* (empty spaces), *Type of Use* (activities performed by citizens), and *Network* (networks of different modalities) [2]. To describe the interaction and the synergy among these components and guide the intervention choices, IMM introduces seven key categories (*e.g.*, Urban Porosity – the spatial relationship between urban built-ups and voids – and Diversity – the structural relationship derived from the different kinds of land uses) and, for each key category, six quantitative measures, named *Metrics*, expressing morphological characteristics. An example of a metric is the Building Density (BD), defined as the total number of buildings in an area divided by the entire area and associated with the Porosity key category.

The *Indicators* serves as tools for performance evaluation(*e.g.*, the density of the Public Transportation Stops), and are organised into *Design Ordering Principle* (DOP) families, which are sets of actions that designers can take to improve the system's behaviour [9].

The IMM methodology comprises four main phases: 1. Investigation, 2. Assessment and Formulation, 3. Intervention and Modification, and 4. Optimisation [16]. For instance, when applied to Milan's Porto di Mare neighbourhood [13], urbanists aimed to promote walkability and cycling while enhancing integration with public transportation. They calculated metrics and identified the weakest key category (*e.g.*, the street network's characteristics influencing overall connectivity and volumes). Based on this, they provided a site-specific DOP ranking with design suggestions. The identified indicators were optimised and used to evaluate and compare different projects in *phase 3*. By comparing indicator values, architects selected the best project and optimised it in *phase 4* to achieve the optimal modification plan.

3 Motivation and Approach

While the IMM methodology has been adopted in numerous case studies, it still presents some limitations. Most of the data are acquired from automatic surveying systems – open geographic databases such as OpenStreetMap[1] or processing through Q-GIS software[2] –, and organised into a spreadsheet without a fixed scheme or methodology. The Q-GIS software, in particular, allows the users to compute several measures on a map. Each measure, or set of measures, creates a project level; the architects use Excel files produced for each project level. This procedure requires the architects to work with huge files representing components on multiple levels.

We started our analysis with a technical inspection of the provided data to help the domain experts better understand such limitations before approaching and formalising them. Several data quality issues emerged, and it was therefore clear that the data must be reorganised, and the domain experts had to be involved in the modelling process to ensure that the new design was usable and effective for them. We then decided to proceed with a *Human-centered Analysis*, based on interviews and focus groups, which we describe in the following sections. Our main goal was to identify the challenges and needs of domain experts and understand how better to support the IMM digitisation process and the *Conceptual Data Design*, which allowed us to represent the output of the human-centred analysis into a conceptual representation of the involved data using an ER diagram.

4 Human-Centred Analysis

As illustrated in Fig. 1, in our research, we actively engaged domain experts in the co-design of the Entity-Relationship (ER) model itself. Unlike traditional practices where domain experts primarily provide requirements and the technical team subsequently crafts the ER model, our approach prioritised collaborative efforts from the outset. We first explored stakeholders' needs and elicited their requirements.

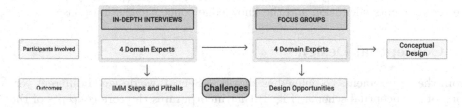

Fig. 1. The process for the human-centred conceptual design.

---

[1] https://www.openstreetmap.org/.
[2] https://qgis.org/it/site/.

First round of Interviews. We conducted in-depth unstructured interviews with 4 domain experts [7], focusing on the IMM phases (investigation, formulation, modification, optimisation). The main investigated aspects were: (i) the workflow of activities performed by each subject; (ii) current challenges that the subjects encounter every day during their work activities and possible solutions; and (iii) future interaction modalities the domain experts imagine when talking about cutting-edge technologies.

Challenge Elicitation. The researchers independently analysed the results of the interviews, with a final session focused on reaching a consensus for the challenge identification. They performed an inductive thematic analysis [11] and clustered the identified themes into the challenges referred to difficulties in: *i)* managing and integrating the different data sets; *ii)* identifying the relationship holding among different elements of the analysis (e.g., indicators, key categories); *iii)* selecting the best measure and analysis procedure due to the lack of formalisation of the decision-making process; *iv)* understanding the best possible action in a given analysis scenario and simulating the effect of their actions.

Focus Groups and Identification of Design Opportunities. After challenge elicitation, we invited the domain experts to participate in two focus groups, to identify the most prominent design opportunities. We discussed (i) their expectations on how technology could help them overcome the identified challenges and (ii) how they wanted to interact with the system.

Table 1 summarises the relationships we found between challenges and design opportunities. The process helped us elicit requirements for the knowledge formalisation step described in the next section. For this paper, we mostly focus on challenge 2, *Relationship among different elements*. We opted for an ER schema that could encapsulate all the logical order and operations among the elements used to assess sustainability and performance, with a formalism still understandable by the domain experts, and that led them to identify and correct issues during various interactions of the design process.

Indeed, as mentioned above, the interaction with the domain experts continued further. We directly involved them in the modelling process, fostering a deeper understanding of the domain's intricacies that bridged the gap between the requirements elicitation and the knowledge conceptualisation phases.

5 Conceptual Design

From the requirements co-defined as shown in Sect. 4, we designed different versions of a conceptual schema (Fig. 2) that incorporates the core concepts of the IMM method from Sect. 2.1. In the last version of the schema, the one described in the following, in addition to the static definition of each main concept, the schema allows to represents:

– the concept's values for each case study (e.g., the Bike lane density in different areas of a city or other cities), distinguished by the attribute *Case_Study*;

Table 1. Identified challenges and corresponding design Opportunities

Challenges	Design Opportunities
1. Data Management	a. Support for the selection of meaningful data b. Support during the transformation of raw data into indicators and metrics
2. Relationship among different elements	a. Clarify relationships among different elements b. Clarify why a certain choice impacts elements c. Clarify why certain choices have been made during the previous steps
3. Decision-making	a. Suggestion of actions based on the analysis b. Simulate the effect of a choice on the map
4. Actionability and Performance Evaluation	a. Clarify how to maximise performance b. Tell the users when to stop with the simulation since performance has been maximised

- the concept's value in the specific project over time; and
- the count of each specific built environment transformation represented by the attribute *Step*.

Component. This entity represents city elements, organised in a generalisation hierarchy with five subclasses. Four subclasses correspond to the subsystems, described in Sect. 2.1, to which each Component belongs. Additionally, the subclass **Boundary** represents the extent and perimeter of each case study.

All the Components are identified by a *CompCode*, a *Case_Study*, and a *Step*. The attribute *Step* tracks the built environment's evolution over time, while the boolean attribute *Elected* denotes whether that Component is part of a transformation elected as best at least once. Note that, as we explained earlier, our goal is not only to represent the essential elements of IMM analysis as static ones, but we also want to model their evolution in time and space: the attributes Step and Case Study, in the Component entity, satisfy this need. Further attributes of Components are *Geometry* and *Date*, the latter indicating when the measure was taken. Each Component subentity has its unique attributes that we do not show to preserve Fig. 2's clarity.

Parameter and Timed Parameter. A Parameter is an aggregation of some attribute of the instances of the Component it refers to, serving as an interme-

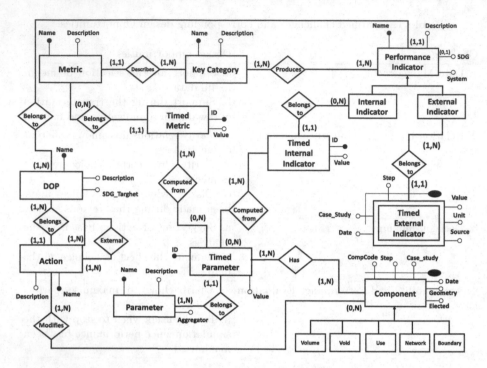

Fig. 2. Final ER diagram. The entities represent the elements of the IMM analysis. *Timed* entities represent the evolution of the built environment over time.

diate step for computing Metrics and Indicators. An example of a Parameter is the total area occupied by the buildings, calculated as SUM (Building.Area), where Building is the component representing the buildings and Area is one of its attributes.

In the ER schema, the entity *Parameter* represents the set of all the possible parameters, each with its description and mathematical expression (*Aggregator*). The entity *Timed Parameter*, instead, represents the *Value* assumed, at a certain time, by the parameter connected to it. This timed parameter represents the different values each Parameter can take, therefore it is in a 1:N relationship with the parameter entity. To identify the component from which that Parameter is calculated and thus also the step and case study it refers to, the entity also features an N: M relationship with the Component entity.

Metric and Timed Metric. Metrics can have different values depending on the case study and transformation step. Not all Metrics are always computable; therefore, as in the case of Parameters, the value of each *Metric* for a specific case study and step is represented by the *Timed Metric* entity, having its own *ID* and *Value* attributes. This entity is also associated with the Timed Parameter allowing connections with the Components, the Step, and the Case study from which its value is derived.

Performance Indicator. Parameters in IMM compute the Indicator values. Similar to Metrics, we distinguish between the entity representing the concept of Indicator (*Performance Indicator*) and its actual value in a specific Case study and Step (*Timed Indicator*).

The entity *Performance Indicator*, contains numerous attributes describing its nature. Metrics and Indicators can be computed from Parameters, but some Indicators may originate from external sources and not through parameters. To address this, we introduce a generalisation in the schema, defining two types of indicators: *Internal Indicator* and *External Indicator*. The first entity, Internal Indicator, represents indicators calculated through parameters and is associated with *Timed Internal Indicator*, which serves a role similar to Timed Metric. On the other branch of the generalisation, *Timed External Indicator* is a weak entity representing values of external indicators. It is identified by *Name*, *Step*, and *Case_study* and contains the *Value* retrieved from an external source. The *Date* attribute is mandatory and indicates when the Indicator was computed according to the source.

Key Category. The *Key Category* entity is identified by its *Name* and the *Describe* relationship indicates the groups of metrics describing the Key Category, while the *Produce* one identifies which Indicators and Parameters must be modified to improve the Key Category's value.

DOP. This entity represents the different families of Design Ordering Principles, each identified by a *Name* and specified by a *Description* and *SDG_Target*.

Action. The *Action* entity represents the actions that modify the Components and, thus, the built environment. They are categorised according to the corresponding DOP but can interact with each other, and this is represented by the *External* relationship. Indeed, actions belonging to the same DOP are defined as internal, but they can interact with actions belonging to different DOPs, which are external ones. The relationship *Modifics* indicates which Component the action acts on.

6 Discussion

The ER schema described above is a support for the solution to the technical problems described in Sect. 3, satisfying the requirements elicited through the adopted human-centred process. Besides representing the main elements of IMM (Components, Metrics, Indicators, Actions and DOPs), the schema supports the architects in the selection of meaningful data and clarifies the relationship among the different elements (Design Opportunities 1.a and 2.a). Table 2 summarises how the different elements of the final schema contribute to meeting the Design Opportunities identified. Specifically, the schema meets the Challenges listed in Table 1 as follows.

Data Management and Relationships among Different Elements. The conceptual schema has already led to the design and implementation of a normalised and unified database. Here, constraints can be imposed to avoid redundancies and inconsistencies among the various transformations of different built environments. The introduced Component generalisation hierachy dramatically reduces the number of null values by assigning each Component a separate entity and eliminating the layer Size resulting from data processing in Q-Gis, which is not optimal for automatic analysis. In addition, the relationships between different elements, particularly Components, Metrics and Indicators, are no longer expressed informally but identified by the relationships between the corresponding entities or Timed entities and the Parameter one, facilitating the transformation from raw data to Indicators and Metrics *cf.*, Table 2 row 1. In addition, the proposed schema allows us to represent the evolution of the built environment through timed entities and different case studies. Thanks to this choice, the schema provides the users with a complete overview of the entire history of the work performed, making it possible to use this information when explaining why a particular change was made and quantifying the effect it had on the built environment *cf.*, Table 2 rows 2 and 3.

Decision-making. Exploiting the connection between an applied change and the specific affected component enables parallel analysis of multiple case studies and eases the recommendation of certain actions or the simulation of certain effects on the urban structure *cf.*, Table 2 row 4.

Actionability and Performance Evaluation. This challenge, although addressed during the process, must be further explored. The main criticality is that, at the moment, there are no benchmarks for optimising the built environments. That said, we firmly believe that the organisation of the data with a model oriented to automatic analysis can also improve this aspect. Future work may try to store the IMM data in other types of databases (*e.g.*, graph db, useful to represent the relationships between a single parameter and numerous components) or using a relational DBMS that supports the preservation of spatial relationships, *e.g.*, Postgres with PostGIS extension. At the moment, representing in the ER diagram how the different actions impacted the different case studies and at what step (due to which actions) each case study turned out to be the best makes it easier for the stakeholders to identify what actions should be taken to optimise the form and consequently suggest when it is appropriate to stop *cf.*, Table 2 row 5.

Table 2. ER elements and Design Opportunities satisfied.

ER elements	Design Opportunities
1. Relationship *Has*, *Computed from* and *Belongs to*, *Parameter* entity	1.a Support in the selection of meaningful data 1.b. Support during the transformation of raw data into indicators and metrics
2. `Timed` entities, relationship *Modifies*	2.b. Clarify why a certain choice impacts elements
3. *Component* entity, `Case_study` attribute	2.c. Clarify why certain choices have been made during the previous steps
4. Relationship *Modify*	3.a. Suggestion of actions based on the analysis 3.b. Simulate the effect of a choice on the map
5. Relationship *Modify* and *Component* entity, `Step` attribute	4.a. Clarify how to maximise performance 4.b. Tell the users when to stop with the simulation since performance has been maximised

7 Conclusions and Future Work

In this paper, we have discussed the effectiveness of the synergy between Human-centred Design and Conceptual Modelling, using an ER schema as a common language defining a bridge across two research groups. The novelty of this process is the human-centred perspective, which involved relevant stakeholders to i) conceptualise the domain experts' knowledge and ii) co-design the system's main features and relative requirements to be met. The process also helped the architects to rediscover and better frame their analysis processes. Moreover, the human-centred design process helped us identify inconsistencies and redundancy in the data and the analysis process as well, which we eliminated during the ER schema design. At the same time, the resulting schema includes temporal and spatial dimensions and appropriate connections between `Timed` and static entities and enables the main requirements to be met. Being the result of a multi-faceted analysis, the defined conceptual schema will be the basis for future work addressing multiple aspects, namely the definition of data integration techniques, the design of a unified knowledge base supporting decision-making, and the design of interactive digital environments supporting the analysis process.

References

1. Ahvenniemi, H., Huovila, A., Pinto-Seppä, I., Airaksinen, M.: What are the differences between sustainable and smart cities? Cities **60**, 234–245 (2017)
2. Biraghi, C.A.: Multi-scale modelling approach for urban optimization: urban compactness environmental implications. Phd thesis, Politecnico di Milano (2019)
3. Carmichael, T., Collins, A.J., Hadžikadić, M. (eds.): Complex Adaptive Systems. UCS, Springer, Cham (2019). https://doi.org/10.1007/978-3-030-20309-2
4. Engin, Z., et al.: Data-driven urban management: mapping the landscape. J. Urban Manag. **9**(2), 140–150 (2020). https://doi.org/10.1016/j.jum.2019.12.001, https://www.sciencedirect.com/science/article/pii/S2226585619301153
5. Gruninger, M.: Methodology for the design and evaluation of ontologies. In: International Joint Conference on Artificial Intelligence (1995)
6. Kotis, K., Vouros, G.A.: Human-centered ontology engineering: the HCOME methodology. Knowl. Inf. Syst. **10**, 109–131 (2006)
7. Locke, K.: Qualitative research and evaluation methods. Organ. Res. Methods **5**(3), 299 (2002)
8. Lopez, M., Gomez-Perez, A., Sierra, J., Sierra, A.: Building a chemical ontology using methontology and the ontology design environment. IEEE Intell. Syst. Appl. **14**(1), 37–46 (1999)
9. Massimo, T.: Integrated modification methodology (IMM): a phasing process for sustainable urban design. WASET World Acad. Sci. Eng. Technol. (2013)
10. Nations, U.: Agenda 2030 (2015). https://unric.org/it/agenda-2030/
11. Nowell, L.S., Norris, J.M., White, D.E., Moules, N.J.: Thematic analysis: striving to meet the trustworthiness criteria. Int J Qual Methods **16**(1), 1609406917733847 (2017)
12. Pinto, H.S., Martins, J.P.: Ontologies: how can they be built? Knowl. Inf. Syst. **6**, 441–464 (2004)
13. Piselli, C., Altan, H., Balaban, O., Kremer, P.: Innovating Strategies and Solutions for Urban Performance and Regeneration. Advances in Science, Technology & Innovation, Springer, Cham (2022). https://doi.org/10.1007/978-3-030-98187-7
14. Richards, D.: The reuse of knowledge: a user-centred approach. Int. J. Hum Comput Stud. **52**(3), 553–579 (2000)
15. Shneiderman, B.: Human-centered artificial intelligence: reliable, safe & trustworthy. Int. J. Hum. Comput. Interact. **36**(6), 495–504 (2020)
16. Tadi, M., Zadeh, M.H.M., Ogut, O.: Measuring the influence of functional proximity on environmental urban performance via integrated modification methodology: four study cases in Milan. Int. J. Urban Civ. Eng. (2020)

Using Knowledge Graphs to Generate SQL Queries from Textual Specifications

Robson A. Campêlo[1]([⊠])[iD], Alberto H. F. Laender[2][iD],
and Altigran S. da Silva[3][iD]

[1] Instituto Federal de Ciência e Tecnologia Goiano, Campos Belos, GO, Brazil
robson.campelo@ifgoiano.edu.br
[2] Departamento de Ciência da Computação, Universidade Federal de Minas Gerais,
Belo Horizonte, MG, Brazil
laender@dcc.ufmg.br
[3] Instituto de Computação, Universidade Federal do Amazonas, Manaus, AM, Brazil
alti@icomp.ufam.edu.br

Abstract. In this paper, we present a tool for querying relational DBs that uses a KG as an approach to generate SQL queries from NL specifications. In this approach, we argue that a KG representation of a relational DB schema can become an auxiliary tool in the translation process. Furthermore, we propose to automate the process of generating such a KG. Our approach to provide an NL interface for relational DBs comprises two major tasks: (1) generation of a KG from a relational DB schema and (2) translation of NL queries to SQL based on the semantics provided by the respective KG. We study the effectiveness of our approach using a benchmark dataset containing 82 NL query examples from the Spider dataset, considering the domain of Formula 1. Our approach is able to correctly translate these queries, which is verified against the expected results provided by our benchmark.

Keywords: Knowledge Graphs · Natural Language Queries · SQL Queries

1 Introduction

Natural language (NL) interfaces for databases (DBs) have attracted increasing interest due to their significant contribution in order to improve user interaction [1]. Such interfaces allow users to interact with a DB using an NL, without requiring complex commands and the knowledge of a specific query language [5,8], thus providing regular users with a more natural way to access a database.

There are several works in the literature that deal with the problem of translating NL sentences into queries in a structured query language such as SQL, by using specific approaches that can be categorized into two distinct groups. The first group concerns approaches that use mapping rules for the elements associated with the database mentioned in the query expressed in natural language.

T. P. Sales et al. (Eds.): ER 2023 Workshops, LNCS 14319, pp. 85–94, 2023.
https://doi.org/10.1007/978-3-031-47112-4_8

Approaches of this type include tools like Templar [2], NaLIR [7], ATHENA [9], ATHENA++ [10] and SQLizer [11]. The second group includes approaches centered on the automatic translation of queries by using machine learning and deep learning techniques, which have as examples tools such as DBPal [3], Type-SQL [13] and SQLNet [14].

One of the main difficulties that are inherent to the task of developing NL interfaces for querying a DB is to understand the semantics of the respective application since it is necessary to establish a mapping strategy between certain query terms and the DB schema. In addition, it is also necessary to understand the intention of the user when submitting a query. In this respect, ATHENA adopts an approach based on a domain-oriented ontology, which describes the semantics of the involved entities and the relationships between them [9]. However, the ontology construction process for ATHENA is not automated and requires the participation of an expert on the respective domain in order to define the entities involved and their respective relationships.

A concept related to ontologies, but with some distinct characteristics for knowledge representation, is that of a knowledge graph (KG). While an ontology is a broader conceptual abstraction, a KG is a more concrete way of organizing and visualizing knowledge in an interconnected way by means of a graph structure [4].

In this paper, we present an approach for querying relational DBs that uses a KG as a kind of an oracle to support the generation of SQL queries from NL specifications. In this approach, we argue that a KG representation of a relational DB schema can become an auxiliary tool in the translation process. Furthermore, we propose to automate the process of generating such a KG from a relational database schema.

The rest of this paper is organized as follows. Section 2 provides an overview of our proposed tool. Then, Sect. 3 presents a short case study of such a tool. Finally, Sect. 4 presents our conclusions and provides some insights for future work.

2   Overview of Our Proposed Approach

In this section, we present an overview of our approach to providing an NL interface for relational DBs. This approach comprises two major tasks: (1) generation of a KG from a relational DB schema and (2) translation of NL queries to SQL based on the semantics provided by the respective KG. For this, we describe some details of its implementation and present an example of the execution flow of a query in natural language.

2.1   Generation of a KG from a Relational Database Schema

The first task of our approach aims to automate the process of generating a KG from a relational DB schema. This task is fundamental for our approach since it synthesizes the basic vocabulary that will be used to automatically generate the relational queries. For achieving this, our first aim is to semantically enrich this graph based on data related to the application domain, in order to expand the knowledge representation present on it. Once the KG is built, the translation process maps the KG into DB elements, allowing the generation of SQL queries structured according to a natural language specification.

To generate a KG from a relational DB, our translation tool performs three steps. In the first step, taking its schema name as input, it submits a query to the DB to retrieve metadata related to the *Information_schema.Tables*[1]. In the second step, it queries the DB to retrieve metadata related to the *Information_schema.Columns*[2]. Then, in the third step, it organizes the metadata retrieved in the previous steps, including examples of data value from the first tuple of each table, to generate a set of RDF[3] triples in the format (*table_name, column_name, value*) in order to define the nodes and edges that comprise the structure of the KG.

It is important to highlight that in our approach the KG is not generated as a full copy of the DB. In other words, it does not need to include all records from the DB to translate the NLQs. Instead, the KG is constructed based only on the first record of each table, aiming to provide a representation of the DB schema with some example data to support the translation strategy in order to provide the semantics of the query terms. Figure 1 shows an excerpt of the KG generated for our example database, which shows a specific view of the Formula 1 race scenery involving the 2008 Monaco Grand Prix. However, due to lack of space we do not present a detailed description of the full schema from which the KG is generated and just list its tables, which are: *circuits, constructorresults, constructors, constructorstandings, drivers, driverstandings, laptimes, pitstops, qualifying, races, results, seasons and status.*

2.2 Translating NLQs to SQL

The next step of our approach aims to perform the translation of the NLQs submitted by the users, based on the KG previously generated to assist the understanding of their semantics considering the terms contained in the sentence. Figure 2 shows the architecture proposed for our NL-SQL translation tool, which is divided into four modules, as described below.

The NLP Module. This module receives the query sent by the user and processes it throughout the following steps: (1) removal of stopwords, (2) tokenization/lemmatization, (3) part-of-speech tagging, (4) filtering of relevant terms based on pos-tagging results, (5) named entity recognition, and (6) identification of dependencies between terms (dependency parsing). The Stanford NLP

---

[1] This is a view that describes the structure of all tables existing in the DB.

[2] This is a view that describes all columns present in the DB tables.

[3] Resource Description Framework (RDF) is a standard model for data interchange on the Web. It serves as data model for generating KG's, enabling the representation of structured, linked, and semantic knowledge.

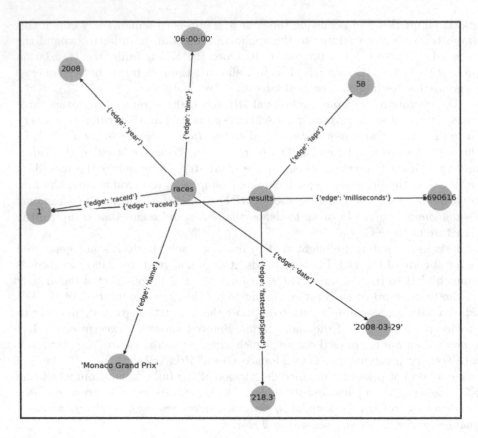

Fig. 1. Excerpt of the generated KG.

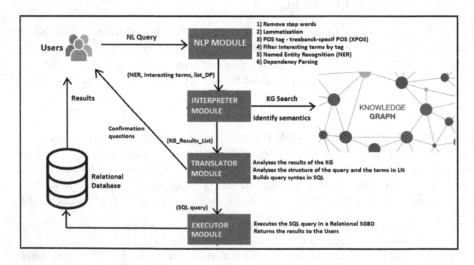

Fig. 2. Architecture of the NL-SQL translation tool.

library[4] was used to implement these techniques. Next, we describe the functioning of the most important steps:

a) Removal of Stopwords: This step consists of removing terms considered irrelevant for translation, based on the process of dividing the sentence into individual words or tokens (tokenization). It is important to highlight that some terms typically considered as stopwords in English were intentionally kept since they are of interest to the translation tool, such as "more", "most", "than", "between", "in", "after", "before", "and","or".

b) Part-of-Speech Tagging (POS Tagging): In this step, each term is labeled according to the grammatical class it belongs to using the POS tag technique. This task aims to perform a mapping from NL to Entity-Relationship (ER) components, aligning these components with the user's parts of speech. We used treebank-specific POS (XPOS) tags because they provide more specific tags, allowing for better filtering to select the terms of interest or keywords in the translation process to SQL.

c) Named Entity Recognition (NER): Here the terms found in a NL sentence are identified and named according to key entity types, allowing for greater efficiency in querying for equivalent terms in the KG (eg. QUANTITY, EVENT, PERSON, DATE, etc.).

d) Dependency Parsing: This technique is an approach used to analyze the grammatical structure of a sentence and identify dependency relationships between its words. For example, certain compound terms used by a user, such as "first name", "last name" and "fastest lap speed", are identified by this technique as being semantically related terms within the context of the user's sentence and, therefore, they should be considered together for querying the KG.

Thus, the output of the NLP Module includes, respectively, three specific lists containing named entities, terms of interest (labeled with POS tags) and terms obtained from a Dependency Parsing. This result is passed on to the Interpreter Module, which is describe next.

The Interpreter Module. This module receives the output from the NLP Module and is responsible for identifying the semantics of the terms present in the user's query, which are related to the DB schema. To achieve this, it queries the KG to find nodes and edges that are related to those natural language terms.

As described in Sect. 2.1, the KG is generated as a set of RDF triples [table_name, column_name, value]. Therefore, the query in the KG is performed by comparing the output of the NLP Module with the nodes and edges of the KG, in order to identify whether a particular term present in the user's query refers to the name of a table, the name of a column or a value existing in the KG. To perform such a kind of comparison, we used the Levenshtein Distance, which is a metric largely used for this purpose.

---

[4] Available at https://stanfordnlp.github.io/stanza/.

The Translator Module. This module receives the list of results from the Interpreter Module and performs an analysis of these results, also considering the structure of the NL query. Therefore, its purpose is to generate a syntactically correct SQL query that can be executed in a relational DBMS, taking into account a specific DB schema. Furthermore, it is essential that the generated query corresponds exactly to the user's data of interest.

To generate a simple SQL query, based on a set of translation rules, the Translator Module analyzes the list of results from the KG, identifies its table and column's names, and defines the condition values required to generate the respective SELECT-FROM-WHERE clauses. For other cases (e.g., distinct, subqueries, grouping, sorting, etc.), it is also necessary to examine the terms from the NL query, particularly those resulting from the POS Tagging, NER and Dependency Parsing processes. This module also utilizes a dictionary of expressions that maps to SQL components certain expressions found among the NL query terms. Finally, in cases where it is not possible to define a valid SQL query due to the existence of multiple possible results, the Translator Module interacts with the user using simple confirmation questions (yes/no) to clarify the response.

Although our set of translation rules was only applied to our benchmark, we argue that it is general enough and can be applied to any other dataset. This means that its rules depend only on the generated KG.

The Executor Module. This module receives an SQL query that comes from the Translator Module and executes it on a relational DBMS. In the implementation described in this paper, we used the MySQL DBMS and the PyMySQL library of the Python programming language[5], which allows establishing a connection with a DB instance by providing the following connection data: *port number*, *username*, *password*, and *DB instance name*. Additionally, it is necessary to specify the character encoding type to represent texts, for example, *utf8mb4*.

The Executor Module is also responsible for returning to the user the result of the query submitted to the relational DBMS, which is done by means of a function called *read_sql_query* from the Pandas library[6] This function takes the SQL query and the database connection data as parameters, and returns the data in the form of a DataFrame[7], which is then printed to the user.

2.3   Example of an NL Query Execution

Figure 3 exemplifies the execution flow for translating an NL query using our tool. Assuming that a user submits the query "What was the fastest lap speed

---

[5] Available at: https://pypi.org/project/pymysql/.

[6] Pandas is a Python library for data analysis and manipulation. Available at: https://pandas.pydata.org/.

[7] A two-dimensional data structure provided by the Pandas library, which is similar to a table, where the data is organized in rows and columns.

in the Monaco Grand Prix in 2008?", the NLP Module identifies the terms of interest by using the POS Tagging technique, which considers the grammatical classes adjective (JJ), superlative (JJS), noun (NN), preposition (IN), and cardinal (CD). Additionally, the NLP Module identified the related composed term "fastest lap speed" by using the Dependency Parsing technique and the terms "Monaco Grand Prix" and "2008" as named entities of types Event and Cardinal, respectively.

The Interpreter Module identifies the semantics of the terms of interest based on the results of a query submitted to the KG and verifies that the translation of that query into SQL will involve the columns *fastestLapSpeed*, *year* and *name*, which belong to the tables *Results* and *Races*. Additionally, this module also identifies that these two tables are related to each other by the column *raceId*, which allows joining them.

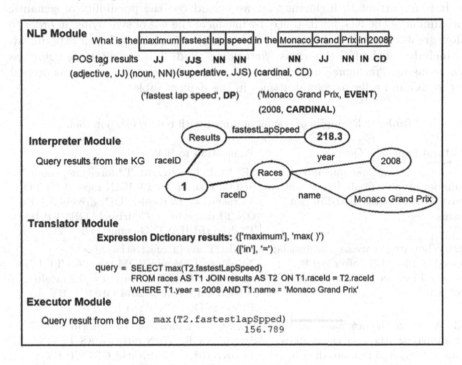

Fig. 3. Execution flow of a query.

Following the execution flow, as we can see in Fig. 3, the Translator Module uses the results from the Interpreter Module to generate the SQL query, also making use of a dictionary of expressions that we developed to map user expressions to specific SQL components. This has been done to define the use of the SQL function *max( )* and the equality operator "=" between the column and value pair in order to set the conditions for the WHERE clause. Finally, the Executor Module executes the SQL query and shows the result to the user (if it is not empty).

3   Short Case Study

In this section, we present a short case study of our tool using 82 NL query examples from the Spider dataset [12], considering the Formula 1 domain. This benchmark dataset provides, in addition to these examples, the equivalent SQL query, which represents the expected result that the translation should achieve. These queries were divided into three groups: Easy (10 queries), Intermediate (24 queries) and Difficult (48 queries), based on the level of complexity involving the SQL syntax of the operations performed, i.e., Selection with Sorting, Join+Selection with Range or Join+Aggregation+Function. Table 1 shows four examples randomly chosen from the 82 queries obtained from the translations performed by our tool, where the correctness of these queries is verified against the expected result provided by the Spider library.

It is important to highlight that we considered the possibility of semantic enrichment of the KG, for instance, by means of the use of synonyms, in order to allow greater flexibility of the tool with the users' vocabulary. In our experiment, we included in the KG the terms *"first name"* and *"last name/family name"* as synonyms for *"forname"* and *"surname"*, respectively, which are terms present in the KG and refer to column names in the *drivers* table.

Table 1. Examples of results obtained with our translation tool.

Natural Language Queries	Translation to SQL
(1) Find the id, forename and number of races of all drivers who have at least participated in two races.	SELECT T3.driverId, T3.forename, count(*) FROM results AS T1 JOIN races AS T2 ON T1.raceId = T2.raceId JOIN drivers AS T3 ON T1.driverId = T3.driverId GROUP BY T3.driverId HAVING count(*) >= 2
(2) What are the average fastest lap speed in races held after 2014 grouped by race name and ordered by year?	SELECT avg(T2.fastestLapSpeed), T1.name, T1.year FROM races AS T1 JOIN results AS T2 ON T1.raceId = T2.raceId WHERE T1.year > 2014 GROUP BY T1.name ORDER BY T1.year
(3) What are the last names and ids of all drivers who had 11 pit stops and participated in more than 5 races results?	SELECT T1.surname, T1.driverId FROM drivers AS T1 JOIN pitstops AS T2 ON T1.driverId = T2.driverId GROUP BY T1.driverId HAVING count(*) = 11 UNION SELECT T1.surname, T1.driverId FROM drivers AS T1 JOIN results AS T2 ON T1.driverId = T2.driverId GROUP BY T1.driverId HAVING count(*) > 5
(4) What is the maximum fastest lap speed in the Monaco Grand Prix in 2008?	SELECT max(T2.fastestLapSpeed) FROM races AS T1 JOIN results AS T2 ON T1.raceId = T2.raceId WHERE T1.year = 2008 AND T1.name = 'Monaco Grand Prix'

In the technological context in which this work is inserted, a tool that has currently attracted enormous attention from both the academic community and the general public is ChatGPT[8] The potential of this tool in interacting with users to provide information has been explored in various fields, among which we mention the translation of NL to SQL queries.

Thus, we compared the translation results of the queries shown in Table 1 with those obtained by ChatGPT version 3.5. To enable ChatGPT to generate accurate SQL results, we provided the DB schema to ensure a fair comparison of our tool with ChatGPT's text-to-SQL translation capabilities [6]. The results showed that out of the four queries shown in Table 1, only query 4 was correctly translated by ChatGPT. In query 2, the translation was partially correct due to the inclusion of an additional column in the GROUP BY clause. Finally, in queries 1 and 3, ChatGPT translated incorrectly. In query 1, the join with the "drivers" table was missing, and in query 3, the execution in the DBMS did not return any results.

4   Conclusions and Future Work

In this paper we presented a natural language interface tool for querying relational databases that uses a knowledge graph as an approach to automatically generate SQL queries from natural language specifications. This tool executes two major tasks: the first automates the generation of a KG from a DB schema and the second executes the NL to SQL translation. For the KG generation, we used a database schema provided by the Spider library, which contains data related to the domain of Formula 1. In our experiment, we implemented this schema in a MySQL version supported by the PyMySQL library and used our tool to generate a representative KG from this schema.

The results of our short case study show the effectiveness of our approach in understanding the semantics of NLQs, since we were able to translate 82 query examples provided by the Spider library. Our results, when compared to translations generated by the ChatGPT, also proved to be more accurate, as we observed that the ChatGPT's translation method relies more heavily on the user's sentence text to generate the SQL query, rather than on knowledge about the DB schema. Furthermore, even when providing the DB Schema to the ChatGPT, we have observed inaccuracies in some of its translation results.

As future work, we plan to expand the experiments with our tool by including the entire Spider dataset and compare its results with those generated by the ChatGPT. Our aim is to verify whether our translation results remain more accurate than those generated by the ChatGPT.

Acknowledgment. This work is supported by the author's research grants from CAPES and CNPq. The authors would like to thank the reviewers for their comments on this work.

---

[8] Available at: https://openai.com/blog/chatgpt.

References

1. Affolter, K., Stockinger, K., Bernstein, A.: A comparative survey of recent natural language interfaces for databases. VLDB J. 793–919 (2019). https://doi.org/10. 48550/arXiv.1906.08990
2. Baik, C., Jagadish, H.V., Li, Y.: Bridging the semantic gap with SQL query logs in natural language interfaces to databases. In: Proceedings of the IEEE 35th International Conference on Data Engineering (ICDE), pp. 374–385 (2019). https:// doi.org/10.48550/arXiv.1902.00031
3. Basik, F., et al.: DBPal: a learned NL-interface for databases. In: Proceedings of the 2018 International Conference on Management of Data, pp. 1765–1768 (2018). https://doi.org/10.1145/3183713.3193562
4. Hogan, A., et al.: Knowledge graphs. ACM Comput. Surv. (Csur) 1–37 (2021). https://doi.org/10.1145/3447772
5. Kim, H., So, B., Han, W., Lee, H.: Natural language to SQL: where are we today?. Proc. VLDB Endow. 13(10), 1737–1750 (2020). https://doi.org/10.14778/3401960. 3401970
6. Liu, A., Hu, X., Wen, L., Yu, Philip, S.: A Comprehensive evaluation of ChatGPT's zero-shot Text-to-SQL capability. arXiv preprint arXiv:2303.13547 (2023). https:// doi.org/10.48550/arXiv.2303.13547
7. Li, F., Jagadish, H.V.: Constructing an interactive natural language interface for relational databases. Proc. VLDB Endow. 8(1), 73–84 (2014)
8. Quamar, A., Efthymiou, V., Lei, C., Özcan, F.: Natural language interfaces to data. Found. Trends Databases 319–414 (2022). https://doi.org/10.48550/arXiv. 2212.13074
9. Saha, D., Floratou, A., Sankaranarayanan, K., Minhas, U.F., Mittal, A.R., Özcan, F.: ATHENA: an ontology-driven system for natural language querying over relational data stores. Proc. VLDB Endow. 9(12), 1209–1220 (2016). https://doi.org/ 10.14778/2994509.2994536
10. Sen, J., et al.: Athena++: natural language querying for complex nested SQL queries. Proc. VLDB Endow. 13(12), 2747–2759 (2020). https://doi.org/10.14778/ 3407790.3407858
11. Yaghmazadeh, N., Wang, Y., Dillig, I., Dillig, T.,: SQLizer: query synthesis from natural language. Proc. ACM Program. Lang. 1(OOPSLA), 1–26 (2017). https:// doi.org/10.1145/3133887
12. Yu, T., et al.: Spider: a large-scale human-labeled dataset for complex and cross-domain semantic parsing and text-to-SQL task. arXiv: Computation and Language (2018). https://doi.org/10.18653/v1/D18-1425
13. Yu, T., Li, Z., Zhang, Z., Zhang, R., Radev, D.: TypeSQL Knowledge-based type-aware neural text-to-SQL generation. arXiv preprint (2018). https://doi.org/10. 48550/arXiv.1804.09769
14. Xu, X., Liu, C., Song, D.: SQLNet generating structured queries from natural language without reinforcement learning. arXiv preprint (2017). https://doi.org/ 10.48550/arXiv.1711.04436

A Systematic Approach to Generate TOGAF Artifacts Founded on Multiple Data Sources and Ontology

Sérgio Guerreiro[1,2,3]($\boxtimes$)(iD) and Pedro Sousa[1,2,3](iD)

[1] Link Consulting SA, Av. Duque de Ávila 23, 1000-138 Lisbon, Portugal
{sergio.guerreiro,pedro.sousa}@linkconsulting.pt
[2] INESC-ID, R. Alves Redol 9, 1000-029 Lisbon, Portugal
[3] Instituto Superior Técnico, University of Lisbon, Av. Rovisco Pais 1, 1049-001 Lisbon, Portugal
{sergio.guerreiro,pedro.manuel.sousa}@tecnico.ulisboa.pt

Abstract. Enterprise Architecture (EA) standards are widely used in industry and recognized as a *lingua franca* between project' stakeholders. However, standards are most of times very extensive, requiring specialized knowledge and consuming a large effort if instantiated manually. Furthermore, the risk of producing inconsistent artifacts increases with project' complexity, *e.g.*, changing the name of a business actor requires propagation to all artifacts using that entity. Moreover, EA generation consumes data from multiple sources, *e.g.*, excel or BPMN files, that need to be normalized, classified, and consistently referenced in the artifacts. This paper proposes a systematic approach where the conceptual understanding of a project is shared using an ontology which in turn supports the entire EA artifacts automatic generation. Results show that there are no similar solution available in the literature. In addition, the usage of our systematic approach in four different EA projects evidences a bounded linear increase in effort as artifacts increase in complexity.

Keywords: Artifact · Concept · Document · Enterprise Architecture · Ontology · TOGAF

1 Introduction

Enterprise Architecture (EA) is defined by TOGAF [7] as *"the structure of components, their inter-relationships, and the principles and guidelines governing their design and evolution over time"*. More recently, as noticed by [17], more thorough definitions can be offered, as the one proposed by Greefhorst and Proper [6] that consider three perspectives for architecture: *"regulation-oriented, design-oriented, and knowledge-oriented, where the first corresponds to the prescriptive perspective, the second corresponds to the descriptive perspective, and the third corresponding to the high-level design decisions of the system"*. The EA way of working are proposed by many standards that are widely used in

T. P. Sales et al. (Eds.): ER 2023 Workshops, LNCS 14319, pp. 95–106, 2023.
https://doi.org/10.1007/978-3-031-47112-4_9

industry, *e.g.* TOGAF, DODAF, MODAF, *etc.*, and are recognized as a *lingua franca* between clients, suppliers, consultants and/or researchers. However, these standards are most of times very extensive, requiring specialized knowledge and consuming large amounts of time and effort to be instantiated. In practice, large EA projects are expected to generate, and maintain, complex artifacts (bottom part of Fig. 1) that require a large effort if they are produced manually. Furthermore, the risk of inconsistencies between artifacts increases with the dimension of projects, *e.g.*, a change in a single business actor requires propagation to all artifacts that are using that entity to enable traceability. In the top of this problem, EA consume data from multiple sources (top part of Fig. 1), *e.g.*, data sources files with modelling elements of the business, application, technology, motivation, or even simple spreadsheets; which need to be normalized, classified, and then, referenced on the artifacts.

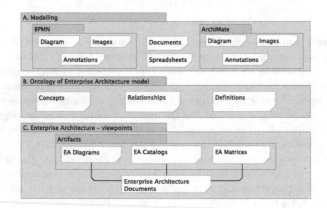

Fig. 1. Conceptual layers to construct EA viewpoints, an example using TOGAF standard.

From a different perspective, an ontology represents the fundamental **concepts**, **relationships** and **definitions** that are used in a specific application domain [8]. It targets the facilitation and dissemination of its understanding between stakeholders with different interpretations of the same reality. Moreover, an ontology is the result of stakeholder's discussions reaching a consensual agreement between them [11]. We aim at integrating ontology representation techniques with EA artifacts construction, where ontology is used as a scale faithful to guide all the concepts that need to be included during EA generation and maintenance (middle part of Fig. 1). In specific, the research question (RQ) addressed by this paper is: *"How to generate, and maintain, the consistency of Enterprise Architecture artifacts in a complex organizational model, with a bounded linear increase in effort as TOGAF artifacts increase in complexity?"*. In short, this paper proposes a systematic approach where the conceptual understanding of a project is shared using an ontology, which is used to classify and to normalize the received data streams, and finally, to automatically generate artifacts that are referenced and synchronized by that understanding. The benefit is to obtain a bounded linear[1] increase in the project' effort as artifacts increase in complexity. TOGAF [7] is used because is a well known standard in the industry, notwithstanding any other standard can be used in our approach. This document is organized as follows. In Sect. 2 the concepts used are introduced and the proposed solution is explained. Afterwards, Sect. 3 discusses the results obtained with the application of this solution in a large EA project. Section 4 identifies, in the literature, the alternative proposals to integrate EA and ontology, and compares them with our proposal. Finally, Sect. 5 concludes the paper and identifies future work.

2   Generation of EA Artifacts Founded on Ontology

This section details the steps proposed in the systematic approach to generate, and maintain, the EA artifacts. It is founded on the data collected from multiple data sources, and on ontology to normalize, classify and relate that data. As depicted in Fig. 1 the collected data relies on previous modelling effort, *e.g.*, business processes using BPMN or any type of available document about the organization' reality.

An **Enterprise Model** is a direct graph $G_t = (A_t, R_t)$ being A a set of artifacts, R a set of relationships, and t is a discrete variable representing time [17]. A **Viewpoint** is *"a specification of the conventions for constructing and using a view; a pattern or template from which to develop individual views by establishing the purposes and audience for a view and the techniques for its creation and analysis"* [9]. An **ontology** is the representation of the essential understanding of a given domain, consisting is a body of knowledge that is recognized by all the stakeholders involved in a EA project, and also considered as the true source for all the artifacts' instantiations [3]. A **project** holds a set of architectural statements that change artefacts states after its successful completion [17].

We use TOGAF 9.2 [7] to illustrate the approach, however, any other representation language can be considered. TOGAF defines that a document is composed by multiple Artifacts [14], and an Artifact is composed by multiple Concepts (*cf.* depicted in Fig. 2). A TOGAF concept is thus the finer grained element requiring common understanding between the stakeholders involved in the EA artifacts generation: suppliers, clients, architects, project managers, *etc.*. The meaning of each TOGAF concept need to be discussed prior to the artifacts' generation in order to align the deliveries expectations and to avoid misunderstandings or biased interpretations.

The solution encompasses six sequential steps, which are described in each one of the following subsections. Next section, the validation, describes how these steps were instantiated in four different EA projects.

---

[1] bounded linear follows the definition from [2]: *"normalized systems are information systems that are stable with respect to a defined set of anticipated changes, which requires that a bounded set of those changes results in a bounded amount of impacts to system primitives".*

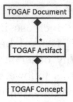

Fig. 2. Decomposition of a TOGAF Document into finer concepts.

2.1   The EA Document Elicitation

The EA document elicitation step must comply to the organization requirements in order to capture correctly its *as-is* or *to-be* situation. By its turn,
related literature refers that organization requirements heavily depend on the
fit for purpose [1,15] of the EA initiative. Therefore, whether the purpose is to
create a new architecture, or planning the transition between architectures, or to
decommission the current architecture, then different documents can be elicited.
Figure 3 depicts an excerpt of the TOGAF documents that are recommended
for each one of the TOGAF Architecture Development Method phase [7], where
each phase concerns a specific purpose, which can be used as a baseline for any
project.

Deliverable	Output from...	Input to...
Architecture Building Blocks (see 32.2.1 Architecture Building Blocks)	F, H	A, B, C, D, E
Architecture Contract (see 32.2.2 Architecture Contract)	-	-
Architecture Definition Document	B, C, D, E, F	C, D, E, F, G, H

Fig. 3. An excerpt of the prescribed documents by TOGAF organized by their location
in the Architecture Development Method [7].

2.2   The EA Artifacts Elicitation

Considering TOGAF as the body of knowledge for EA artifacts, they are
divided into: *diagrams*, *catalogs* and *matrices*, and then, classified by type of target architeture (Business, Data, Application, Technology, Preliminary, Vision,
Requirements Management and Opportunities and Solutions). Knowing that
each EA document contains multiple artifacts, and that the previous step 2.1
already elicited the desired document, the decision is straightforward if TOGAF
template documents are used. Each TOGAF template document suggests what
are the artifacts that should be included. Therefore, this is a stable source to
identify which are the artifacts to be elicited for each document. If needed, any
other artifact can be added. Furthermore, as long as consistency is guaranteed,
an artifact can be placed in more than one document.

Alternatively, the fit for purpose principle can be further extended to elicit
what are the exact artifacts that should be produced avoiding eventual biased
effects of an EA standard. In this circumstances, each EA artifact should be given
a specific and well-defined purpose.

2.3   The EA Concepts Elicitation

This step is the most relevant in terms of EA generation, encompassing the capability to normalize and classify the data that is going to be extracted using an ontology definition. An ontology comprises a set of concepts, relationships and definitions that are stablished for each EA project and that are only valid within that scope. A first iteration to create the ontology is to identify all the concepts that are required in each one of the selected artifacts from step 2.2, and then, map each concept with a known data source. For instance, if the *"Value Stream Catalog"* from Business Architecture has been selected for generation, and there is an ArchiMate model available containing the *"Value"* and *"Business Actor"* elements, then both ArchiMate elements should be considered as key concepts in the ontology to generate that artifact. This is a straightforward approach, but if done iteratively, with validation milestones, has the possibility to converge for a robust ontology. If the project' stakeholders involved are connoisseurs of the domain, then other option is to bootstrap the ontology from that existing knowledge, requesting them to express all the key concepts and design the ontology from scratch. Either way, a table with concepts and definitions, could be produced with the content similar to that exemplified in Table 1. Relationships are inherited from the data source, *e.g.*, the sequence flows existing between tasks of a BPMN model.

Table 1. An illustration of concepts definition.

The concept name	The agreed concept definition	The data source where the concept can be extracted	An optional alias to other concept
Actor	The intervenient of a business process	Excel File	
Pool	The intervenient of a business process	BPMN file	Actor
...	...	...	...

2.4   Concepts Mapped with Artifacts and Documents

Table 2 exemplifies the mapping between concepts, artifacts and documents. Let us consider a universe of 3 concepts: X, Y and Z; 6 artifacts, namely: 3 diagrams, 1 catalog and 2 matrices; and 3 documents. Diagram 1 and 2 only use the concept X, while diagram 3 requires concepts Z and Y. Catalog 1 requires the concept Z. Matrix 1 requires the concepts Y and Z, while Matrix 2 requires the concepts X and Y. Then, document 1 requires the following artifacts: diagram 3 and matrix 2. Document 2 requires the following artifacts: catalog 1 and matrix 1. Finally, document 3 includes all the available artifacts.

Table 2. Example of mapping the concepts, artifacts and documents.

		Diagrams			Catalogs	Matrices		
Artifacts	Diagram 1	Diagram 2	Diagram 3		Catalog 1	Matrix 1		Matrix 2	
Concepts	X	X	Z	Y	Z	Y	Z	X	Y
Documents — Document1			•	•				•	•
Documents — Document2					•	•	•		
Documents — Document3	•	•	•	•	•	•	•	•	•

2.5 Concepts Extraction from Multiple Data Sources

After the previous steps consisting in the ontological construction, a software tool to consume data from the multiple data sources and to compute it, is preferrable than processing manually. Each data source needs to be normalized, we suggest adding annotations, where each piece of data is classified into a single concept, and then stored in a repository. Therefore, the classification part of this step is currently human dependent. AI techniques can be considered as future improvement for this step.

2.6 Automatic Documents Generation

After provisioning all data annotated in the repository, each document is programmed with a definition of the artifacts required. By its turn, each artifact is constructed with the class of concept and relationships required at each position in a viewpoint. A language for representation need to be chosen here, *e.g.*, BPMN, ArchiMate, or any other; and the tool need to have visualization support for that language. Finally, the documents are automatically generated iteratively selecting the instances of each concept class from repository data, and filling them in the artifacts. This a very repeating task if done manually.

3   Validation

The systematic approach is validated using *(i)* explanations and *(ii)* results' discussion based on the execution of four EA consultancy projects as listed in Table 3. Each project is characterized by application domain, project' purpose, approximate time employed (for comparison consider that similar man/month allocation was used), and then a set of approximate metrics to understand the complexity involved, namely, the number of concepts found, the data sources, the number of documents produced and the number of artifacts produced.

The EA document elicitation: distinct elicitation actions were performed. The exact document list to produce in *Project 1* was defined since the initial specifications. In *Project 2* and *3* the documents were proposed (and accepted by the client) as an initiative of the consultancy team. While in *Project 4* the initial high-level specifications were used to specify a document list baseline that required adjustment during project execution time.

Table 3. Sample of projects where the systematic approach has been applied.

Project ID	Application Domain	Purpose of the EA generation	Approximate time employed	Quantity of Concepts	Quantity of Data Sources and format	Quantity of Documents	Quantity of Artifacts
1	Public administration	IT systems' portfolio to support financial services	6 months	25	5 in excel and BPMN	7	50
2	Public banking	IT services dependencies to increase its resilience	<6 months	50	1 in workshop	2	9
3	Private banking	The application components portfolio to align business and technology	<6 months	6	2 in workshop and word	2	5
4	Justice	IT technologies' portfolio to normalize usage in suborganizations	>1 year	25	20 in excel	3	50

The EA Artifacts elicitation: is a finer definition when compared with documents. In *Project 1, 2* and *3* artifacts were elicited by the consultancy team at project kick-off, while *Project 4* defined them from specifications some time after project start. Again iterative approach is also possible in complex projects where the specification is not closed at project' beginning.

The EA Concepts elicitation: has been designed by the internal development team in all projects, using abstraction of the provided documentation and by organizing meetings to elicit the core concepts. Communication is of uptmost importance at this stage.

Concepts mapped with Artifacts and Documents: for a better visualization, a full example of mapping *Concept → Artifact → Document* that has been used in one of the projects is depicted in Fig. 4. It is noticed that concepts are repeated in many artifacts increasing the need to be consistent in their usage and being able to trace the concept' changes in all artifacts. Moreover, the 3 types of artifacts (Diagram, catalogs and matrices) refer to the same concepts. It is also usual that some artifacts are specialized referring to a small set of artifacts, while other refer many of them (or even all of them).

Concepts extraction from multiple data sources and automatic documents generation: concepts from *project 1* were extracted from BPMN sources, and excel files containing BPMN annotations. *Project 2* and *4* used excel files as data sources. In *Project 3*, due to the fact that the organization is in an initial maturity phase, the data source extraction was skipped and the available data was injected directly in the concepts. For all projects, documents were produced using the ATLAS[2] tool. The artifacts (to be included in the documents) were produced using the notation of ArchiMate and BPMN for the diagrams, and excel files for the catalogs and matrices.

Concerning the RQ posed in introduction, we propose to generate, and maintain, the EA using a core ontology of *Concept → Artifact → Document* that requires alignment between the project specifications, the organization' client expectations and the consultancy team. This alignment is produced by a matrix defined with the client' involvement as examplified in Fig. 4, and then programmed once in the ATLAS tool. Whenever the data sources change the

---

[2] Available for consultation at https://atlas.linkconsulting.com.

Documents: vii. Architecture Roadmap	vi. Architecture Requirements Specification	v. Architecture Definition Document	iv. Architecture Vision	iii. Business Principles, Business Goals, and Business Drivers	ii. Architecture Repository	i. Architecture Principles	Concepts	Artefacts	
		x					Business Process	i. Business Model diagram*	Diagrams
		x					Business Process	ii. Value Chain diagram	
		x					Value		
		x					Application	iii. Solution Concept diagram	
		x					Actor		
		x					Objective		
		x					Requirement		
		x					Business Capability	v. Business Capability Map	
		x					Business Process	vi. Value Stream Map	
		x					Value		
		x					Business Process	vii. Business Footprint diagram	
		x					Goal		
		x					Actor		
		x					Application		
		x					Business Object	viii. Business Service/Information diagram;	
		x					Business Service		
		x					Business Process	ix. Functional Decomposition diagram	
		x					Business Service	x. Product Lifecycle diagram*	
		x					Product		
		x		x			Goal	xi. Goal/Objective/Service diagram	
		x		x			Objective		
		x					Business Service		
		x					Business Service	xii. Business Use-Case diagram	
		x					Actor		
		x	x				Actor	xiii. Organization Decomposition diagram	
		x	x				Role		
		x	x				Location		
		x					Business Process	xiv. Process Flow diagram	
		x					Business Event	xv. Event diagram	
		x					Actor		
		x					Business Process		
		x					Business Object	xvi. Conceptual Data diagram	
		x					Data Object	xvii. Logical Data diagram	
		x					Actor	xviii. Data Security diagram	
		x					Business Object		
		x					Application	xix. Application Communication diagram	
		x					Application Service		
		x					Application Service	xx. Application Use-Case diagram	
		x					Actor		
		x					Business Object		
		x					Objective	xxi. Benefits diagram*	
		x					Outcome		
		x		x		x	Principle	i. Principles catalog	Catalogs
		x					Business Process	ii. Value Stream catalog	
		x					Value		
		x					Business Capability	iii. Business Capabilities catalog	
		x					Business Process	iv. Value Stream Stages catalog	
		x					Value		
		x	x				Actor	v. Organization/Actor catalog	
		x	x				Organization Unit		
		x		x			Goal	vi. Driver/Goal/Objective catalog	
		x		x			Objective		
		x		x			Driver		
		x					Role	vii. Role catalog	
		x					Business Service	viii. Business Service/Function catalog	
		x					Business Process		
		x	x				Location	ix. Location catalog	
		x					Business Process	x. Process/Event/Control/Product catalog*	
		x					Business Event		
		x					Product		
		x					Business Service	xi. Contract/Measure catalog	
		x					Application Service		
	x	x					Business Service Contract		
	x	x					Application Service Contract		
		x	x				Application	xii. Application Portfolio catalog	
		x					Application Service	xiii. Interface catalog	
		x					Actor	i. Actor/Role matrix	Matrices
		x					Role		
		x					Application	ii. Application/Data matrix	
		x					Data Object		
		x	x				Application	iii. Application/Function matrix	
		x	x				Principle		
		x	x				Business Process		

Fig. 4. Example of mapping concepts with artifacts and documents to a given domain.

documentation is recreate again. Therefore, ensuring consistency between all artifacts.

Some questions still remain in this alignment process: *Which documents are relevant? Considering some selected documents, what are the artifacts that are recommended? Are they enough, or are others needed? Considering the available project' concepts, which one can be used within each artifact?* Furthermore, we noticed that keeping the ontology as simple as possible, with many alias, is easier for human understanding and discussion. Visualization has also been identified as a positive asset to understand the ontology.

4 Related Work

This section reviews the knowledge available in the literature regarding EA and ontology topics. To that end, a search has been conducted using Web of Science, Google Scholar, IEEE and ACM databases, considering the following topic searches: *"enterprise architecture" AND "ontology"*, until 2022 (included). The returned number of hits for each database is N = {170, 33, 93, 115}, with a total number of 378 papers, where 2 are technical reports, 13 are books, 1 are chapters in books, 288 are conference papers, 72 are journal articles, and 2 are miscelaneous. The yearly distribution reveals a slight decrease in number of publications in recent years when compared with the period: 2007...2018. Yet, many contributions are noticed by this search. The following procedure for the literature review has been used. Firstly, the references are collected from each database, then the duplicates are excluded. Secondly, this selection of papers are screened by title and abstract, where the most referred terms are used to score each paper in a scale [0...1]. The considered inclusion score for title is the top 30% and for abstract is the top 5%. A final collection of 48 papers is considered to a fine-grained analysis. The most relevant and referred concepts, by descending order, in the 48 abstracts are: *"enterprise; architecture; models; information; EA; business; management; Ontology; ontologies; process; support; order; framework; modeling; elements; development; role; representation; environment; integration; issues; case; study; knowledge; language; context; concepts; ontological; problem; implementation; holistic; system; tool; maintain; view; engineering; technologies; description; formal; part; communication; components; goal; data; functions; specific; application; benefits; consistent; integrated; domain; sources; comprehensive; evaluate; artifacts; rules; changes; terms; ontology-based; challenges"*. While the concerns related with architecture exist in this set, also the ontologies, languages, concepts, are present; and the aspects of integrating its concepts along with the consistency. Table 4 organizes the selected papers by interest core[3]. The most referred interest core is located in EA, while a smaller amount is located in Ontology. 10 papers uses a specific ontology: the enterprise ontology. Enterprise and Knowledge Engineering have a minority interest. However, Enterprise Engineering is represented in this search with books, which is not directly comparable with papers. Similarly, with our systematic approach,

---

[3] The full citations are not included in this paper due to length limitations.

Table 4. Distribution of papers by interest core classification.

Interest core	Number of papers found
Enterprise Architecture	43
Ontology	20
Enterprise Ontology	10
Enterprise Engineering	4
Knowledge Engineering	1

18 papers are both interest in EA and ontologies, which is a strong indicator about the community interest in this integration. A large diversity of domain are identified in the literature, what corroborates our position that each specific project requires an own ontology: Services computing, Security and/or Privacy, Goal modelling, Repositories, Access management, Concern modelling, Systems' Integration, Government, Semantic Architecture, Libraries, Semantic web, Decision making, and Governance. A different approach than ours, is presented in [11,12] that propose an integration between an ontology of the business terms with an ontology for EA components and EA relationships, aiming a common understanding between humans and systems to support integrations in enterprises and collaborations between enterprises. However, no practical applications were attempted. [13,18] propose ontologies to extract knowledge from EA models, what is the opposite that we are proposing. Similarly with our proposal, [5,10,14,16] present approaches to generate EA using an EO ontology: DEMO [3] or OWL. However, prior knowledge to these ontologies is required, and few validation in industry exist. The common understanding between stakeholders, as referred as an hard requirement in our approach is also corroborated by [4] that exemplifies how to construct it using a wiki tool.

5   Conclusion and Future Work

The uncontrolled effort that is required to manually construct artifacts in complex EA projects triggered the need for this research. It is impracticable to start an EA project without knowing the exact effort that is required. Using a metaphor, it is the same as designing a building without knowing the associated cost growth factors. An initial review of the related literature showed us that no full solution is available for this problem. Therefore, we researched an approach and tested it in practice, with four EA complex consultancy projects, to show its usefulness. As reported, our systematic approach evidences a linear, and thus controlled, increase in effort as TOGAF artifacts increase in complexity. The positive impact of our approach applied to a specific domain is given by the following aspects: (i) the network of dependencies between architectural concepts is explicitly presented to all project' stakeholders, (ii) the unique identification of each concept instance allows navigation between them, and thus, the traceability of the dependencies between concepts is easier, and *(iii)* the alignment between architectural layers can generate complementary views of the same model, and thus achieve different documentation purposes.

As future work, we identify the following threads: application of the approach to more case studies or industrial projects to further explore its social and technical implications, and automate the extraction of concepts types from the multiple data sources using AI techniques assisted by human classifications.

Acknowledgement. This work was supported by a pre-registered project, named as eProcess, which is under national funds with reference C669314338-00003137 (LINK CONSULTING - TECNOLOGIAS DE INFORMAÇÃO S.A.).

References

1. Currie, A.: From models-as-fictions to models-as-tools (2017)
2. De Bruyn, P., Mannaert, H., Verelst, J., Huysmans, P.: Enabling normalized systems in practice-exploring a modeling approach. Bus. Inf. Syst. Eng. **60**, 55–67 (2018)
3. Dietz, J.L., Mulder, H.B.: Enterprise Ontology: A Human-Centric Approach to Understanding the Essence of Organisation. Springer, Cham (2020). https://doi.org/10.1007/978-3-030-38854-6
4. Fuchs-Kittowski, F., Faust, D.: The semantic architecture tool (SemAT) for collaborative enterprise architecture development. In: Briggs, R.O., Antunes, P., de Vreede, G.-J., Read, A.S. (eds.) CRIWG 2008. LNCS, vol. 5411, pp. 151–163. Springer, Heidelberg (2008). https://doi.org/10.1007/978-3-540-92831-7_13
5. Gomes, B., Vasconcelos, A., Sousa, P.: An enterprise ontology approach for defining enterprise information architecture. In: Proceedings of the 5th ECIME, pp. 609–615 (2011)
6. Greefhorst, D., Proper, E.: Architecture Principles. Theory and Practice. Springer, Berlin, Heidelberg (2010). https://doi.org/10.1007/978-3-642-20279-7
7. Group, O.: Togaf® standard, version 9.2 (2023). Accessed 18 Apr 2023
8. Guerreiro, S.: Conceptualizing on dynamically stable business processes operation: a literature review on existing concepts. BPMJ **27**(1), 24–54 (2021)
9. ISO/IEC/IEEE: Systems and software engineering - architecture description 42010:2011 (2011). Accessed 19 Apr 2023
10. Jabloun, M., Sayeb, Y., Ben Ghezala, H.: Enterprise ontology oriented competence: a support for enterprise architecture. In: 2013 3rd ISKO-MAGHREB. ISKO, IEEE, 345 E 47th St, New York, NY, 10017, USA (2013)
11. Kang, D., Lee, J., Choi, S., Kim, K.: An ontology-based enterprise architecture. Expert Syst. Appl. **37**(2), 1456–1464 (2010)
12. Minaei, B.: Enterprise Architecture Development Based on Enterprise Ontology. academia.edu (2013). query date: 2022-11-03 10:52:59
13. Montecchiari, D.: Ontology-based validation of enterprise architecture principles in enterprise models. In: BIR Workshops (2021). http://ceur-ws.org/Vol-2991/paper17.pdf, query date: 2022-11-03 10:52:59
14. Nogueira Santos, F.J., Santoro, F.M., Cappelli, C.: Addressing crosscutting concerns in enterprise architecture. In: IEEE 18th EDOC, pp. 254–263 (2014)
15. Parker, W.S.: Model evaluation: an adequacy-for-purpose view. Philos. Sci. **87**(3), 457–477 (2020)

16. Silva, N., Mira da Silva, M., Sousa, P.M.M.V.A.D.: Modelling the evolution of enterprise architectures using ontologies. In: 2017 IEEE 19th CBI, vol. 01, pp. 79–88 (2017)
17. Sousa, P., Vasconcelos, A.: Enterprise Architecture and Cartography: From Practice to Theory; From Representation to Design. Springer, Cham (2022). https://doi.org/10.1007/978-3-030-96264-7D
18. Sunkle, S., Kulkarni, V., Roychoudhury, S.: Analyzing enterprise models using enterprise architecture-based ontology. In: Moreira, A., Schätz, B., Gray, J., Vallecillo, A., Clarke, P. (eds.) MODELS 2013. LNCS, vol. 8107, pp. 622–638. Springer, Heidelberg (2013). https://doi.org/10.1007/978-3-642-41533-3_38

Bridging the Gap: Conceptual Modeling and Machine Learning for Web Portals

Dadhichi Shukla[1]([✉]), Eugen Lindorfer[1], Sebastian Eresheim[2], and Alexander Buchelt[2]

[1] STRG GmbH, Vienna, Austria
{Dadhichi.Shukla,Eugen.Lindorfer}@strg.at
[2] UAS Saint Poelten, Saint Pölten, Austria
{Sebastian.Eresheim,Alexander.Buchelt}@fhstp.ac.at

Abstract. In recent years e-commerce has become a significant part of the economy by turning into a multi-billion dollar industry. Modern web-tools and services enable business owners to create online portals within a short period. In most cases, the business owners design their online shops based on existing user journey templates offered by webpage service providers. As a result, it limits their analysis of the user journey of their shop portal prior to the launch (go online). It becomes extremely challenging to understand what the experience of the visitors following the template user journey could be. In this paper, we propose a system to leverage the fundamentals of conceptual modeling and reinforcement learning to model web portals and analyze their structure. We employ two reinforcement learning methods, namely, Q-learning and SARSA, to train agents and navigate in a simulated environment representing the web portal. The paper is an attempt at creating a bridge between conceptual modeling and reinforcement learning by taking an empirical study approach.

Keywords: Conceptual modeling · Reinforcement learning · Agents · Online Shop Simulation · Web Portals · Empirical Study

1 Introduction

The field of conceptual modeling has vastly advanced in understanding complex system by applying levels of abstraction. It facilitates fragmenting a system with complex functions in a way to make it human-interpretable. Despite the well-known merits associated with conceptual modeling (CM), it has yet to see extensive adoption in the field of machine learning (ML). Both of the domains have seen limited overlap with each other. It is perhaps due to the applicative nature of modern-day machine learning methodologies.

Today, machine learning is used in every aspect of our digital presence. Thanks to the vast amount of digital data available, the ML models can be trained to identify user patterns while browsing the net. The training data consists of user interactions such as tracking data, sharing pictures, likes and comments on articles, viewed videos, online shopping, etc. The ML models attempt to understand patterns of user behavior, primarily used for personalized content recommendation.

T. P. Sales et al. (Eds.): ER 2023 Workshops, LNCS 14319, pp. 107–116, 2023.
https://doi.org/10.1007/978-3-031-47112-4_10

Works by Terragni et. al [17] and Sarirah et al. [14] propose approaches by apply process mining techniques to web server logs of user journey in order to investigate user behaviors, identify clusters of different users, and improve user journey. In light of online portals adhering to particular templates for the user journey (or sitemap) during webpage design, an interesting question arises: How can we analyze a website's user journey prior to its launch? Given that web server logs help identify various user behaviors, there is no-way to gather real user data prior to the launch. To address the question, we propose to employ reinforcement learning (RL), a subfield of machine learning, to analyze complex web portals, particularly online shopping portals, characterized by a multitude of interconnected pages.

We can utilize fundamentals of conceptual modelling to describe specific objectives to describe web portals. Successively, implement RL techniques, incorporating the objectives, and then performing an empirical study. We designed a simulated environment consisting of nodes and edges representing an online shopping portal. The RL agents were trained to navigate through the environment. One of the challenges in the analysis of web portals is identifying if there is a correlation with users online behavior and user journey. However, user's online behavior is complex in nature comprising various web actions. The forthcoming sections explain the ability of our current approach to emulate user behaviors.

Next, Sect. 2 describes various research efforts aiming to bridge the two vast domains of conceptual modeling and machine learning. Additionally, we highlight papers applying reinforcement learning methods for web-based challenges. In Sect. 3, we explain our framework to model web portals and how to apply reinforcement learning to the analysis of the portals. Section 4 illustrates results from our experiments, where an agent learns to navigate the simulated environment. Lastly, in Sect. 5 we present our concluding arguments, emphasizing how conceptual modeling and machine learning can complement each other.

2 Related Work

Maass et al. [7] provide an extensive study addressing how conceptual modeling and machine learning can support each other in the development of better solutions. Within the context of the proposed paper, we attempt to address the above by using virtual RL agents to navigate a simulated online store to emulate user behavior. The agent's behavior is established based on RL methods such as Q-learning [18] and SARSA (State-action-reward-state-action) algorithm [13].

Previous research has applied RL in online environments where agents operate on human-created artifacts. Shi et al. [16] presented the World of Bits (WoB) platform, where agents accomplish tasks on an online page by simulating actions such as key presses and mouse clicks. Other study proposed by Gur et al. [6] was along the same lines, where an agent learns to navigate a webpage. A RL agent learns to solve the tasks by mastering a set of primitive skills, such as filling up appropriate text in a form or deciding a date, and combining those to form and solve complex, compositional tasks such as booking an airline ticket, logging in to a website, etc. The key similarity of our work and the aforementioned studies is the modeling of web portals and having a RL agent to perform various tasks. It aligns with the fundamentals of: (i) conceptual modeling, by having a web portal model to describe aspects of the real-world for the purpose of simplifying its understanding, and (ii) reinforcement learning, by implementing an agent to perform real-world tasks, analyzes of empirical results, and drawing conclusions about the modeled web portals.

Often, it appears the fields of conceptual modeling and machine learning are weakly connected [8], however, numerous research works has exhibited otherwise. A RL-based tool called MORTAL proposed by Yuan et al. [19] to learn a relational schema by interacting with a relational database management system (RDBMS). In a recent study, Bork et al. [3] focused on systematic mapping between the two interdisciplinary fields i.e., conceptual modeling (CM) and artificial intelligence (AI), and the advantages of strategically utilizing one another. Reinhard et al. [12] propose a conceptual model for interactive labeling with the human-in-the-loop method using RL.

A study by Feltus et al. [5] investigated the role of AI in helping with domain conceptualization. The authors provide a review of symbolic and subsymbolic AI techniques. Reinforcement learning is categorized as a subsymbolic approach since the agent learns to act and react in an environment with the objective of solving a task by accumulating rewards. Conceptual modeling of web portals, intertwined with training RL agents to perform human-style tasks, enables a new way of analyzing user journeys on webpages.

3   Modeling

If we look at nature, we will observe animals explore and interact with their environment to survive by traveling around, gathering food, etc. Similarly, the paradigm of reinforcement learning captures an agent's behavior in the environment in search of a reward. Barreto et al. [1] provide an extensive explanation of model-based and model-free agents. The authors propose an algorithmic model that strikes a balance between model-free and model-based approaches, resembling human-like methods of describing the world and selecting strategies for interaction. Comparably, when it comes to web portals, we have to consider the human factor in the design of the user journeys of the web pages. Our work primarily focuses on online shopping portals.

Consider a retail business owner who wishes to start an e-commerce portal. While a physical store entails numerous expenses like rent and maintenance, its benefit lies in enabling firsthand customer experience through direct observation of their interaction with products. Take, for instance, a boutique clothing store. Despite the costs, the physical store permits customers to engage with fabrics, try on outfits, and evaluate their appearance in mirrors, offering a tactile shopping experience. Having the experience as a predicate, the owner can exercise changes in the shop, such as redesigning aisles, placement of mirrors, speedy customer service, enhancing the shop aesthetics, etc.

In comparison to a physical store, an online shop deploys existing user journeys. A user journey template is analogous to aisle design in a physical shop. Having an online shop enables the owner to scale the customer base far wider than the reach of the physical shop [2,9]. Although the owner benefits from cost reduction and scalability by having an online shop, it is difficult to gauge the shopping experience of an online customer. Most web portals use an existing user journey template to create an online shop, thereby leaving minimal room for experimenting and understanding the customer experience. Unbeknownst of its merits and demerits, an existing user journey template is used frequently. Moreover, given today's tools, the owner can analyze a user journey only after the online shop goes online and not during the offline development phase. We attempt at solving this problem by having virtual RL agents, emulated as real users with various behaviors, generate empirical data prior to the launch of a website.

Conceptual modeling plays a crucial role in designing an empirical approach by providing a framework to represent the various components and objectives of the user journey. It aids in breaking down the complex system of an online shopping portal into manageable parts, making it easier to analyze and understand. Through conceptual modeling, we can identify the different stages and touchpoints in the user journey, such as product browsing, adding items to the cart, and completing a purchase. We propose to model web portals, consisting of webpages, as a graph structure, with a webpage as a node and edges representing all links accessible from the webpage. Modeling web portals with a graph structure provides a simplistic overview of all the connections among pages.

The graph can be exported in any of the standard formats, like XML, JSON, etc. We use the JSON format as it enables seamless integration with popular RL libraries [4, 11]. The ease of integration with RL-libraries facilitates the process of experimenting with new environments into the training pipeline, making experimentation more efficient. The JSON-based RL environment enables the creation of diverse and customizable scenarios for training an agent. Overall, a structure consisting of nodes and edges, can be exported from all kinds of web pages, especially e-commerce shopping portals where each node consists of elements such as add to cart, review, like, comment, favorites, etc.

Reinforcement learning, allows us to train an agent to make decisions and take actions within the online shopping environment. The agent's objective is to maximize a reward, defined according to desired outcomes. The RL agent learns through trial and error, adjusting its actions based on the observed rewards and the environment's responses. In the context of analyzing an online shopping portal, we use reinforcement learning as an empirical approach, where a user is modeled as an agent and the online shopping portal as the environment.

By combining the principles of reinforcement learning and conceptual modeling, we can simulate user behaviors, study the actions of an agent on a user journey, and eventually optimize the web portals. To implement the empirical approach towards conceptual modeling of online portals and to put reinforcement learning to use, we propose the following steps:

1. **Conceptual Model**: Develop a conceptual model to represent the user journey of the online shopping portal. The model should capture the different stages, touchpoints, and possible actions a user can take. It may include components such as product discovery, search functionality, product details, reviews, cart management, and checkout.
2. **State Representation**: Define the states of the environment where an RL agent can observe and interact within the conceptual model. States are represented by variables

to capture relevant information, such as the current page, items in the cart, search history, time allowed for shopping, amount of money to spend, or user preferences. The state representation should reflect the context of a task and enable the agent to make informed decisions.

3. **Action Space**: Define the possible actions the agent can take in each state. Actions can include browsing different products, adding items to the cart, applying filters, using search functionalities, or proceeding to checkout. The action space of the agent should encompass a range of choices reflecting interactions by real-world users in the online shopping portal.

4. **Reward Design**: Design a reward function providing feedback to the agent based on its actions. The reward function can consider metrics such as successful purchases, time and money spent on the portal, etc. Its role is to incentivize actions leading to desired task outcomes and discourage actions diverging from task completion.

5. **Training and Evaluation**: To train the agent based on the specified conceptual model, use RL algorithms such as Q-learning, SARSA, Deep Q-Networks (DQN) [10], Proximal Policy Optimization (PPO) [15], etc. The agent interacts with the environment by selecting actions based on its current state, receiving rewards, and updating its policy to improve future decision-making. Iteratively exploring the state space and adjusting the agent's behavior are part of the training process.

6. **Empirical Study**: Conduct experiments using the trained agent to simulate user behaviors and study their impact on the online shopping experience. Vary the agent's behavior by adjusting parameters related to the state space, reward design, and training RL model. Evaluate the effectiveness of different user journey variations.

7. **Analysis and Optimization**: Analyze the empirical study results to gain insights into the relationship between the user journey and user behavior. Identify patterns, trends, and influential factors contributing to positive user experiences and desired outcomes. The results would provide insights into suboptimal user journey designs, diminishing the conversion rate of a shopping portal. Use these insights to optimize the conceptual model and refine the user journey design, improving the online shopping portal's performance and user satisfaction.

By combining the empirical approach of reinforcement learning with the principles of conceptual modeling, we can iteratively improve the user journey design in an online shopping portal. This method facilitates data-driven decision-making and optimization, leading to enhanced user experiences, increased conversion rates, and improved business outcomes, where the data is generated by RL agents simulated as real users instead of using the data from a real user. Moreover, the method also provides a framework for understanding the complex dynamics of user behavior and enables targeted interventions to guide users towards desired actions within the online shopping environment free from collecting real users web log data.

4   Experiments and Results

We simulated an online shopping portal consisting of webpages such as "jeans", "shoes", and "shirts". The simulated shopping portal starts with the *homepage* node, with edges connected to three level-1 nodes (or main category): *bestsellers*, *the essentials*, *trending*. Each level-1 node has three level-2 nodes namely *jeans*, *shoes*, and *shirts*, also called as wardrobe capsules. Lastly, each wardrobe capsule has two articles or level-3 nodes, for example, the *shoes* capsule consists of *loafers* and *trainers*. A partial view of the simulated shopping portal modeled as a JSON structure is described in listing 1, where ellipsis (...) indicate the rest of the nodes.

```
{
    "nodes":
    [
        {
            "node_label": "homepage",
            "product_page": "False",
            "nodes_list": ["homepage", "bestsellers",
            "the essentials", "trending"],
            "action_elems": ["jump_to"]

        },
        {
            "node_label": "the essentials",
            "product_page": "False",
            "nodes_list": ["homepage", "jeans", "shoes", "shirts"],
            "action_elems": ["jump_to"]
        },
        {
            "node_label": "shoes",
            "product_page": "False",
            "nodes_list": ["homepage", "loafers", "trainers"],
            "action_elems": ["jump_to"]
        },
        {
            "node_label": "loafers",
            "product_page": "True",
            "nodes_list": ["homepage", "shoes"],
            "action_elems": ["jump_to"]
        },
        {
            "node_label": "trainers",
            "product_page": "True",
            "nodes_list": ["homepage", "shoes"],
            "action_elems": ["jump_to"]
        },
        {
            "...": "..."
        }
    ]
}
```

Listing 1. Shopping portal JSON representation.

In total there are 31 nodes modeled as the environment for the RL agent. All the nodes have an edge connecting to the *homepage* node. Each node consists of four keys:

1. **node_label**: Name of a node.
2. **product_page**: To identify whether a node consists of a product/article to be bought.
3. **nodes_list**: List of connected nodes, namely, nodes an agent can jump to.
4. **action_elems**: List of actions an agent can perform.

Currently, the agent can execute only one type of action, i.e., *jump_to*, to go to one of the connected nodes. We are working towards incorporating other types of action such as *add_to_cart*, *review*, *like*, *comment*, etc. Introducing additional types of actions would also involve a substantial action space, manageable through reinforcement learning methods like DQN, PPO, etc.

The state space of the agent is a two-dimensional tuple consisting of discrete values. The first dimension represents the node label where the agent is currently present, whereas the second dimension tells the agent where it is supposed to go. The total number of states is computed by N^s; where N is the number of nodes and s is the dimension of the state space. Therefore, in our particular shopping environment, there are $31^2 = 961$ states.

The action space of the agent is a one-dimensional discrete value, representing the node label where the agent decides to *jump_to*. Since the environment consists of 31 nodes, the action space consists of 31 actions. However, not all actions are permitted in each state. The number of actions the agent can choose from is based on the actions stated in *action_elems* and *nodes_list* of the node where the agent is currently present.

In the navigation task, the agent is provided with a node label where it has to start and find its way to a target node. It selects one of the nodes from the *nodes_list* of its current node using ε-greedy method and executes the *jump_to* action. The selection of the start and target nodes was made in two ways: (i) **randomized**: the start and the target node are randomly selected, and (ii) **pre-determined**: the start and the target nodes are selected manually ensuring a minimum of 2 edges between the nodes. The reward function of the agent is defined as follows: the agent receives a reward of (i) **1** if the targeted node is reached, (ii) **-1** if the maximum amount of steps during a training episode is reached, or (iii) **0** when an agent jumps to a node other than the targeted node.

The results for the navigation task are shown in Fig. 1. The y-axis of the plots represents the length of an episode, i.e., the number of steps the agent takes to navigate to the target node, whereas the x-axis is the number of episodes i.e., the number iterations to train an agent. The dark blue line shows the average episode length per episode and the light blue background area shows the standard deviation, over 100 different seeds. Figure 1a and Fig. 1b illustrate results from randomized and pre-determined start and target nodes, respectively. A noticeable difference between the two node selections ways is the number of episodes required by an agent to converge. The randomized method begins to converge at around 650 episodes, whereas the pre-determined method requires less than 20 episodes to reach the target node. Both the RL approaches continue to drop their episode length of the training process as the number of episodes increase.

If we observe closely the result of the pre-determined nodes Fig. 1b, the SARSA agent converges comparably faster with respect to Q-learning. In other words, SARSA

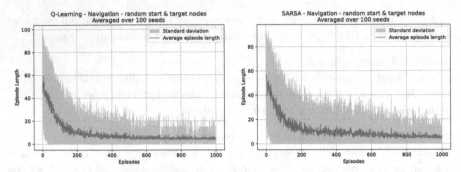

(a) Agent navigating from a randomly selected start node and a target node. **(l)** Agent trained with Q-learning, **(r)** Agent trained with SARSA.

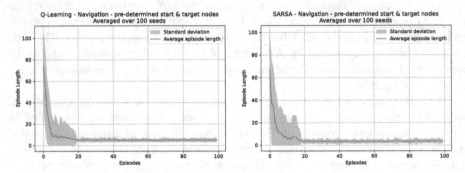

(b) Agent navigating from a pre-determined manually selected start node and a target node. **(l)** Agent trained with Q-learning, **(r)** Agent trained with SARSA.

Fig. 1. Comparing Q-learning with SARSA for the navigation task. Dark blue line indicates average episode length and light blue background shows standard deviation over these episodes each with 100 seeds. (Color figure online)

helps the agent solve the navigation task relatively quickly. When dealing with randomized nodes, SARSA exhibits slower convergence in contrast to Q-learning, as a result of less knowledge gained during each iteration due to its non-exploratory nature.

One can deduce, through observation, an agent using the SARSA approach displays risk-averse tendencies or is focussed, consequently limiting its exploratory behavior. The Q-learning agent, however, demonstrates risk-taking and exploratory behavior. Although the differences in the convergence values are comparably close, they will eventually be significant with addition of newer nodes, edges, and agent actions to the environment.

5 Conclusion and Future Work

Given the extensive amount of research in the fields of conceptual modeling and machine learning, it is reasonable to expect they would complement each other. We addressed creating a bridge between the two domains by taking an empirical approach due to the applicative nature of machine learning, especially reinforcement learning. We chose the problem of analyzing whether user journeys of online portals, specially the ones yet to be launched, engender specific user behaviors online. Using a JSON structure to model web portals in our conceptual modeling phase undeniably facilitates easy addition and removal of nodes, connections among the nodes, and available actions for an agent on a node. The simplicity and expressiveness of the environment as JSON, makes the graph's topology, node attributes, and edge properties a step towards AI explainability.

Expanding upon the work presented in this paper, we intend to introduce a broader range of types of action an agent could perform at each node. Consequently, we would have to exercise sophisticated learning methods such as DQN, PPO, masked-PPO, etc., due to their ability to handle large action space. By employing various RL methods and changing reward functions, we could potentially simulate different types of complex user behaviors on a web portal. Consequently, the analysis of user journeys of a web portal can be completely simulated to identify emphasizing and de-emphasizing segments of the portal. The integration of empirical methods and conceptual modeling with machine learning not only enhances the validity of the models but also facilitates a deeper understanding of the underlying processes. In conclusion, empirically-informed conceptual modeling practice in conjunction with reinforcement learning represents a powerful and rigorous approach to understanding complex systems such as web portals.

Acknowledgement. The research leading to these results are produced under the STRG.Agents project funded by the Austrian Research Promotion Agency (Die Österreichische Forschungsförderungsgesellschaft FFG) under the grant agreement number 45196538 and the FFG number 898104.

References

1. Barreto, A., Hou, S., Borsa, D., Silver, D., Precup, D.: Fast reinforcement learning with generalized policy updates. Proc. Nat. Acad. Sci. **117**(48), 30079–30087 (2020)
2. Bjerkan, K.Y., Bjørgen, A., Hjelkrem, O.A.: E-commerce and prevalence of last mile practices. Transp. Res. Proc. **46**, 293–300 (2020)
3. Bork, D., Ali, S.J., Roelens, B.: Conceptual modeling and artificial intelligence: a systematic mapping study. arXiv preprint arXiv:2303.06758 (2023)
4. Brockman, G., et al.: OpenAI gym (2016)
5. Feltus, C., Ma, Q., Proper, H.A., Kelsen, P.: Towards AI assisted domain modeling. In: Advances in Conceptual Modeling: ER 2021 Workshops CoMoNoS, EmpER (2021)
6. Gur, I., et al.: Environment generation for zero-shot compositional reinforcement learning. In: Advances in Neural Information Processing Systems (2021)
7. Maass, W., Storey, V.C.: Pairing conceptual modeling with machine learning. Data Knowl. Eng. **134**, 101909 (2021)
8. Maass, W., Storey, V.C.: Why should machine learning require conceptual models? (2021)
9. Maat, K., Konings, R.: Accessibility or innovation? Store shopping trips versus online shopping. Transp. Res. Rec. **2672**, 1–10 (2018)
10. Mnih, V., et al.: Human-level control through deep reinforcement learning. Nature **518**, 529–533 (2015)

11. Raffin, A., Hill, A., Gleave, A., Kanervisto, A., Ernestus, M., Dormann, N.: Stable-baselines3: reliable reinforcement learning implementations. J. Mach. Learn. Res. **22**, 12348–12355 (2021)
12. Reinhard, P., Li, M.M., Dickhaut, E., Reh, C., Peters, C., Leimeister, J.M.: A conceptual model for labeling in reinforcement learning systems: a value co-creation perspective. In: International Conference on Design Science Research in Information Systems and Technology (2023)
13. Rummery, G.A., Niranjan, M.: On-line Q-learning using connectionist systems. University of Cambridge, Department of Engineering Cambridge, UK (1994)
14. Husin, H.S., Ismail, S.: Process mining approach to analyze user navigation behavior of a news website. In: Proceedings of the 4th International Conference on Information Science and Systems (2021)
15. Schulman, J., Wolski, F., Dhariwal, P., Radford, A., Klimov, O.: Proximal policy optimization algorithms. arXiv preprint arXiv:1707.06347 (2017)
16. Shi, T., Karpathy, A., Fan, L., Hernandez, J., Liang, P.: World of bits: an open-domain platform for web-based agents. In: International Conference on Machine Learning (2017)
17. Terragni, A., Hassani, M.: Analyzing customer journey with process mining: from discovery to recommendations. In: 2018 IEEE 6th International Conference on Future Internet of Things and Cloud (FiCloud) (2018)
18. Watkins, C.J., Dayan, P.: Q-learning. Mach. Learn. **8**, 279–292 (1992)
19. Yuan, G., Lu, J.: MORTAL: a tool of automatically designing relational storage schemas for multi-model data through reinforcement learning. In: Proceedings of the ER Demos and Posters 2021, International Conference on Conceptual Modeling (ER 2021) (2021)

JUSMOD

JUSMOD – 2nd International Workshop on Digital Justice, Digital Law and Conceptual Modeling

Silvana Castano[1], Mattia Falduti[2], Cristine Griffo[3], and Stefano Montanelli[1]

[1] Department of Computer Science, Università degli Studi di Milano
silvana.castano@unimi.it, stefano.montanelli@unimi.it
[2] The Square Center, Italy
mattia.falduti@thesquarecentre.org
[3] Eurac Research, Italy
cristine.griffo@eurac.edu

Law plays a crucial role in almost every aspect of our life, both public and private. Thousands of legal documents are constantly produced by institutional bodies, such as Parliaments and Courts, which constitute a prominent source of information and knowledge for judges, lawyers, and other professionals involved in legal decision-making. To cope with the growing volume, complexity, and articulation of legal documents as well as to foster digital justice and digital law, increasing effort is being devoted to digital transformation processes in the legal domain.

Conceptual modeling plays a crucial role in this scenario, to formalize features and the nature of terminology used in legal documents and to promote the development and the adoption of legal ontologies, shared vocabularies, and open linked data about legislation, case law, and other relevant legal information. Furthermore, advanced functionalities for legal data and process modeling and management are advocated, embracing modern technologies like Semantic Web, NLP, and AI, to enable semantic text search and exploration, legal knowledge extraction and formalization, legal decision-making and legal analytics.

JUSMOD 2023, the 2nd International Workshop on Digital JUStice, digital law, and conceptual MODeling, constituted a meeting venue for a variety of researchers involved in digital justice and digital law crossing different disciplines besides computer science, like law, legal informatics, management, economics, and social sciences.

Launched in 2022, also with the push of the *Next Generation UPP* national multidisciplinary project of the Italian Ministry of Justice, JUSMOD represents an opportunity to share, discuss, and identify new approaches and solutions for modeling, analysis, formalization, and interpretation of legal data and related processes. The program of its second successful edition included 6 long papers and 2 short papers selected by the Program Committee after the peer-review process.

JUSMOD Organization

Workshop Chairs

Silvana Castano   University of Milan, Italy
Mattia Falduti   The Square Center, Italy
Cristine Griffo   Eurac Research, Italy
Stefano Montanelli   University of Milan, Italy

Program Committee

Tommaso Agnoloni   IGSG, CNR, Italy
Jean-Rémi Bourguet   University of Vila Velha, Brazil
Samuel M. Brasil Jr.   National School of the Judiciary (ENFAM), Brazil
Daniel Braun   University of Twente, The Netherlands
Juliana Justo Castello   Vitória Law School, Brazil
Maria das G. Teixeira   Federal University of Espírito Santo, Brazil
Mirna El Ghosh   Sorbonne University, France
Matthias Grabmair   Technische Universität München, Germany
João O. Lima   Prodasen-Brazilian Senate, Brazil
Zoran Milosevic   Deontik, Australia
Matteo Palmonari   Università di Milano-Bicocca, Italy
Anca Radu   European University Institute, Italy
Davide Riva   Università degli Studi di Milano, Italy
Giovanni Sileno   University of Amsterdam, The Netherlands
Andrea Tagarelli   Università della Calabria, Italy

New-Generation Templates Facilitating the Shift from Documents to Data in the Italian Judiciary

Amedeo Santosuosso, Stefano D'Ancona, and Emanuela Furiosi[✉]

IUSS Pavia, Palazzo del Broletto, Piazza della Vittoria, n.15, Pavia, Italy
{amedeo.santosuosso,stefano.dancona,
emanuela.furiosi}@iusspavia.it

Abstract. The Italian judicial system may be about to make a historic transition: from the mere digitization of its documents to an activity intrinsically shaped by the possibilities that the digital environment offers (digitalization as controlled flow of data) and that allows, for the first time operationally, the use of artificial intelligence techniques.

The paper describes the research carried out by the authors (The paper was jointly conceived by the authors. However, the introduction of Sect. 3 and Sect. 4 were written by Amedeo Santosuosso, Sects. 1 and 2 were written by Stefano D'Ancona and Sects. 3.1 and 3.2 were written by Emanuela Furiosi.) as part of the NEXT GEN UPP research project, launched in 2021 by the Italian Ministry of Justice, with the aim, among others, of enhancing the digitalization of judicial activities as a basis for the improvement of organizational processes and the application of Legal Analytics (LA) tools to legal texts.

The activity described in this paper relates to the design of a "new-generation template" that can be used by judges and lawyers in carrying out their activities.

These new templates can maximize the efficiency of judicial activity while creating an enabling IT environment for the use of the most advanced artificial intelligence technologies, both current and future.

Keywords: Digitalization of Judicial Proceedings · AI and Law · Documents vs. Data · Segmentation · Law as Data · Italian Judicial System

1 Introduction

The Italian judicial system may be about to make the historic transition from the mere digitization of its documents (that is, from the mere use of digital tools) to an activity intrinsically shaped by the possibilities that the digital environment offers and that also uses, for the first time operationally, artificial intelligence techniques.

In the previous phase, the content and mode of acquisition of documents, as well as the lawyers and judges' deeds in the process, were essentially unchanged from the past, and the only difference was the transfer in digital format of information expressed as signs impressed on paper (with pen, typewriter or computer printing makes little difference: analog mode), producing digital "documents". In the phase now opening, information

T. P. Sales et al. (Eds.): ER 2023 Workshops, LNCS 14319, pp. 121–130, 2023.
https://doi.org/10.1007/978-3-031-47112-4_11

can be directly produced and organized in such a way as to be machine-readable and have characteristics that make it directly explorable with artificial intelligence techniques (legal analytics), which by definition operate on "data" and not on documents. It is the shift *from documents to data*, long known and investigated at the theoretical level [1], that now becomes the cornerstone of a major innovation (i.e., implementation) in the judicial field.

This paper describes the research the authors carried out as part of the NEXT GEN UPP research project, that has focused on the development of templates for court deeds, with defined structural requirements, that can enable judges and lawyers to draft structurally homogeneous "digital entities". The Italian Ministry of Justice launched the project in 2021, with the aim, among others, of enhancing the digitalization process of judicial activities as a basis for the improvement of organizational processes and, in particular, the application of Legal Analytics (LA) tools. In other words, to create a technical environment in which different disciplines, including data science, artificial intelligence (AI), machine learning (ML) and natural language processing (NLP), can converge.

The activity described in the present paper, carried out by the working group of the Scuola Universitaria Superiore, IUSS Pavia, was first aimed at designing a "conceptual template" of judicial decision making, and then a "new generation template", able to maximize the efficiency of judicial activity, while at the same time creating an enabling computer environment for the use of the most advanced artificial intelligence technologies, current and forthcoming.

The path taken by the research group was to design a new "technological object" that can be used for the drafting of judicial deeds (judgments in the first place, but also defenders' briefs).

2   The State of the Art in the Use of Templates

In a recent review of the publications dedicated at legal prediction the authors [2] stress the importance of a sufficient knowledge about the exact dataset the studies are based on and the awareness of researchers of the type of data they are analyzing, and conclude "unfortunately, this is frequently not the case" (p. 207). This very comprehensive review to the studies that have used machine learning techniques and claimed to be predicting court decisions, published from 2015 to 2021, shows very clearly, among other very interesting aspects of legal prediction, the importance of homogeneity of judgements and the way they are divided in sections.

It is interesting the case of ECHR judgments whose structural homogeneity derives from Rule 74 of the Rules of Court, which, precisely imposes specific obligations on the Court in the way the judgments are drafted [3–7]. It is precisely from this acquired knowledge, the Authors of this paper have moved in an attempt to provide, within the Italian court system, a template of judgment and party deed (the deeds of lawyers)[1] that can be used by legal practitioners in the context of civil judgments. The theoretical assumption is that the creation of templates of judgments makes possible to have an accurate textual analysis, reducing if not avoiding prior tagging by domain experts [8–12]. Studies in this field in Italy have not yet been provided, unless we have unintentionally missed some research.

In paragraph 3 the bidirectional path from a mere conceptual template toward a new-generation template is described and the basic structure of new generation template is designed in a way which might be workable by technicians. Finally, in paragraph 4 further research steps and practical applications are outlined.

3 From Conceptual Templates to New Generation Templates

The design of "new-generation templates" for use in civil judgments (of first and second instance) has been conceived as a tool having a twofold nature: a fundamental junction of the various aspects of the proceedings and a catalyst for broader change in the transition from the analog to the fully digital era.

Just to give a few examples, a process conceived as a flow of data and not as a mere accumulation of documents produced by judges and the parties' lawyers implies significant technological support, which concretely means a reengineering of the current Italian Telematic Civil Process (PCT), not to mention the necessary changes in the tools the lawyers use in their judicial and non-judicial activity. In addition, in the background, several prejudices against technology are still present among both judges and lawyers and need to be faced.

As it is often the case in information technology, technical activities require prior clarification of the logic and sequence of the process. Thus, the first step for designing "new-generation templates" is a clear understanding of the logical and legal ideas the currently used templates are based on: i.e., a conceptual template reproducing in an ordered way what legal practitioners assume when they use them simply through a word processor (such as Word or Google Docs and similar). Of course, this is a bidirectional path, starting from a mere conceptual template (which could also be totally analog) and moving toward a new-generation template (i.e., a fully digital entity), which then can retroact on the same basic conceptual structure if a way emerges that is better suited to the digital context and is therefore clearer and more efficient. In some sense the new generation templates can be considered as the technological and conceptual evolution of the templates that already exist in a basic form within the library of the Judge's and

---

[1] In Computer science, *template* is: «A document or file having a preset format, used as a starting point for a particular application so that the format does not have to be recreated each time it is used» (American Heritage® Dictionary of the English Language, Fifth Edition. Copyright © 2016 by Houghton Mifflin Harcourt Publishing Company. Published by Houghton Mifflin Harcourt Publishing Company), at https://www.thefreedictionary.com/template.

Assistant's Consolle, accessible through the PCT. The difference stands in technology and, mostly, in the mindset: rather than trying to adapt the digital environment to the old analogic structure of the documents, we have now the possibility to shape procedural rules according to what the digitalization of the legal environment makes possible.

3.1 Conceptual Templates

In designing the conceptual models for both court documents and pleadings, two aspects were considered. Firstly, the minimum content required by the Code of Civil Procedure (Article 132 for judgment and Articles 163 and 167 for pleadings) was taken into account. Secondly, inputs from judges and lawyers who collaborated on the project, including their best practices and reported needs, were also considered [13]. The work started from the conceptualization of the civil judgment template of first instance. Fifty judgments from first-instance courts, covering various domains of civil law, have been gathered, selected for their structural interest and quality.

These judgments were subjected to meticulous analysis and a process of reduction to a scheme, in which essential elements were saved and semantic contents (i.e., legal concepts and arguments) were removed. Figure 1 shows the transition from a *17-pages judgment* (first column on the left) to an essential *1-page scheme* where only sections of the judgement are highlighted (second column). The third column shows the same *1-page scheme populated* with cross-references to other documents within the legal proceeding, in compliance with the provisions of Article 132 of the Code of Civil Procedure.

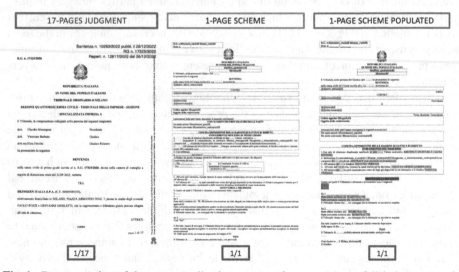

Fig. 1. Representation of the conceptualization process: from a common full judgement to the conceptual template. The picture is intended to show the logical development and not to be readable in the content.

3.2   New Generation Templates

The new generation templates have been designed as native digital objects capable of collecting structured data and organizing them in dedicated repositories (such as data warehouse -DWH- and data lake -DL-), which allow the creation of datasets that can be exploited with retrieval tools and artificial intelligence techniques. The design of new generation templates assumes a proper digitalization of the process and the centrality of data, in a way which modifies how judges and lawyers write and structure their documents and reshape how data and information interact with each other.

In this context, the judge essentially has the task of organizing data and information within the decision according to his/her institutional position. Figure 2 represents the sources of data and the place that AI may take within the judgement (document builder).

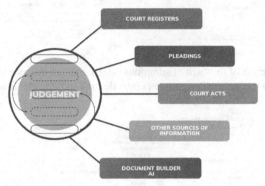

Fig. 2. Representation of the judgement as a space where information and data from multiple sources are organized.

New generation templates are rooted in previous research and experience that has undergone significant acceleration due to the innovations introduced by the so called Cartabia Reform [14] for the digitalization of civil processes. This reform follows the direction already undertaken by the experiment addressed in this paper, explicitly mandating the electronic filing of documents, the use of native digital files, electronic signatures, and the adoption, precisely, of computerized templates for legal documents.

Hereinafter we present the main acts of the parties' lawyers and how they are related to the final judgement. In Fig. 3 are shown the writ of summons and the response of the opposing lawyer: in green are the sections that are automatically filled in using the available technology, in yellow are abstracts derived from plaintiff pleadings, in pink are abstracts derived from defendant pleadings.

Figure 4 aims to give an idea of the combination of data having different sources and origins in the creation of final judgment. In green are sections that are automatically filled in using the available technology, in yellow are sections derived from plaintiff pleadings, in pink are sections derived from defendant pleadings, in light blue are sections derived from the court minutes.

New-generation templates operate on two fronts: on one hand, they simplify the drafting of individual decisions; on the other hand, they contribute to the creation of

Fig. 3. Visual comparison of the "writ of summons" (left) and "appearance and defense/answer" (right) new generation templates. The picture is intended not to be readable in the content.

Fig. 4. Judgement new-generation template. Scheme readable only for the color code, the picture is intended not to be readable in the content.

datasets, enabling the use of artificial intelligence systems to further help of judges' work.

Regarding the first aspect, new-generation templates are designed considering the complex structural and content-related connections between the exchanged documents throughout the legal process and the judgment. The focus was on maximizing those interconnections between information and data contained in those documents, organizing them into interoperable databases, to be reorganized by the judge within the judgment. Figure 5 represents the section of judgement dedicated to the case facts and their legal ground.

1.   By writ of summons ritually served on ...... the plaintiff sued $$defendant$$ claiming that **ABSTRACT art. 163, c. 3, n. 4 c.p.c._claim1 + claim2 + claim***n*
2.   After the cross-examination was established, $$defendant$$ constituted himself by appearance of ___ requesting ABSTRACT
With supplementary brief $$plaintiff$$: **ABSTRACT ex art. 171-ter c.p.c. n. 1**, $$defendant$$ ABSTRACT ex art. 171-ter c.p.c. n. 1
With supplementary brief $$plaintiff$$ **ABSTRACT ex art. 171-ter c.p.c. n. 2**, $$defendant$$ ABSTRACT ex art. 171-ter c.p.c. n. 2
With supplementary brief $$plaintiff$$ **ABSTRACT ex art. 171-ter c.p.c. n. 3**, $$defendant$$ ABSTRACT ex art. 171-ter c.p.c. n. 3
3.   At the hearing of ...., Judge **MINUTES ART. 183 cc. 1,2,3 c.p.c.**
4. By Order No. _____ filed on _____ the Judge **ORDER art. 183 co. 4 c.p.c.**
5. Considered the case ready for decision, **MINUTES ART. 189 c.p.c.**
6. The parties filed briefs: $$plaintiff$$: **ABSTRACT ART. 189, c.1, n. 2** $$defendant$$ ABSTRACT ART. 189, c.1, n. 2; $$plaintiff$$: **ABSTRACT ART. 189, c.1, n. 3** $$defendant$$ ABSTRACT ART. 189, c.1, n. 3
7. At the hearing of ...., Judge **MINUTES-189, c. 3 c.p.c.**

Fig. 5. Section of the judgement new-generation template regarding the concise statement of the case facts. The close interconnection of the judgement with the pleadings is highlighted. In green are sections that the existing system automatically fills in; in yellow are sections derived from plaintiff pleadings; in pink are sections derived from defendant pleadings; in light blue are sections derived from the court minutes.

Using new-generation templates the judgement is almost automatically compiled in those sections related to events and facts (see Fig. 5), and in sections where the dialogue between the parties is summarized (see Fig. 6). This is also made possible by the elaborate structuring of the pleadings, which involves the drafting of specific abstracts that are designed to be transposed (unaltered) into the judgment.

This structure reduces the laborious task of reconstructing the procedural facts and provides the judge with a clear synoptic overview where the parties' procedural positions are faithfully reproduced. Of course, the judge is still responsible for the decision-making part of the judgement, determining the order of issues to be addressed (see Fig. 6), making decisions on individual issues and drafting the corresponding reasoning (see Fig. 7). The content of party briefs drafted based on the template should also achieve appreciable levels of clarity.

New generation templates are both a model for the construction of individual court deeds and a guide to their drafting at the various stages of the trial. In the progress of each specific judgment, new data are inserted and combined with existing data, shaped according to the different trial position of the writer (lawyer, judge and more). For example, an appointed expert writes his report by taking data provided by the parties,

REASONS FOR DECISION

The issues on which the court is asked to rule are as follows:

1. .....
2. .....

About Issue 1.
The plaintiff argues that **$$ABSTRACT ART. 189, c.1, nn. 1 e 2$$**.
The defendant argues that $$ABSTRACT ART. 189, c.1, nn. 1 e 2$$.
The Court considers that ...REASONING... as a result, the claim is to be granted/dismissed.

About Issue 2.
The plaintiff argues that **$$ABSTRACT ART. 189, c.1, nn. 1 e 2$$**.
The defendant argues that $$ABSTRACT ART. 189, c.1, nn. 1 e 2$$.
The Court considers that ...REASONING... as a result, the claim is to be granted/dismissed.

*

Fig. 6. Excerpt from the judgement new-generation template regarding the reasoning section.

CLAIM STATEMENT	DEFENSE	JUDGEMENT
Claim 1. $$ABSTRACT_ Claim1$$ plus a clear and specific presentation of the legal elements constituting the grounds of the claim.	**Against Claim 1.** The defendant challenged the admissibility and merits of the claim $$ABSTRACTagainst claim1$$	**About Claim 1.** The plaintiff argues that $$ABSTRACT_ Claim1$$ The defendant argues that $$ABSTRACTagainst claim1$$ The Court considers that ...REASONING..., the claim is to be granted/dismissed.

Fig. 7. A Representation of the procedural dialogue structured by the new generation template, which includes AI support in the 'reasoning' section.

organizing them in a way that is functional to the task he is called upon to perform, and adding the technical data. At the end of the process, the judge will similarly proceed by adding the evaluations proper to his decisional function. And so on. The proceeding as a flow in which data are added along the various stages of the trial is shown in Fig. 8.

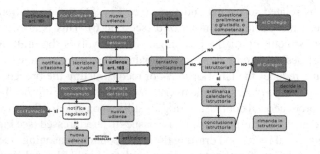

Fig. 8. Diachronic representation of the phases of a specific judgment and related data flow.

From a different point of view (and this is the *second aspect*), the use of new-generation templates, as previously mentioned, enables the collection of good quality data and the creation of datasets that facilitate the use of cutting-edge technologies to further support the judge's and lawyers' activities.

On this repository can work an AI system as a document builder [15]. Specifically, the same new-generation template includes the contribution of the so-called document builder. The document builder is an AI evolving technology, capable of querying the well-organized dataset fed with good quality data, created and organized based on the new-generation templates, assisting the judge in making decisions and motivating them by suggesting the most relevant legal precedents for the specific case at hand. The same dataset can also be queried by artificial intelligence (AI) to make visible those pieces of information that may not be immediately seen by the judge.

4 Final Considerations. Achievements and Prospects

At the beginning of the research, the Authors set the goal of designing a new generation template to be provided to the judiciary for two essential purposes.

The first was to maximize the efficiency of the justice service. The second, to be achieved in the medium to long term, was to create an enabling IT object for the use of the most advanced AI technologies. At the conclusion of the research, it is necessary to compare the aims with the achieved results, as well as any identified limitations.

From a theoretical point of view, it is possible to deepen the analysis of best practices among lawyers and judges, trying to reduce prejudices about the use of technology in the legal field. The potential introduction of judgment templates through the Judge's Consolle and the Assistant (especially for judgments) would imply that judges would have to adapt to new working methods. Currently, judges have complete discretion in structuring the judgment as they see fit, including merging sections that, as shown in the model, would be better kept distinct.

From a practical point of view the Authors offer this work to the attention of the Ministry of justice hoping in a decision to start a close interaction with the technical offices and experiment some templates within the Italian Telematic Civil Process (PCT). This would also be an opportunity to overcome the difficulties and misunderstandings occurred in the first phase of application of the Article 46 Civil Procedural Code (Implementing provisions). This rule establishes "*The Minister of Justice, in consultation with the Superior Council of the Judiciary and the National Bar Council, defines by decree the computerized templates of judicial acts with the structure of the necessary fields for entering information in the process registers*". This would be the best development moving from the research published in this paper.

References

1. Livermore, M.A., Rockmore, D.N. (eds.): Law as Data: Computation, Text, & the Future of Legal Analysis. Santa Fe Institute Press (2018)
2. Medvedeva, M., Wieling, M., Vols, M.: Rethinking the field of automatic prediction of court decisions. Artif. Intell. Law **31**, 195–212 (2023). https://doi.org/10.1007/s10506-021-09306-3
3. Aletras, N., Tsarapatsanis, D., Preoţiuc-Pietro, D., Lampos, V.: Predicting judicial decisions of the European court of human rights: a natural language processing perspective. PeerJ Comput. Sci. **2**, e93 (2016)
4. Medvedeva, M., Vols, M., Wieling, M.: Using machine learning to predict decisions of the European court of human rights. In: Artificial Intelligence and Law, pp. 1–30 (2019)

5. Pinotti, G., Santosuosso, A., Fazio, F.: A rule 74 for Italian judges and lawyers. In: Guizzardi, R., Neumayr, B. (eds.) Advances in Conceptual Modeling. ER 2022. Lecture Notes in Computer Science, vol. 13650, pp. 112–121. Springer, Cham (2022). https://doi.org/10.1007/978-3-031-22036-4_11

6. Liu, Z., Chen, H.: A predictive performance comparison of machine learning models for judicial cases. In: 2017 IEEE Symposium Series on Computational Intelligence (SSCI), pp. 1–6 (2017)

7. Visentin, A., Nardotto, A., O'Sullivan, B.: Predicting judicial decisions: a statistically rigorous approach and a new ensemble classifier. In: 2019 IEEE 31st International Conference on Tools with Artificial Intelligence (ICTAI), 13 February 2020 (2020)

8. Zhong, L., Zhong, Z., Zhao, Z., Wang, S., Ashley, K., Grabmair, M.: Automatic summarization of legal decisions using iterative masking of predictive sentences. In: Proceedings of the Seventeenth International Conference on Artificial Intelligence and Law, Association for Computing Machinery, New York, NY, USA, 17–21 June 2019, pp. 163–172 (2019)

9. Santosuosso, A., Pinotti, G.: Bottleneck or crossroad? problems of legal sources annotation and some theoretical thoughts. Stats 3(3), 376–395 (2020). https://doi.org/10.3390/stats3030024

10. Wyner, A.Z.: Towards annotating and extracting textual legal case elements. Informatica e Diritto: special issue on legal ontologies and artificial intelligent techniques 19(1–2), 9–18 (2010)

11. Zanoli, E., et al.: Annotators-in-the-loop: testing a novel annotation procedure on Italian case law. In: Proceedings of the 17th Linguistic Annotation Workshop (LAW-XVII) , Toronto, Canada, pp. 118–128. Association for Computational Linguistics (2023)

12. Regarding the U.S.A. see Matt Perez, Why A Stanford Law Project Aims To Standardize Court Forms, in Law360, referring to Stanford's Law and Policy Lab's Filing Fairness Project (2022)

13. Several Judges from the Specialized Section for Business Matters of the Milan Court and some civil sections of the Monza Court were involved. For the legal profession, meetings were held with groups of lawyers from highly qualified law firms in Milan and at a national level, such as those from the Milan office of LCA Law Firm and those from Grimaldi Alliance (with both in-person and remote participation from Rome, Parma, Verona, Treviso, Padua, Naples, Bari, and Turin)

14. Santosuosso, A.: From documents to data, in JuLIA Handbook. Artificial Intelligence, Judicial Decision-Making and Fundamental Rights, Scuola della Magistratura (ed.) (2024, in preparation)

15. The development of the document builder is being carried out by the Department of Computer Science at the University of Milan, led by a working group headed by Professor Silvana Castano, as part of the NEXT GEN UPP research project, launched in 2021 by the Italian Ministry of Justice (2021)

Supervised Learning, Explanation and Interpretation from Pretrial Detention Decisions by Italian and Brazilian Supreme Courts

Marco Billi[3]([⊠])(iD), Thiago Raulino Dal Pont[2](iD), Isabela Cristina Sabo[1](iD), Francesca Lagioia[3,4](iD), Giovanni Sartor[3,4](iD), and Aires José Rover[1](iD)

[1] Department of Law, Federal University of Santa Catarina, Florianópolis, Brazil
[2] Department of Automation and Systems Engineering, Federal University of Santa Catarina, Florianópolis, Brazil
[3] CIRSFID - Alma AI, Alma Mater Studiorum, University of Bologna, Bologna, Italy
marco.billi3@unibo.it
[4] Department of Law, European University Institute, Florence, Italy

Abstract. Pre-trial detention is a debated measure in different legal systems since it deprives defendants of their liberty prior at the initial stage of proceedings. To order this measure, the judge must justify it by highlighting the risks that the arrested person presents to society and to the criminal procedure itself. An example of a factor related to preventive custody, in countries such as Italy and Brazil, is involvement in criminal organizations. The paper presents the results of experimental research with supervised learning, in particular using XAI techniques, such as decision trees and Shapley Additive Explanations. Our corpora are composed of unstructured data (texts of judicial decisions) and structured data (factors extracted from such judicial decisions), from the case law of Italian and Brazilian Supreme Courts. As a result, we have identified a collection of factors that play an important role in the reasoning of the judge and in predicting outcomes, including common factors between the two countries. In particular, we have verified that involvement in criminal organizations consistently leads to the decision to maintain imprisonment in the Brazilian scenario, while in the Italian context, this is unclear. Finally, we conclude that data structuring based on the extraction of factors from the decision texts not only increases the prediction's quality but also allows for their interpretation and explanation.

Keywords: E-justice · Machine learning · Classification · Decision trees · XAI · Criminal Law · Pretrial detention

This research has been supported by Brazilian National Council for Scientific and Technological Development (CNPq) and Coordination for the Improvement of Higher Education Personnel - Institutional Program for Internationalisation (CAPES/PrInt); ADELE (Analytics for Decision of Legal Cases, EU Justice program Grant (2014–2020); COMPULAW (Computable law), ERC Advanced Grant (2019-2024); LAILA (Legal Analytics for Italian Law), MIUR PRIN Programme (2017), the European Commission under the NextGeneration EU programme, PNRR - M4C2 - Investimento 1.3, Partenariato Esteso PE00000013 - "FAIR - Future Artificial Intelligence Research" - Spoke 8 "Pervasive AI".

1   Introduction

Pre-trial detention is a debated measure in different legal systems, since it limits the fundamental principle of the presumption of innocence, by depriving defendants of their liberty at the initial stage of proceedings. Thus, it needs to be justified by the necessity to prevent defendants from absconding or committing further offence(s) or interfering with the investigations. Being part of a criminal organization, for example, has been considered a relevant risk-factor allowing judges to adopt such a measure, especially due to the expansion of organized crimes in several countries, including Italy and Brazil [6]. These countries have specific legal provisions to fight criminal organizations.[1]

Aiming to obtain an interpretation and explanation of judicial decisions that address pre-trial detention, this work applies supervised learning and XAI techniques to the case law of the Supreme Courts of the mentioned countries. Both unstructured data (texts of the decisions) and structured data (factors extracted from such decisions) are considered.

Specifically, we intend to evaluate whether the use of structured data (i) improves the quality of predictions and (ii) enables such predictions to be explained. In particular, we investigate whether participation in criminal organizations is relevant to confirming pre-trial detention.

To this end, we built four different datasets of Italian and Brazilian judicial decisions, as detailed in Sect. 2. Section 3 describes the techniques applied, in particular for classification (supervised learning) and XAI. Section 4 reports the experiments and discusses the results, delineating commonalities and differences between the two legal systems. Section 5 concludes and outlines possible future research.

This work follows recent attempts at explaining decision-making systems through relevant factors and at extracting factors [3,4,9,10,14].

2   Datasets

We built four different datasets. Two of them are composed of judicial decisions (unstructured data) collected from the Brazilian and Italian Supreme Courts (982 and 718 documents, respectively). The remaining two datasets (one Brazilian and one Italian) are based on structured data, prepared by using clustering techniques for extracting relevant factors (F) [14]. In summary, after applying clustering techniques to the documents, the legal experts analysed the groups obtained and observed which legal factors were present. After this analysis, the experts extracted from each document the factors identified as relevant in the previous analysis. Thus, each document is characterized according to the following categorical variables (features).

(A) Decision reasons (binary variables). Whether the decision addresses:

– $F1$: excess of time in prison.

---

[1] Italian Criminal Code, article 416-bis and Brazilian Law no. 12.850/2013.

- $F2$: suspension of time in prison or suspension of the proceedings.
- $F3$: nullity of the interrogation or hearing of the accused.
- $F4$: connection between different crimes or proceedings.
- $F5$: sending back the case to a previous stage or to a judge of another instance (remittance of proceedings)
- $F6$: presence of wiretaps.
- $F7$: the complexity of the proceedings or existing risks such as of the prisoner's flight.
- $F8$: facts inferred and not proven (i.e., the judge understands that a person is aggressive from the evidence of the case).
- $F9$: the victim's statement.
- $F10$: a prisoner caught in *flagrante delicto*.
- $F11$: defence restriction (i.e., the defendant did not have access to prosecution documents).

(B) Crime categories (C) (binary variables). Whether the committed crime was:

- $C1$: against a person (including sexual crime).
- $C2$: against a property.
- $C3$: public safety (including crime provided in the firearms law).
- $C4$: against the government, justice administration or public economy.
- $C5$: provided in special laws.
- $C6$: related to criminal organizations.
- $C7$: provided in the drug law.

(C) Others (categorical variables):

- Location: State or regional capital where the crime occurred.
- Judge rapporteur: Judge's name who was responsible for reporting the case.
- Date: Year in which the decision was issued.

(D) Prisoner status (binary variable): whether the decision stated that the accused must be released. This is our target variable, according to which we split each corpus into two subsets, one containing the decisions in favour of the defendant (i.e., ordering release), the other containing the judgements in favour of prosecution (i.e., ordering detention). In the Italian corpus, the first subset contains 104 judgments and the second 614; in the Brazilian corpus 282 and 700.

3 Methodology

We approach the research problems with two goals: (i) comparison of learning performance from predictions obtained with structured and unstructured data, by using traditional classification techniques; (ii) observing whether the extracted variables (classification with structured data) contribute to explain and interpret the predictions. For the latter purpose, we relied on XAI techniques and decision trees. We reused existing implementations and standard methods to perform the experiments, including Orange 3 [5], and open-source libraries.

3.1   Classification

Classification in supervised learning consists in inferring a function based on labelled data (the training set) and in using that function to map new unlabelled data. Supervised classification requires a sufficient number of labelled records to learn the model [12].

We employed these approaches for the classification task $[1, 2, 8, 15, 16]^2$:

(A) Linear models: Support Vector Machines (SVM) with Radial Basis Function (RBF) kernel, Logistic Regression, and Naïve Bayes;
(B) Tree-based/Ensemble models: Adaboost, Decision Tree, Gradient Boosting, and Random Forest;
(C) Instance-based model: k-Nearest Neighbours (kNN);
(D) Neural-based model: Multi-Layer Perceptron (MLP).

For the structured datasets, preprocessing was unnecessary, as the variables used for the experiments (crimes and factors) are binary. For the unstructured datasets, on the other hand, some preprocessing was necessary: conversion to lowercase, lemmatizing, stopwords removal, and feature extraction using Term-Frequency Inverse-Document-Frequency (TF-IDF) [1].

For training the models, we used k-fold Cross-validation with five stratified folds. The final metrics were the average over the metrics achieved in the five steps.

Accuracy and F1 Score metrics were used to evaluate prediction quality. [1].

Following the literature [11], we designated the majority classifier –always predicting the majority class– as the baseline.

We applied some XAI methods to explain the outcomes of our model.

In particular, we employed the Shapley Additive Explanations (SHAP) to explain individual predictions [13] and Decision Trees to interpret decision rules through their visualization [7].

4   Experiments and Results

In this section we present the evaluation of the classifications by our model and then consider the corresponding explanations.

4.1   Predictions Based on Structure and Unstructured Data

To evaluate the extracted features, i.e., factors and crime categories, we applied classification techniques to the four datasets to predict whether the court will reform or maintain the imprisonment. In the classification of the unstructured judgments, the input was the opinion of the judges, with the exclusion of its final decision part (the verdict). The results are presented in Tables 1 and 2.

---

^{2} For all models, the hyperparameters were kept to the default settings from Orange Data Mining.

Table 1. Accuracy results for the four datasets.

Dataset	Factors BR	Text BR No Verdict	Factors IT	Text IT No Verdict
BASELINE	71,3	**71,3**	**85,4**	85,4
AdaBoost	75,2	66,3	84,5	81,9
Gradient Boosting	75,2	71,1	84,8	84,7
kNN	73,7	70,0	85,1	**85,8**
Logistic Regression	74,3	67,4	85,2	85,7
Naive Bayes	74,7	55,7	84,5	20,3
Neural Network	75,1	65,8	83,8	85,7
Random Forest	74,9	70,7	85,1	85,5
SVM	67,4	37,9	**85,4**	85,5
Tree	**75,9**	63,7	84,9	81,6

Table 2. F1-Score results for the four datasets.

Dataset	Factors BR	Text BR No verdict	Factors IT	Text IT No Verdict
BASELINE	0,416	0,416	0,460	0,460
AdaBoost	0,755	**0,651**	**0,817**	0,812
Gradient Boosting	0,752	0,629	0,805	0,810
kNN	0,745	0,607	0,792	0,802
Logistic Regression	0,731	0,632	0,788	**0,825**
Naive Bayes	0,755	0,578	0,787	0,154
Neural Network	0,752	0,636	0,808	0,813
Random Forest	0,751	0,624	0,809	0,809
SVM	0,683	0,371	0,786	0,788
Tree	**0,758**	0,643	**0,817**	0,803

Comparing the results, we can see that for the Brazilian dataset, predictions based on structured data perform significantly better than those based on textual input for all methods. For the Italian dataset, factor-based predictions are slightly better for some methods and slightly worse for others. Overall, Accuracy and F1 Score for the Italian corpora are slightly higher than those of the Brazilian.

For both datasets, we obtained the best F1 score when applying Decision Trees (in the Italian dataset this approach tied with AdaBoost).

We must remark that our datasets are highly unbalanced, for both the BR and IT, in favour of the "Not Released" outcome. Therefore, if we considered as baseline for comparison a model always guessing a "Not Released" decision, we would obtain a Macro F1-score of 0.416 for the Brazilian dataset and of 0.46 for the Italian one. The accuracy would be 71% for the Brazilian dataset and 85% for the Italian.

This highlights the difficulty of predicting outcomes in this domain, as the Supreme Courts of both countries are reluctant to overturn the result of previous decisions, especially in cases of pretrial detention. Italy is a particularly difficult country to approach, as the number of "Not Released" decisions is much higher than its counterpart.

4.2   Explanation Based on SHAP

To evaluate the impact of each factor (categorical variable) on the best model (Decision Tree), we applied classical XAI techniques, such as SHAP, presented in Figs. 1a and 1b. By way of interpretation of the results, if the impact is positive, that factor will favour a "Not Released" decision. In contrast, if the impact is negative, the factor will favour a "Released" decision.

For the Brazilian dataset, we verify that the features with the highest impact in favour of the "Not Released" outcome are Factors 5 (remand, sending back the case), 6 (presence of wiretaps), and 9 (victim's statement). The crime category with the highest effect favouring the same outcome is Crime 2 (crime against property), while Crime 6 (criminal organization) has a high impact on both outcomes, although with a slight edge in favour of the "Released" decision. The only feature with a higher impact on the "Released" outcome is Factor 2, (suspension of time in prison or proceedings).

For the Italian dataset, we observe a clear distinction from the Brazilian results. First, we verify that Factor 8 (inferred facts) clearly favours the "Not Released" outcome. In the Italian case, most features had a higher impact, comparatively to the Brazilian results, in favour of the "Not Released" outcome. Only three factors favour the "Released" outcome, i.e., Factors 4 (connection between proceedings) and 6 (presence of wiretaps), and Crime 2 (crime against property).

These results, partially go against those obtained by the Brazilian dataset, where Factor 6 and Crime 2 are in favour of the opposite outcome (Not Released). In the Brazilian case, we also have more factors which do not clearly favour one of the two outcomes. In particular, Factors 1 (excess time in prison), 2 (suspension of time in prison), 5 (sending back the case), and 7 (complexity of the proceedings), as well as Crime 6 (criminal organization), do not have a clear edge or impact, although their effect is slightly in favour of the "Not Released" outcome. Factor 1 does not play a crucial role in the Brazilian dataset too.

4.3   Interpretation Based on Decision Trees

As a final experiment, we generated small decision trees from the structured datasets, presented in Figs. 2 and 3. This technique allows us to interpret the results of the model, by examining the correlation between the nodes representing the categorical variables.

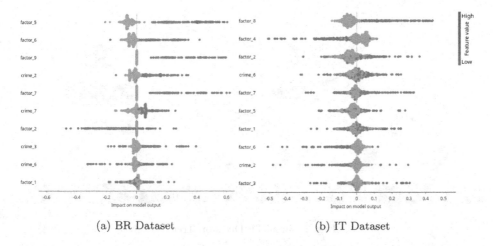

(a) BR Dataset   (b) IT Dataset

Fig. 1. SHAP analysis over trained Decision Trees on the two structured datasets.

Regarding the Italian decision tree, we can see that the presence of Crime 6 (criminal organizations) and Factor 8 (inferred facts), is the shortest path to a "Not Released" decision, containing 70 samples in the dataset. Similarly, the absence of Factor 8 (inferred facts) and the presence of Factor 11 (defence restriction) is also one of the most common paths for the "Not Released" dataset, with 37 instances.

The vast majority of "Released" decisions (253 data instances) occurs through the absence of Factors 8 (inferred facts), 11 (defence restriction), 2 (suspension of time in prison or proceedings), 4 (connection between crimes) and 9 (victim's statement). If we look for positive correlations starting from node #2, we can see that the absence of Factors 8 (inferred facts), 11 (defence restriction) and 2 (suspension of time in prison or proceedings), and the presence of Factor 7 (risk of the prisoner) and Crime 7 (drug law), lead to other 13 instances of a "Released" decisions. From this path, we can infer that when the committed crime is related to drugs, and there are no inferred facts (for example, there is no information that the prisoner is aggressive), it is possible that the defendant will be released since there is low risk for public order. In addition, it can be noted that when the prisoner is suspected of being part of a criminal organization, he can be released from pre-trial detention if he argues that he is not a risk to the proceedings and his detention time has been wrongly suspended. This corresponds to many of the "Released" instances we have analysed, and it is one of the results we expected.

Regarding the Brazilian small decision tree (3), we can see that the absence of Factor 5 (remittance of proceedings) with the presence of Crime 4 (against the government) and 7 (drug law) or Crime 6 (criminal organizations) is a path to a "Not Released" decision, containing respectively 125 and 21 samples in the dataset. As for crime against the government ("white collar" crime) in particular,

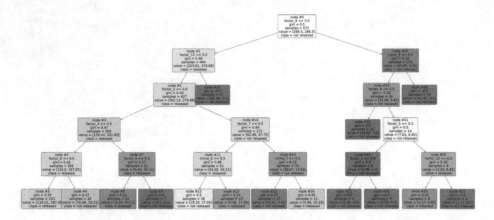

Fig. 2. Small IT Decision Tree

contrary to our expectations, leads to decisions to maintain imprisonment in the majority of the scenarios.

The almost totality of "Released" instances occurs with the absence of Factors 5 (remittance of proceedings), 7 (risk of the prisoner), 9 (victim's statement), 10 (*in flagrante delicto*), and 2 (suspension of time in prison or proceedings), while there are only 3 instances that are not part of the leftmost tree path. This path shows clearly which factors a case should not contain for the prisoner to achieve his or her freedom.

Comparing the two legal systems, we can observe that participation in criminal organizations in Brazil is a factor favouring imprisonment in both datasets. However, it is not necessarily clear in Italy. Also, there are certain factors that, when not addressed in the case, can lead to the prisoner's release in both countries, such as when the case allows for a statement or testimony from the victim.

Comparing the results provided by decision tree with those obtained through SHAP, we can observe that the two complement each other in the Italian case, although with inconsistencies. First, 8 out of the 10 factors detected by SHAP appear in the small decision tree, meaning that the importance of these 8 is confirmed. The scale to which each factor impacts the outcome, and which outcome in particular, is subject to discussion. For example, according to SHAP, Factor 6 is supposed to favour "Released" outcomes, while in the tree (node #19) it leads mainly to a "Not Released" decision. A similar result is obtained with the Brazilian dataset. 9 out of 10 factors appear in both decision tree and SHAP, sometimes even more than once. On the other hand, Factors 7 and 9 (nodes #1 and #2) lead to the result "Released", although according to SHAP they clearly favour the opposite outcome.

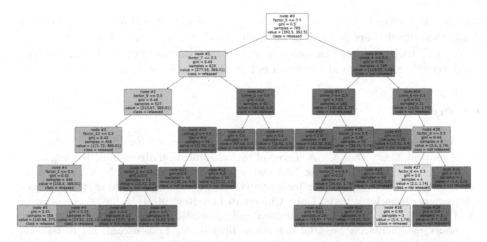

Fig. 3. Small BR Decision Tree

5   Conclusion and Future Works

Our results show that data structuring based on the extraction of factors from
the decision texts not only increases the prediction's quality but also allows for
their interpretation and explanation through XAI techniques.

Regarding the explanation based on SHAP and the interpretation based on
Decision Trees, we can conclude that our dataset does not contain any single
factor decisive on its own, even if the total number of favoured outcomes differs
highly between certain features. Through our analysis, we elicited connections
between factors that we did not expect in advance.

In this respect, we emphasize that both SHAP and Decision Trees showed
the same factors have the greatest impact in both legal systems. We observed
that, in the Brazilian scenario, Factor 5 (sending back the case) has a significant
impact on decisions about the release of prisoners. And in the Italian context,
we noticed that Factor 8 (inferred facts) is significant in decisions that maintain
imprisonment.

As for the crime category of involvement in criminal organizations, we found
that in the Italian dataset, it is not clear whether this category motivates the
Supreme Court to maintain the detention. This emerges from the SHAP and
the decision tree results. On the other hand, for the Brazilian dataset, this crime
category is clearly in favour of the "Not Released" outcome.

However, this crime category contributed to a "Not Released" outcome in
combination with additional factors that provide further information on the
case.

Future research in this area will focus on extracting minimal abducible and
contrastive explanations from Decision Trees. We will also look at models that
could help us in dealing with non-binary values and unbalanced datasets, to check
if we could improve our accuracy and F1 score. Furthermore, we will compare

the results obtained here with experiments from classification on unstructured data, to verify if there are any missing factors that our model does not take into account. Finally, further discussion is needed on whether decision trees, together with SHAP values, provide a global explanation of the model.

References

1. Aggarwal, C.C.: Machine Learning for Text. Springer, Cham (2018)
2. Aggarwal, C.C., Zhai, C.: A survey of text classification algorithms. In: Aggarwal, C.C., Zhai, C. (eds.) Mining Text Data, pp. 163–222. Springer, New York (2012)
3. Ashley, K.D.: Prospects for legal analytics: some approaches to extracting more meaning from legal texts. Univ. Cincinnati Law Rev. **90**(4), 5 (2022)
4. Dal Pont, T.R., et al.: Classification and association rules in Brazilian supreme court judgments on pre-trial detention. In: Kö, A., Francesconi, E., Kotsis, G., Tjoa, A.M., Khalil, I. (eds.) EGOVIS 2021. LNCS, vol. 12926, pp. 131–142. Springer, Cham (2021). https://doi.org/10.1007/978-3-030-86611-2_10
5. Demšar, J., et al.: Orange: data mining toolbox in Python. J. Mach. Learn. Res. **14**(1), 2349–2353 (2013)
6. Dipoppa, G.: How criminal organizations expand to strong states: migrant exploitation and political brokerage in northern Italy (2021)
7. Došilović, F.K., Brčić, M., Hlupić, N.: Explainable artificial intelligence: a survey. In: 2018 41st International Convention on Information and Communication Technology, Electronics and Microelectronics (MIPRO), pp. 0210–0215. IEEE (2018)
8. Hair, J.F., Black, B., Black, W.C., Babin, B.J., Anderson, R.E.: Multivariate Data Analysis. Cengage Learning, Andover (2019)
9. Horty, J.: Reasoning with dimensions and magnitudes. Artif. Intell. Law **27**(3), 309–345 (2019)
10. Horty, J.F., Bench-Capon, T.J.: A factor-based definition of precedential constraint. Artif. Intell. Law **20**(2), 181–214 (2012)
11. Katz, D.M., Bommarito, M.J., Blackman, J.: A general approach for predicting the behavior of the supreme court of the United States. PLoS ONE **12**(4), e0174698 (2017). https://doi.org/10.1371/journal.pone.0174698
12. Kotu, V., Deshpande, B.: Data Science, 2nd edn. Morgan Kaufmann (Elsevier Science), Cambridge, MA (2019)
13. Lundberg, S.M., Lee, S.I.: A unified approach to interpreting model predictions. In: Advances in Neural Information Processing Systems, vol. 30 (2017)
14. Sabo, I.C., Billi, M., Lagioia, F., Sartor, G., Rover, A.J.: Unsupervised factor extraction from pretrial detention decisions by Italian and Brazilian supreme courts. In: Guizzardi, R., Neumayr, B. (eds.) ER 2022. Lecture Notes in Computer Science, vol. 13650, pp. 69–80. Springer, Cham (2023). https://doi.org/10.1007/978-3-031-22036-4_7
15. Silva, I.N.D., Spatti, D.H., Flauzino, R.A., Liboni, L.H.B., Alves, S.F.D.R.: Artificial Neural Networks. Springer, Switzerland (2018)
16. Zhang, Y., Haghani, A.: A gradient boosting method to improve travel time prediction. Transp. Res. Part C: Emerg. Technol. **58**, 308–324 (2015)

Few-Shot Legal Text Segmentation via Rewiring Conditional Random Fields: A Preliminary Study

Alfio Ferrara[ID], Sergio Picascia[ID], and Davide Riva[(✉)][ID]

Department of Computer Science, Universitá degli Studi di Milano,
Via Celoria, 18, 20133 Milano, Italy
davide.riva1@unimi.it

Abstract. Functional Text Segmentation is the task of partitioning a textual document in segments that play a certain function. In the legal domain, this is important to support downstream tasks, but it faces also challenges of segment discontinuity, few-shot scenario, and domain specificity. We propose an approach that, revisiting the underlying graph structure of a Conditional Random Field and relying on a combination of neural embeddings and engineered features, is capable of addressing these challenges. Evaluation on a dataset of Italian case law decisions yields promising results.

Keywords: Text Segmentation · Legal Document Processing · Conditional Random Fields

1 Introduction

Several legal systems around the world are undergoing a complex process of digital transformation, in which a pivotal role is played by the digitization and automated processing of legal documents. Court decisions and law codes are known to be long and articulated documents, lacking of standard rules and conventions defining their structure.

Text Segmentation is concerned with the ex-post recognition of the structure of a textual document, which may be related to the discussed topics, *topical segmentation*, or to the functions that each segment plays in the text, *functional segmentation*. Here we look particularly at the latter, as an important contribution in the legal domain to provide valuable information to several downstream tasks, thus becoming a critical component of a complete Information Extraction pipeline [2]. However, we acknowledge the need to overcome 3 key challenges that concern the legal domain: (a) the articulated structure of documents comprises the possibility that parts of text which perform a single function are discontinuous within the text; (c) models need to operate in a *few-shot* scenario, characterized by the scarcity of annotated data and strong label imbalance; (d)

T. P. Sales et al. (Eds.): ER 2023 Workshops, LNCS 14319, pp. 141–150, 2023.
https://doi.org/10.1007/978-3-031-47112-4_13

specificity of the legal jargon, with its own semantics and syntax deviating from common language, requires expert knowledge for proper understanding.

Given the scarce availability of annotated data, in this study we explore 3 solutions that modify a Linear-Chain Conditional Random Field (CRF) model to address the 3 challenges outlined above. In particular, we propose adding connections between non-consecutive portions of text to deal with (a) while keeping the number of parameters feasible. To tackle (b), we inject prior statistical information into the model. Finally, we measure the impact of domain-specific feature extraction models to confront (c). The resulting model is evaluated on a dataset of Italian court decisions which underwent manual segmentation by part of legal experts in the context of the *Next Generation UPP* Project, funded by the Italian Ministry of Justice.

The paper is organized as follows: Sect. 2 describes previous work on Text Segmentation with a focus on the legal domain; Sect. 3 presents the proposed approach; Sect. 4 contains an evaluation of the approach on the dataset of Italian court decisions; and Sect. 5 discusses the conclusions and future work.

2   Related Work

Text Segmentation (TS) consists in partitioning a text into coherent portions called *segments*. Coherence may be interpreted from a topical, semantic, structural, or functional perspective, and criteria for partitioning are highly dependent on the given interpretation. TS approaches can be categorized as *linear* or *hierarchical* [5], where a linear approach sees a textual document as a sequence of segments [7], while a hierarchical approach splits segments at several levels, down to a predefined granularity [4]. From another point of view, we can distinguish between *region-oriented* approaches, that aim at detecting segment boundary position [12], and *class-oriented* approaches that classify each text unit (be it a paragraph, sentence, clause, or even a single word) into a segment type [11].

In the legal domain, Aumiller et al. [1] advocated for topical TS as a way to improve downstream applications, such as information retrieval and document summarization. However, we argue that *functional* TS may often be more relevant than topical TS in the legal domain, since the same information may be more valuable when found in certain parts of the document (e.g. highly argumentative parts) rather than others (e.g. introductory parts). Functional TS of legal documents has been addressed in the context of judgements by the US Security and Exchange Commission, using CRFs [11], as well as Italian court decisions, adding a sentence-level mean-pooling layer and feed-forward neural network on top of a BERT model fine-tuned for Italian legal language [9]. Both approaches, as well as ours, fall in the *linear*, *class-oriented* category; however, it must be noticed that approaches are hardly portable from one application to another, due to the development of different segmentation schemas for different legal documents and the subsequent need to annotate a sufficient amount of documents to re-train models. Furthermore, supervised approaches like the ones in this category often have to deal with a few-shot scenario with significant label imbalance.

Inspired by GRAPHSEG [5], our model builds a sentence relatedness graph and trains a classifier on it. However, to capture functional instead of semantic relatedness, we adopt a supervised approach to graph construction. In the same way, we train a CRF to classify graph nodes with manually defined labels instead of relying solely on an unsupervised clustering algorithm, which would produce an unlabeled classification. Thus, our work is, to the best of our knowledge, the first to explore functional TS of legal documents in a few-shot scenario, simultaneously addressing domain specificity, segment discontinuity, as well as class imbalance.

3 Methodology

The proposed approach to Text Segmentation is displayed in Fig. 1.[1]

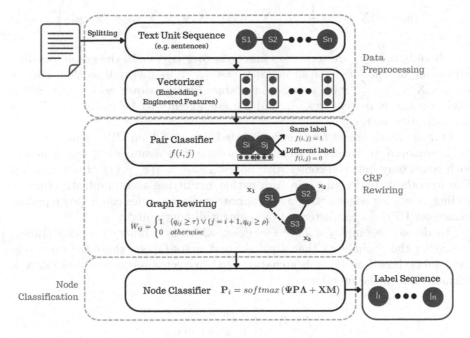

Fig. 1. Workflow of the proposed approach.

3.1 Data Preprocessing

TS models typically operate on sequences of text units, which can be single words, phrases, clauses, sequences or even paragraphs according to the granularity level that is appropriate for the application at hand.

---

[1] A Python package will be made available at https://github.com/umilISLab/ TextSegmentation.

After being split, each text unit s_i is represented as a d-dimensional vector $\mathbf{x}_i$ using a Vectorizer. The Vectorizer concatenates the embedding vectors generated by a neural embedding model (e.g. BERT [3]) with additional, engineered features that have been found to be informative in legal text structure [6]. Such variables are the counts of verb moods and tenses, the number of references to law articles and previous judgements, the sentence length, and the sentence position in the document[2].

3.2 CRF Rewiring

Given a graph $G = (V, E)$ with N nodes where each node v has an associated feature vector $\mathbf{x}_v$ and a label y_v with value in label set $L = \{l_1, \dots, l_K\}$, a CRF [8] is a discriminative graphical model whose general form is written as:

$$\mathbb{P}(y_i = l_k | \mathbf{X}, G) = \frac{1}{Z} exp \left(\sum_{e \in E} \lambda_k t_k(e, \mathbf{X}, \mathbf{y}) + \sum_{v \in V} \mu_k s_k(\mathbf{x}_v, y_v) \right) \qquad (1)$$

where $t_k(e, \mathbf{X}, \mathbf{y})$ are *transition functions* over edges e of the graph, possibly depending on the labels of all nodes $\mathbf{y} = (y_1, \dots, y_N)$ as well as their feature vectors $\mathbf{X} = (\mathbf{x}_1, \dots, \mathbf{x}_N)$, $s_k(\mathbf{x}_v, y_v)$ are *state functions* over node v, depending solely on the node features, λ_k and μ_k are weights to be learnt and Z is a normalization factor.

Our approach starts from an undirected, Linear-Chain CRF, which is typically employed in sequential labelling problems as it assumes a graph structure with edges only between consecutive nodes, i.e. $E = \{(v_i, v_{i+1}) : i = 1, \dots, N\}$. The hypothesis underlying our work is that modifying such graph structure by adding/removing edges between appropriately chosen nodes can improve performance on TS tasks characterized by segment discontinuity.

To do so, we employ a *Pair Classifier*, which is a binary classifier trained to predict the probability that a pair of text units (s_i, s_j) share the same segment label, based on the concatenation of their vector representations $(\mathbf{x}_i, \mathbf{x}_j)$. Formally, given the function:

$$f(i, j) = \begin{cases} 1 & y_i = y_j \\ 0 & otherwise \end{cases} \qquad (2)$$

we estimate the probability $q_{ij} = \mathbb{P}(f(i, j) = 1 | \mathbf{x}_i, \mathbf{x}_j)$.

A natural advantage of employing such a model rather than a model that directly aims at predicting segment labels is that the Pair Classifier has at its disposal $\binom{N}{2}$ data for training from a single document made of N text units, thus mitigating possible overfitting issues.

The probabilities q_{ij} predicted by the Pair Classifier are then exploited to rewire the linear-chain graph structure based on the following two rules:

---

[2] Notice that the proposed model is capable of working with any vector representation of the analyzed text units.

– an edge is added between two non-consecutive nodes (v_i, v_j) if $q_{ij} \geq \tau$;
– an edge between two consecutive nodes (v_i, v_{i+1}) is removed if $q_{i,i+1} < \rho$.

An example of the procedure is provided in Fig. 2.

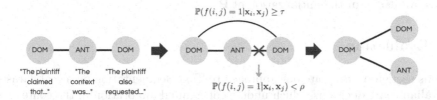

Fig. 2. Example of rewiring with 3 text units having labels DOM, ANT and DOM again (see Sect. 4.1 for details). First, we have a linear-chain structure, then the Pair Classifier evaluates the probability of each connection, and finally one new connection is added between the two DOM-labelled units and an existing one is removed, complying with probability thresholds.

In order to favor the flow of information in the sequence of units $(s_1, \ldots, s_N)$ with respect to "residual" connections between non-consecutive ones, we recommend $\rho \leq \tau$. The larger the difference, the more sequential connections are favored. A stricter constraint on residual connections makes passage of information possible between text units that belong to the same segment with a high degree of certainty. Finally, self-loops are added by default for each node.

The graph $\tilde{G}$ resulting from this rewiring operation will have an adjacency matrix $\mathbf{W}$ such that:

$$W_{ij} = \begin{cases} 1 & (q_{ij} \geq \tau) \vee (j = i+1, q_{ij} \geq \rho) \\ 0 & otherwise \end{cases} \quad (3)$$

Fringe cases are $\tau = \rho = 0$, which yields a fully connected graph, and $\tau = \rho = 1$, for which a link appears only for completely certain connections.

3.3 Node Classification

Once the rewired graph $\tilde{G}$ has been constructed, a CRF classifier is applied for node classification:

$$\mathbf{P}_i = softmax\left(\mathbf{\Psi P \Lambda + XM}\right) \quad (4)$$

where P_{ik} denotes the probability of label l_k for node v_i, $\mathbf{\Psi} \in [0,1]^{N \times N}$ is the transition matrix of the rewired graph, obtained by row-normalization of the adjacency matrix, $\mathbf{X} \in \mathbb{R}^{N \times d}$ is the feature matrix of all nodes, and $\mathbf{\Lambda} \in \mathbb{R}^{K \times K}$ and $\mathbf{M} \in \mathbb{R}^{d \times K}$ are the weight matrices to be learnt. Since Eq. 4 is implicit, having $\mathbf{P}$ on both the left-hand and right-hand sides, we initialize $\mathbf{P}$ and compute it iteratively relying on row-stochasticity of $\mathbf{\Psi}$ to ensure convergence.

To address the problem of class imbalance, we propose a combination of two solutions. The first is *loss weighting*, which consists in weighting each segment label l_k in the training phase by a factor $\frac{N}{N_k K}$, where K denotes the cardinality of the label set and N_k is the number of text units in the training set labelled with l_k. The second solution is to inject prior *positional knowledge* of labels, where available, in the initialization of **P**.

4   Evaluation

In this section we evaluate our approach to Text Segmentation on a real dataset of Italian court decisions, which underwent manual annotation in the context of the *Next Generation UPP* Project, funded by the Italian Ministry of Justice.

4.1   Dataset

The dataset employed for the evaluation consists of 50 Italian case law decisions, retrieved from 12 Courts in Northern Italy and concerning first degree civil law judgements on the matter of unfair competition.[3] The documents have been manually annotated with functional segments by a group of 9 legal experts through an interactive annotation activity. We refer to [13] for further details. The annotation schema, presented in Table 1, comprises five segment labels, representing the functional role a segment plays in the legal document. A NULL label is automatically added to indicate segments with no specific function.

Table 1. The annotation schema for Italian court judgements of civil proceedings.

Labels	Italian	Explanation
COR	Corte e Parti	Court, judicial panel, parties
ANT	Antefatto	Background information
DOM	Domande	Claim(s) and argumentation of the parties
MOT	Motivazione	Reason(s) for the final decision(s)
DEC	Decisione	Final decision(s)

The dataset contains annotated segments in the form of quintuples (document ID, start, end, text, label), where start and end are integers indicating a character position in the document, while label belongs to the set of labels. Segments cannot overlap with one another. Moreover, since at the present moment there is no rule strictly imposing the structure of a court decision in Italian judicial system, segment labels have no semantic nor order relationship with one another. The only two types of segments for which we have reasonable a-priori

---

[3] The dataset cannot be made public at the moment due to privacy-related constraints. An anonymization process to make the data public is ongoing.

positional knowledge are COR, expected at the beginning of a document, and DEC, expected at the end.

For the purpose of our experiments, among the annotated 50 documents, we filter the 38 court decisions with a complete annotation. Each document is split in sentences, which are the text units considered by our model. Since segment boundaries resulting from the manual annotation activity do not always match sentence boundaries, we adopt a heuristic approach to label sentences by assigning label y to a sentence s, if s overlaps by at least $1/2$ of its length (as number of characters) with a segment labeled with y. No sentence in the dataset is equally shared between two segments with different labels, and, in case a sentence overlaps with two such segments, the difference in overlapping lengths is never under 20% of the sentence length. The resulting dataset, which undergoes no further preprocessing, contains documents having an average of ~ 121 sentences (s.d. ~ 54) and ~ 4114 words (s.d. ~ 1821). Label imbalance is evident in that an average of ~ 56 sentences per document are labelled as MOT, while only ~ 7 with DOM, with counts of other labels in between.

Since Inter-Annotator Agreement metrics showed the heterogeneity of annotations, hinting at its intrinsic complexity and subjectivity, we regard a perspectivist approach as preferable. Perspectivist approaches to ground-truthing consist in refraining from the definition of a univocal ground truth so to capture the perspective of different annotators, insofar as they are deemed reliable. As a consequence, in case a single segment is assigned two different labels by two different annotators, we do not aim at solving discrepancies. As an example, the sentence *"All issues are outweighed by the current term of protection (70 years after the author's death)"*[4] is labelled by one annotator as ANT and by another as MOT, thus the two labels are assigned 50% probability each.

4.2 Setting

We implement our model adopting a feed-forward neural network with a single hidden layer as Pair Classifier, and train both the Pair Classifier and the Node Classifier with the AdamW algorithm [10] to minimize cross-entropy loss.

We study the impact of domain-specific features by experimenting with 3 different embedding models: a transformer model fine-tuned on Italian legal documents called Italian-Legal-BERT[5] [9] with mean pooling (ILB), a Sentence-BERT model pre-trained on Italian language (SBERT)[6], and a non-neural model based on TFIDF with ICA dimension reduction so to match the output dimensionality of the other models, i.e. 768. We experiment with a sparse ($\tau = 0.95$), a dense ($\tau = 0.65$), and an intermediate rewiring ($\tau = 0.80$), keeping in all cases $\rho = \frac{\tau}{2}$. Positional knowledge injection for addressing label imbalance was obtained by initializing $\mathbf{P}$ so that the probability of label COR for the first sentence and that of label DEC for the last sentence of each document is set to 1.

---

[4] Traslated from Italian.
[5] Model available at https://huggingface.co/dlicari/Italian-Legal-BERT.
[6] Model available at https://huggingface.co/nickprock/sentence-bert-base-italian-xxl -uncased.

To evaluate how our approach deals with segment discontinuity and class imbalance, we compare it against two baseline models in a 4-fold Cross Validation framework:

- a ONE-BY-ONE classifier, which takes the vector representation of each sentence individually and trains a shallow classifier (e.g. Random Forest) to predict its label;
- a SEQUENTIAL model, equivalent to our model before rewiring, i.e. a Linear-Chain CRF[7].

4.3 Results

Results, in terms of average F_1 scores, are presented in Table 2.

Table 2. 4-fold Cross Validation average F_1 scores.

		Embedding			Embedding + Features		
		ILB	SBERT	TFIDF	ILB	SBERT	TFIDF
ONE-BY-ONE		0.547	0.523	0.505	0.582	0.558	0.599
SEQUENTIAL		0.622	0.547	0.555	**0.656**	0.607	0.601
REWIRED(Ours)	Sparse	**0.650**	0.572	**0.556**	**0.656**	0.559	**0.616**
	Intermediate	0.627	0.560	0.521	0.530	**0.621**	0.591
	Dense	0.620	**0.587**	0.518	0.554	0.543	0.587

F_1 scores are naturally informative for what concerns (i) the capacity of a model to deal with segment discontinuity and (ii) the impact of domain-specific features.

For (i), we notice little improvement from the rewiring operation, whose outcome is closely comparable with the one of the SEQUENTIAL model. In general, rewiring that produces a more sparse graph is preferable, while a dense structure is even detrimental, up to a 0.1 difference in F_1 scores, and performs worse than ONE-BY-ONE models.

For (ii), results show a consistent improvement when an model fine-tuned on domain data (Italian-Legal-BERT) is used. Nevertheless, including the additional, engineered features discussed in 4.2 seem to provide an equally consistent improvement, contributing with information that may not be captured by the embedding model. Indeed, while transition functions $\mathbf{\Psi P \Lambda}$ have higher coefficients (in absolute value) with respect to state functions $\mathbf{X M}$, additional features play an important role. For instance, verbs at future tense and verbs at subjunctive mood are always among the 10 features with the strongest influence on the prediction of DOM.

---

[7] We employed CRFSuite implementation for this model, available at www.chokkan. org/software/crfsuite.

As error analysis, Fig. 3 shows the confusion matrix between predicted and ground truth labels.

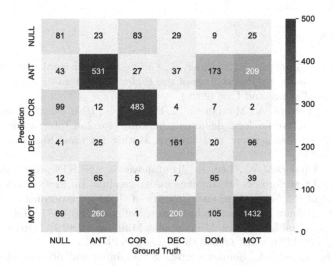

Fig. 3. Confusion matrix.

Besides the label imbalance, it can be appreciated that confusion between segments labelled with COR and DEC, which are expected at the opposite ends of a document, is minimal. Conversely, ANT, MOT and DEC-labelled segments tend to intertwine the most, since final decisions are often introduced and background information is often recalled within the reasoning section. Indeed, even annotators had the highest disagreement on label pairs (MOT, DEC) and (DOM, ANT). [13]

5   Conclusion and Future Work

In this paper we addressed the Text Segmentation task proposing a model that, revisiting the underlying graph structure of a Conditional Random Field, aims at handling segment discontinuity and operates in a few-shot scenario with scarcity of data. Moreover, we experimented with domain-specific embedding models and engineered features in order to capture more information than we would have with a general embedding model. The preliminary results confirm our hypothesis that domain-specific, engineered features provide useful information to the model, and that our methods to tackle label imbalance are effective.

This work will serve as basis for the integration of a segmentation module in a complete information retrieval pipeline tailored to the legal domain, for instance for document retrieval and building (see [2] for more). To achieve such objective, we aim at constructing a model which has generalization, scalability and domain specificity as its key characteristics.

Acknowledgment. This work is partially supported by the Next Generation UPP project within the PON programme of the Italian Ministry of Justice and by project SERICS (PE00000014) under the MUR NRRP funded by the EU - NextGenerationEU.

References

1. Aumiller, D., Almasian, S., Lackner, S., Gertz, M.: Structural text segmentation of legal documents. In: Proceedings of the Eighteenth International Conference on Artificial Intelligence and Law (2020)
2. Castano, S., Ferrara, A., Montanelli, S., Picascia, S., Riva, D.: A knowledge-based service architecture for legal document building. In: 2nd International Workshop on Knowledge Management and Process Mining for Law. Sherbrooke, Quebec, Canada (2023)
3. Devlin, J., Chang, M.W., Lee, K., Toutanova, K.: BERT: pre-training of deep bidirectional transformers for language understanding. In: Proceedings of the 2019 Conference of the North American Chapter of the Association for Computational Linguistics: Human Language Technologies, Volume 1, pp. 4171–4186. Association for Computational Linguistics, Minneapolis, Minnesota (2019). https://doi.org/10.18653/v1/N19-1423, https://aclanthology.org/N19-1423
4. Glavas, G., Ganesh, A., Somasundaran, S.: Training and domain adaptation for supervised text segmentation. In: Workshop on Innovative Use of NLP for Building Educational Applications (2021)
5. Glavas, G., Nanni, F., Ponzetto, S.P.: Unsupervised text segmentation using semantic relatedness graphs. In: International Workshop on Semantic Evaluation (2016)
6. Grover, C., Hachey, B., Korycinski, C.: Summarising legal texts: sentential tense and argumentative roles. In: HLT-NAACL 2003 (2003)
7. Koshorek, O., Cohen, A., Mor, N., Rotman, M., Berant, J.: Text segmentation as a supervised learning task. In: North American Chapter of the Association for Computational Linguistics (2018)
8. Lafferty, J.D., McCallum, A., Pereira, F.: Conditional random fields: probabilistic models for segmenting and labeling sequence data. In: International Conference on Machine Learning (2001)
9. Licari, D., Comandè, G.: Italian-legal-BERT: a pre-trained transformer language model for Italian law. In: CEUR Workshop Proceedings (Ed.), The Knowledge Management for Law Workshop (KM4LAW) (2022)
10. Loshchilov, I., Hutter, F.: Decoupled weight decay regularization. In: International Conference on Learning Representations (2017)
11. Savelka, J., Ashley, K.D.: Segmenting US court decisions into functional and issue specific parts. In: International Conference on Legal Knowledge and Information Systems (2018)
12. Solbiati, A., Heffernan, K., Damaskinos, G., Poddar, S., Modi, S., Calì, J.: Unsupervised topic segmentation of meetings with BERT embeddings. ArXiv abs/2106.12978 (2021)
13. Zanoli, E., et al.: Annotators-in-the-loop: testing a novel annotation procedure on Italian case law. In: Proceedings of the 17th Linguistic Annotation Workshop (LAW-XVII), pp. 118–128. Association for Computational Linguistics, Toronto, Canada (2023). https://aclanthology.org/2023.law-1.12

Identification and Visualization of Legal Definitions and Legal Term Relations

Catherine Sai[1][✉], Anastasiya Damaratskaya[1], Karolin Winter[2],
and Stefanie Rinderle-Ma[1]

[1] TUM School of Computation, Information and Technology,
Technical University of Munich, Garching, Germany
{catherine.sai,anastasiya.damaratskaya,stefanie.rinderle-ma}@tum.de
[2] Department of Industrial Engineering and Innovation Sciences,
Eindhoven University of Technology, Eindhoven, The Netherlands
k.m.winter@tue.nl

Abstract. Reading, analyzing, and implementing regulatory documents
are cumbersome and still mostly manual tasks. The constantly increasing
amount of regulatory documents causes an equally increasing need for
supporting those tasks through information systems. One key aspect for
accomplishing those tasks is to acquire knowledge of the different legal
definitions, i.e., legal terms accompanied by their explanations, in order
to build a legal vocabulary or an ontology. This paper proposes an approach taking European regulations as input and i) automatically extracting legal definitions, ii) determining semantic relations such as hyponyms,
meronyms, and synonyms between legal terms, and iii) visualizing the
results in form of a knowledge graph and statistics. The approach is
evaluated on European regulations and made accessible particularly for
non-technical users through an easy-to-use web service.

Keywords: Legal Definitions · Legal Knowledge Extraction · Legal
Relations Graph · Natural Language Processing · Regulatory
Documents

1 Introduction

Understanding regulatory documents like legal texts is an important basis for
any business in order to ensure its compliance with the therein outlined requirements. Unfortunately, due to the complexity and ambiguity of such regulatory
documents, their analysis is still performed mostly manually resulting in a time-
consuming, cumbersome, and error-prone task. The rising number of regulatory
documents aggravates this, leading to an increasing need for automated support through information systems. Identifying legal definitions and determining
their semantic relations constitutes one core challenge toward automated analysis of regulatory documents. The presented approach takes up the challenge
of automating this task for European regulations from the EUR-Lex database[1].

---

[1] https://eur-lex.europa.eu.

T. P. Sales et al. (Eds.): ER 2023 Workshops, LNCS 14319, pp. 151–161, 2023.
https://doi.org/10.1007/978-3-031-47112-4_14

We automatically extract legal definitions consisting of the legal term and its corresponding explanation, i.e., "'personal data' means any information relating to an identified or identifiable natural person ('data subject'); [...]" [8]. Moreover, relations among legal terms like hyponyms, meronyms, and synonyms are determined. To ensure accessibility for non-technical users, the approach, similar to tools like [24], is provided as a web service taking the EUR-Lex document's CELEX number, i.e., a unique identifier, as single input. The results are depicted in the form of i) text files containing extracted information, e.g. all definitions, ii) an annotated document allowing for quick look-up of explanations, iii) a frequency diagram showing how often a legal term appears in a paragraph or article, and iv) a knowledge graph (KG) representing the relations among legal terms. This publication is based on and extends the work of one of the author's bachelor thesis [7].

2   Related Work

This work approaches legal definition extraction, semantic relation determination as well as visualizations of the information and insights drawn from them. Similar to our approach, existing works use a combination of NLP techniques like POS tagging, tokenization, and pattern matching through, e.g., regular expressions. However, each of those approaches considers a different input in terms of document type and language, i.e., [2] focus on Indonesian laws, [12] on Brazilian laws, [15] on Chinese laws, [18] on Japanese acts, [22] on German laws and [17,23] on Dutch laws. Our approach differs from theirs as we consider not a single country's laws but European regulations. Though this makes no difference in terms of methodology, from a technical perspective it ensures broader applicability in practice as these documents typically apply to multiple countries and affect a broader range of companies and people.

Furthermore, the mentioned approaches differ in their output and goal. [12] mainly focus on extracting the definitions while [23] also aim for identifying related concepts, [15] construct an ontology, [18] build a legal terminology consisting of legal terms and their explanations and [2] use density-based clustering to determine groups of similar definitions in order to identify inconsistencies. The output of our approach incorporates a combination of these goals.

Besides extracting legal definitions, [22] also determine further relevant semantic information, such as the year of dispute, and [17] investigated the meaning of definitions in regulatory documents deeper, further pointing out the similarity of text patterns by describing the legal term. Compared to this, our approach presents statistical information on the usage of definitions throughout the document. As legal experts might already know which definitions are relevant in their case, by providing statistics per, e.g., term frequency in each paragraph of a regulatory document, those parts of the document containing no or only a few mentions of the desired legal definition can be identified and excluded quickly. [13] applies a zero-shot learning classifier to retrieve a documents relevant terminology and provide concept-term-relations in form of a graph. In contrast to our approach, the users have to provide an initial set of concepts and the evaluation was performed on a US state case law.

Visualization of legal definitions and their relations is, e.g., addressed in [6]. Similar to our approach, a graph-based view is selected, but in contrast, the relations are pre-established. Also, outside of the legal domain, there has been work on definition extraction, i.e. the works of [3,21] are based on English textbooks, fine-tuning pre-trained language models, which require large amounts of labeled data and deliver less reliable results compared to our approach.

In addition, we implemented a publicly available web service using the EUR-Lex API, which allows also non-technical users to easily access all functionalities.

3 Definition Extraction, Relation Determination and Visualization

The overall approach for extracting legal definitions, their relations, and visualizing them is depicted in Fig. 1 and consists of four components. While the *pre-processing*, Sect. 3.1, ensures a valid input is provided, the main contribution of this paper is located in the *processing* component, Sect. 3.2, containing the *Legal Definition Extraction* and *Semantic Relation Extraction* stages. In order to convey the findings also to non-technical users, in the *post-processing* component, Sect. 3.3, we represent the results in various visual manners.

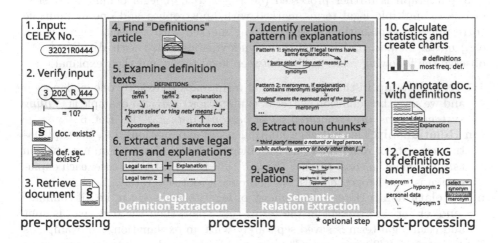

Fig. 1. Overview of the Presented Approach

3.1 Pre-processing

The pre-processing component of the approach is built to handle European regulations. The user defines a document for analysis by entering its CELEX number (cf. Fig. 1, **1.**). The provided CELEX number is then verified in terms of existence, validity, and applicability to our approach (cf. Fig. 1, **2.**), i.e. the regulatory document is a regulation containing an article *"Definitions"*. Finally, the document is retrieved from EUR-Lex (cf. Fig. 1, **3.**) and the processing starts.

3.2 Processing

This component consists of two main stages.

Legal Definition Extraction. After retrieving the desired regulation, the document is parsed in order to identify the section containing the legal definitions (cf. Fig. 1, **4.**). For this, we consider the joint practical guide [4] outlining principles for writing European legislative documents, to identify the parts of the document containing definitions. It highlights that, to avoid faulty interpretation of legal acts and improve general perception, ambiguous terms should be defined in a single article called *"Definitions"*. This article should be at the beginning of the document, and the definitions respected throughout the act. If the separate article containing defined terms does not exist, then regulatory documents include them in the first article together with the scope. Other than that, legal definitions are required to not contain autonomous normative provisions since it can lead to misinterpretation. This is a step that is tailored towards the types of documents we consider, but can be adapted to accommodate other documents than European regulations. The further steps of this component are kept generic.

After the parts of the document containing definitions are retrieved, each text paragraph is further processed (cf. Fig. 1, **5.**). A legal definition consists of one or multiple legal terms marked with apostrophes and one or multiple corresponding explanations of the meaning marked with a specific signal word, e.g., *mean*, *include*, and *be* and their synonyms identified with WordNet[2]. We observe the following relationships between legal terms and their explanations.

1:1 Relationship In this case, one legal term maps to exactly one explanation and we save the definition as (term, explanation) pair. An example is *"'filing system' means any structured set of personal data [...]"* [8], Def. 6

1:n Relationship This reflects a multiplicity of explanations, i.e., the explanation consists of more than one text block and includes a colon, e.g., *"'main establishment' means: (a) [...]; (b) [...];"* [8], Def. 16. In this case, each explanation is saved together with the sentence's root separately.

m:1 Relationship Here, a multiplicity of legal terms is given, i.e., if commas, 'and' or 'or' are present in the terms text span, then the text span is split and each legal term is saved separately with an explanation. An example is *"'C1 tyres', 'C2 tyres' and 'C3 tyres' means tyres belonging to the respective classes set out in Article 8(1) of Regulation (EC) No 661/2009;"* [10] Def. 1

m:n Relationship In this case, we have a multiplicity of legal terms and explanations. Both of them are split and each combination of term and explanation is saved as a pair. For example *" 'term 1' or 'term 2' means: 'explanation a'; 'explanation b' "* would be split into the pairs *'term 1 - explanation a'; 'term 1 - explanation b'; 'term 2 - explanation a'; 'term 2 - explanation b'*

---

[2] https://wordnet.princeton.edu/.

Those four relationships are essential for the Semantic Relation Extraction component, e.g., m:1 and m:n demonstrate the synonymy between legal terms, while 1:1 and 1:n show the possibility of at least one other semantic relation. Additionally to the mentioned relations, we also take into account that one legal term can be subsumed by another legal term but their explanations fundamentally differ, e.g., *personal data* and *personal data breach* [8].

The legal definition extraction algorithm (details cf. [7]) first identifies the *"Definitions"* article, and then searches the first clause of each text paragraph for apostrophes indicating a legal definition. In order to make the algorithm more flexible and capable of dealing with different phrasings, in contrast to regular expression-based approaches like [2,12,19], it makes use of NLP techniques such as tokenization, lemmatization, dependency parsing, and POS tagging. However, we face the same challenge as mentioned in [19], i.e., needing to define where the legal term ends and where its explanation starts. The sentence's root is taken as a leading separator of the legal term and explanation parts. It is identified by applying tokenization and dependency parsing and saved if it is a verb or auxiliary, and its lemma is a synonym of one of the most probable signal verbs. Subsequently, the definition is separated into the legal term and explanation parts and saved according to a relationship type (cf. Fig. 1, **6.**).

Semantic Relation Extraction. This stage of the approach focuses on resolving semantic relations among legal terms such as hyponyms, meronyms, and synonyms. A hyponym is a semantic relationship between words in which the meaning of one word is included in the meaning of another word [5,16]. An example is: "'static nets' means any type of gillnet[...]" [9], Def. 23, here "static nets" is a hyponym of "gillnet". A meronym is a semantic relationship between words in which the meaning of one word is part of another word [16]. An example is: "'codend' means the rearmost part of the trawl[...]" [9], Def. 33, here "codend" is a meronym of "trawl". A synonym is a semantic relationship between words in which words have a similar meaning. An example is: "'purse seine' or 'ring nets' means any surrounding net [...]" [9], Def. 21, here "purse seine" and "ring nets" are synonyms. The classification of semantic relations between legal terms within the regulation is required for various semantic interpreting tasks, such as textual entailment and inquiry answering. However, in most circumstances, identifying a semantic relation between terms is rather challenging [1]. Therefore, we identify a set of patterns in the legal definitions explanation to classify relations as hyponyms, meronyms, and synonyms (cf. Fig. 1, **7.**). The patterns depend i.a. on noun chunks, certain signal-words, punctuation marks, and their combinations, they are described in the following. The legal definition's explanation

1. **refers to another legal document**, e.g. "'product database' means the product database established pursuant to Article 12 of Regulation (EU) 2017/1369;" [10], Def. 10.
2. **is identical for multiple legal terms**, which indicates synonymy.
3. **contains signal-word** e.g. "type of" indicates a hyponym, while "part of" indicates a meronym relation

4. **contains specific punctuation mark + signal-word pattern**, e.g. "'good status' means: (a) for surface water, [...]; (b) for groundwater, [...]" [10], Def. 22. Here the pattern colon and "for" applies, resulting in considering only the main clause.
5. **has a signal-word as a first noun chunk**, such as "following". In this case, the approach continues with the iteration over noun chunks.
6. **contains enumeration after first noun chunk**, e.g. "'end-user' means a consumer, fleet manager or road transport undertaking [...]" [10], Def. 19.
7. **contains multi-word noun chunk**, e.g. "'regional indicative programme' means a multi-country indicative programme [...]" [11] Def. 3.

Various sets of signal words were identified for the different patterns and semantic relationship categories. If none of the above patterns is found, a hyponymy relation is assumed since the explanation typically clarifies the term by using more general terminology with specific details. For semantic relation patterns 3. – 7., the noun chunks are extracted from the explanation (cf. Fig. 1, **8.**) before saving the detected relations (cf. Fig. 1, **9.**).

3.3   Post-processing

In order to convey the results of the two previous components to non-technical users, graphical representations are provided through a web service (cf. Fig. 2).

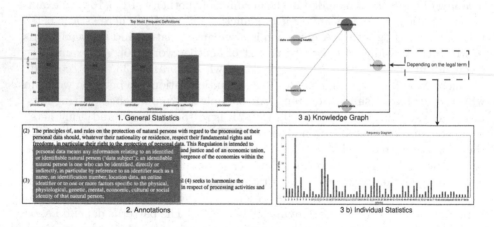

Fig. 2. Visualizations of Results

One of these representations are statistics on the number of occurrences of a legal term in the document (cf. Fig. 1, **10.** and Fig. 2, 1., 3b)). This helps users to gain an overview of the most common terms and the overall intention and objective of the document, e.g., if a user is interested only in specific legal terms, the reading effort can be reduced as only those articles containing the desired legal term can be immediately identified (cf. [25]).

Furthermore, we annotate the regulatory document with the extracted definitions (cf. Fig. 1, **11.** and Fig. 2,2.) by providing the explanations when a user hovers over a legal term that is defined in another part of the document.

For the semantic relations visualization (cf. Fig. 1, **12.** and Fig. 2,3 a)), we follow work on representation of citations in legal texts, e.g., [20]. We opt for a knowledge graph that depicts a legal term together with its related terms as nodes and the semantic relations such as meronymy, synonymy, hyponymy as edges. The visualization of all relations and legal terms would create a too complex graph since on average a regulation contains around 25 definitions (calculated based on the average of the number of definitions taken from the evaluation, cf. Sect. 4). Based on this, and the assumption of the hyponymy relation per default, we can conclude there will be at least 25 relations between two terms, which results in at least 50 nodes, leading to a too complex and indecipherable graph. Therefore, users can analyze each legal term separately by selecting the legal term and a relation type from a list. Afterward, the corresponding knowledge graph is displayed. The node colors indicate different term relations, i.e., red represents the selected legal term, pink other related legal terms, and blue related terms that were not identified as legal terms within this document.

4   Evaluation

Sect. 4.1 describes the data set used for the evaluation, the results are provided in Sect. 4.2. Precision, recall, and F1-score are measured for a) identification of legal definitions, their separation into legal terms and corresponding explanations, b) relation extraction among legal terms, and c) annotating legal terms in the document. Limitations of the approach are discussed in Sect. 4.3.

The approach was prototypically implemented using Python 3, Django, BeautifulSoup, spaCy, WordNet, NLTK, Matplotlib, NetworkX in the backend and core part of the approach. For the frontend HTML, CSS, and Bootstrap were used. The developed web service is publicly accessible at https://demos.bpm.cit. tum.de/legal_definitions/ and the source code and data set used for the evaluation are also made available[3].

4.1   Data Set

The evaluation data set was gathered by searching the EUR-Lex database in March 2023 using the keyword *regulation* and randomly selecting 17 regulations containing articles with definitions. Those 17 regulations contain on average 25 definitions per document and are used to evaluate the ability to extract a) legal definitions and b) semantic relations among legal terms. In order to evaluate c) the ability to extract text segments containing definitions and attach annotations 3 of the 17 regulations are selected. This subset is selected since the gold standards in form of CSV files containing the expected results were derived manually and it is infeasible to check annotations for all those documents.

---

[3] https://github.com/AnastasiyaDmrsk/Legal-Definitions-and-their-Relations.

4.2   Evaluation Results

Due to the fact that the extraction of semantic relations is directly dependent on the correctness of extracted definitions, both tasks of definition and relation extraction are evaluated in conjunction. To calculate all the measures, the gold standards are compared, and the resulting values depicted in Tab. 1.

Table 1. Results for a) definition extraction, b) relation extraction, c) annotations

	precision	recall	F1
a) 452 definitions	1.00	0.97	1.00
b) 568 relations	0.97	0.97	0.97
c) 167 sentences	1.00	1.00	1.00

Both the Definition and Relation Extraction components deliver almost perfect precision, recall, and F1 scores (cf. Table 1, a) and b)). Only one definition is not extracted correctly, which is due to the word "encompass" being used between the legal term and its explanation, which is not included in our signal word list (mean, include, be) and their synonyms. In the semantic relation extraction, some inaccurate behavior originates from the *noun_chucks* function provided by spaCy, as it sometimes faulty recognized first occurred noun(s) by legal definitions, especially, if a hyperonym represented a verbal noun. Another erroneous behavior in relation extraction is caused by a listing pattern, where the noun chunks do not directly follow each other but are distributed along the whole text segment. Applying dependency parsing and using the structural properties of compound words, as presented in [14], did not solve any of the listed issues. Concerning the post-processing component, the representation of the definitions through document annotation shows perfect performance (cf. Table 1, c)).

The web service was tested qualitatively. For example, as the approach takes Regulations of the European Parliament and of the Council as input, other regulatory document types, as well as nonexistent CELEX numbers, should be rejected by the system. In this case, a customized error message is displayed. In terms of statistics and visualizations, it would be beyond the scope of the paper to evaluate those in-depth together with legal experts. However, evaluating the concrete advantages for legal experts will be envisioned in future work.

4.3   Limitations

Limitations of the presented approach lay in its generalizability. Concerning the **pre-processing**, the approach was developed for and tested with the Regulation type of documents within the legislation sector of EUR-Lex. Thus, the presented approach most likely requires extensions when applied to EUR-Lex documents from other sectors, e.g., treaties or case-law, documents without a *"Defini-tions"* Article, or regulatory documents outside EUR-Lex. For the **processing of legal definition extraction**, the algorithm would e.g. need an extension of the allowed sentence root words indicating a valid legal term. Currently, the sentence root must belong to the synonyms of words *mean*, *include*, and *be*, or no legal definition is extracted. However legal definitions from a wider variety of input sources may contain other sentence roots or follow another sentence structure. Similarly, for the **processing of semantic relation extraction**, an extension of the relation patterns is needed to further improve the relation identification and increase its applicability to other regulatory documents.

5 Conclusion

This paper presents an approach to automatically extract legal terms along with their explanations and relations. The results are complemented by relevant statistics and visualizations and made accessible via a web service enhancing the approach's utility, particularly for non-technical experts. The evaluation demonstrated the capability of correctly identifying legal definitions which makes the approach highly reliable when working with sensitive compliance relevant information. As future work we plan to evaluate the usefulness of the statistics and visualizations with legal experts and aim to combine our approach with efforts towards, i.a., extracting references between compliance documents.

References

1. Al-Talib, G.A., Atiyah, A.A.: Extraction and classification of semantic relations from news recommendation. In: International Congress on Human-Computer Interaction, Optimization and Robotic Applications (2022). https://doi.org/10.1109/HORA55278.2022.9799885
2. Amaludin, B., Wardika, F.R., Putra, P.J.M., Paramartha, I.G.Y.: Analyze the usage of legal definitions in Indonesian regulation using text mining case study: treasury and budget law. In: Legal Knowledge and Information Systems (2021). https://doi.org/10.3233/FAIA210324
3. Avram, A., Cercel, D., Chiru, C.: UPB at SemEval-2020 task 6: pretrained language models for definition extraction. In: Workshop on Semantic Evaluation (2020). https://doi.org/10.18653/v1/2020.semeval-1.97
4. Commission, E., service, L.: Joint practical guide of the European Parliament, the Council and the Commission for persons involved in the drafting of European Union legislation. Publications Office of the European Union (2016). https://doi.org/10.2880/5575
5. Cruse, D.A.: Hyponymy and its varieties. Semant. Relationsh. Interdiscip. Perspect. (2002). https://doi.org/10.1007/978-94-017-0073-3_1
6. Culy, C., Chiocchetti, E., Ralli, N.: Visualizing conceptual relations in legal terminology. In: International Conference on Information Visualisation (2013). https://doi.org/10.1109/IV.2013.42

7. Damaratskaya, A.: Identification and Visualization of Legal Definitions and their Relations Based on European Regulatory Documents. Bachelor's thesis, Technical University Munich, May 2023. https://mediatum.ub.tum.de/doc/1715461/document.pdf
8. European Parl., Council of the EU: Regulation (EU) 2016/679 of the European Parliament and of the Council. https://data.europa.eu/eli/reg/2016/679/oj
9. European Parl., Council of the EU: Regulation (EU) 2019/1241 of the European Parliament and of the Council. https://eur-lex.europa.eu/eli/reg/2019/1241/oj
10. European Parl., Council of the EU: Regulation (EU) 2020/740 of the European Parliament and of the Council. http://data.europa.eu/eli/reg/2020/740/oj
11. European Parl., Council of the EU: Regulation (EU) 2021/947 of the European Parliament and of the Council. http://data.europa.eu/eli/reg/2021/947/oj
12. Ferneda, E., do Prado, H.A., Batista, A.H., Pinheiro, M.S.: Extracting definitions from Brazilian legal texts. In: Murgante, B., et al. (eds.) ICCSA 2012. LNCS, vol. 7335, pp. 631–646. Springer, Heidelberg (2012). https://doi.org/10.1007/978-3-642-31137-6_48
13. Ferrara, A., Picascia, S., Riva, D.: Context-aware knowledge extraction from legal documents through zero-shot classification. In: Guizzardi, R., Neumayr, B. (eds.) Advances in Conceptual Modeling. ER 2022. LNCS, vol. 13650, pp. 81–90. Springer, Cham (2022). https://doi.org/10.1007/978-3-031-22036-4_8
14. Hippisley, A., Cheng, D., Ahmad, K.: The head-modifier principle and multilingual term extraction. Nat. Lang. Eng. 11(2), 129–157 (2005). https://doi.org/10.1017/S1351324904003535
15. Hwang, R.H., Hsueh, Y.L., Chang, Y.T.: Building a Taiwan law ontology based on automatic legal definition extraction. Appl. Syst. Innov. (2018). https://doi.org/10.3390/asi1030022
16. Khoo, C.S., Na, J.C.: Semantic relations in information science. Annu. Rev. Inf. Sci. Technol. (2006). https://doi.org/10.1002/aris.1440400112
17. de Maat, E., Winkels, R.: Automated classification of norms in sources of law. In: Francesconi, E., Montemagni, S., Peters, W., Tiscornia, D. (eds.) Semantic Processing of Legal Texts. LNCS (LNAI), vol. 6036, pp. 170–191. Springer, Heidelberg (2010). https://doi.org/10.1007/978-3-642-12837-0_10
18. Nakamura, M., Ogawa, Y., Toyama, K.: Extraction of legal definitions and their explanations with accessible citations. In: Casanovas, P., Pagallo, U., Palmirani, M., Sartor, G. (eds.) AICOL -2013. LNCS (LNAI), vol. 8929, pp. 157–171. Springer, Heidelberg (2014). https://doi.org/10.1007/978-3-662-45960-7_12
19. Nakamura, M., Ogawa, Y., Toyama, K.: Extraction of legal definitions from a Japanese statutory corpus-toward construction of a legal term ontology. In: Law via the Internet Conference (2013)
20. Sartor, G., Santin, P., Audrito, D., Sulis, E., Caro, L.D.: Automated extraction and representation of citation network: a CJEU case-study. In: Guizzardi, R., Neumayr, B. (eds.) Advances in Conceptual Modeling. ER 2022. LNCS, vol. 13650, pp. 102–111. Springer, Cham (2022). https://doi.org/10.1007/978-3-031-22036-4_10
21. Singh, A., Kumar, P., Sinha, A.: DSC IIT-ISM at semeval-2020 task 6: boosting BERT with dependencies for definition extraction. In: Workshop on Semantic Evaluation (2020). https://doi.org/10.18653/v1/2020.semeval-1.93
22. Waltl, B., et al.: Automated extraction of semantic information from German legal documents (2017)
23. Winkels, R., Hoekstra, R.: Automatic extraction of legal concepts and definitions. In: Legal Knowledge and Information Systems (2012). https://doi.org/10.3233/978-1-61499-167-0-157

24. Winter, K., Gall, M., Rinderle-Ma, S.: Regminer: taming the complexity of regulatory documents for digitalized compliance management. In: Proceedings of the Best Dissertation Award, Doctoral Consortium, and Demonstration & Resources Track at BPM. CEUR Workshop Proceedings, vol. 2673, pp. 112–116. CEUR-WS.org (2020). https://ceur-ws.org/Vol-2673/paperDR10.pdf
25. Winter, K., Rinderle-Ma, S.: Untangling the GDPR using conrelminer. CoRR (2018). http://arxiv.org/abs/1811.03399

Comparative Analysis of Disinformation Regulations: A Preliminary Analysis

Antonella Calò[✉], Antonella Longo, and Marco Zappatore

Department of Engineering for Innovation, University of Salento, 73100 Lecce, Italy
{antonella.calo,antonella.longo,
marcosalvatore.zappatore}@unisalento.it

Abstract. The spread of Fake News poses a significant threat to democracy and public discourse. Instances of disinformation have had serious consequences, such as undermining election integrity and reducing vaccine trust over the recent years. Moreover, terminological variations and digital neologisms hinder consensus among scholars. The Internet proliferation has intensified the challenge of managing false content, demanding technological tools and regulations. This study explores the complexities of Fake News and the need for regulatory measures. A comparative law methodology is used to analyze international and European regulations concerning Fake News. Social media policies and reporting methods are also investigated, aiming to find ways to combat misinformation effectively. On the one hand, a fragmented regulatory framework both at a European and International level is revealed. To cope with this scenario, a multilingual ontology to harmonize definitions and facilitate compliance is proposed. On the other hand, the crucial role of Social Media policies, their algorithms' transparency, and educating roles are considered. This leads to the need for an enhanced regulation of social media, educational initiatives of digital media literacy, and AI-driven news apps to provide trusted sources and manage misinformation in a better way.

Keywords: Disinformation · Fake News Regulation · Digital Media Literacy · International Law · European Law · Social Media Policies · Reporting Systems

1 The Threat of Fake News: Growing Challenges and the Need for Regulation

The plurality of communication channels and the spread of Fake News are widespread phenomena today [1]. The more this phenomenon increases, the more it represents a threat to democracy and to the public sphere. Indeed, there is a significant number of examples of disinformation, and Fake News that caused serious issues, such as the most famous one during Donald Trump's 2016 elections, according to which Russian hackers were attacking the vote-system software to track and manipulate election results [2] or the 600% increase of disinformation about vaccines after the decision of the Court of Rimini to recognize that MMT (Measles, Mumps, and Rubella) vaccine caused autism in a child. This caused low trust in the health systems and vaccines, thus leading to a reduction in child immunization [3]. Finally, the huge infodemic during the COVID-19 pandemic was another noteworthy source of disinformation.

T. P. Sales et al. (Eds.): ER 2023 Workshops, LNCS 14319, pp. 162–171, 2023.
https://doi.org/10.1007/978-3-031-47112-4_15

The terms related to Fake News have different definitions, and the phenomena assume different forms, various authors, and multiple motives, such as commercial sensation contents, ideologically driven news, or state sponsored misinformation to influence users [4]. The phenomenon of misinformation/disinformation has always existed, but with the rise of the Internet, it has now increased significantly, causing a more difficult management of false contents. This led to the need of technological tools capable of detecting Fake News, and to the creation of regulations and measures. The latter is the topic of the present paper.

The research questions this work will address are:

1. Which are the regulations at an international and European Level related to the dissemination of Fake News?
2. How to ensure compliance, and harmonization among those regulations also from a terminological standpoint?
3. Is it possible to report Fake News on social media? And how can social networks educate their users?

The paper is organized as follows: the second section will focus on the state of the art and legal/regulatory framework. In the third section we will provide the methodology followed to check the possibility to report Fake News on social media and the application of the methodology. The fourth section will present the results. Finally, conclusions and the future work will be drawn.

2   State of the Art: Legal, and Regulatory Framework in Public/Private Sectors

The study of legal literature concerning misinformation, disinformation, and Fake News reveals a fundamental challenge rooted in the use of terms and language. The evolution of these terms, particularly the adoption of "Fake News" as a label for what was once called "false information," has not brought consensus among scholars and experts [5]. Claire Wardle and Hossein Derakhshan call these challenges "Information disorder" [6]. They showed that the term "Fake News" has been used to refer to different phenomena and results inadequate, therefore the authors proposed three notions:

- **Dis-information**: Information that is false and deliberately created to harm a person, social group, organization, or country.
- **Mis-information:** Information that is false, but not created with the intention of causing harm.
- **Mal-information**: Information that is based on reality, used to inflict harm on a person, organization, or country.

Different interpretations of these terms have been described in [7], while the European definitions can be checked in [8].

The phenomenon of Fake News in the current "post-truth" era is multifaceted and complex, and it often exploits emotions to polarize readers and elicit strong reactions. However, its implications go beyond emotional manipulation. Fake News frequently intertwines with criminal behaviors, such as hate speech and revenge porn facilitated by technologies like DeepFake and DeepNudes. Semantic challenges arise when addressing cybercrimes and developing regulations, particularly regarding terminology and the discrepancies among different legal systems. Supranational cybercrime cases that transcend national jurisdictions pose additional difficulties, as legislative differences hinder comprehension and interoperability [9]. Fake News, and all the phenomenon of disinformation, and misinformation can be considered an example of cybercrime; and, because of the Internet, they certainly circulate out of national jurisdictions. Cybercrimes, and these types of contents shared online, share new abstract terminologies, or neologisms such as: "DeepFake"; "eco-chambers"; "Clickbait", and so on. The fragmented terminology, and their continuous development also cause problems in translation. Furthermore, because of the lack of digitalization, and the lack of trust in digital tools, Artificial Intelligence and Machine Learning, legal professional translators refuse to use the so-called CAT (Computer-Aided translation Tools) tools, leading to a low interoperability between the legal terms [10], and mistranslations of regulations, and lack of compliance. Some examples of mistranslations of the European GDPR regulation in the Italian, Icelandic, Finnish, and Slovak languages can be found in [10].

This is why, regulations appear fragmented, and ontologies have been created to face these problems [9, 11, 12]. In the case of revenge porn, or other types of abstract concepts such as in [11, 13], different European Member States have tried to face the issue with different regulations, which yet do not seem to be harmonized at European level. The situation is even more fragmented when it comes to regulating Fake News. Indeed, the lack of proper definitions, and interpretations of the terms, causes issues when regulating these phenomena, resulting in a lack of regulations (or not harmonized ones). Furthermore, concerns about public and private censorship risk violating Article 19 of the United Nations' right to freedom of speech and expression, as well as other directives and regulations safeguarding freedom of expression [5, 14].

To mitigate these risks and promote effective regulation, collaboration with social media platforms has become crucial. Social media algorithms, designed to recognize users' preferences, often reinforce confirmation bias, and create echo chambers. The opacity of algorithms, often referred to as "black boxes", contributes to the erosion of trust in technology and data sharing practices. The adoption of Explainable AI (XAI) can enhance transparency and address biases by providing insights into the involvement of personal data [15]. Research on XAI in societal challenges has shown promising results in enhancing transparency, mitigating biases, and combating disinformation [16].

In the subsequent section, we will delve into a comparative analysis of existing regulations at both public and private levels, examining the strategies employed by social media platforms to report and combat disinformation. By exploring these approaches, we aim to shed light on the complex landscape of Fake News regulation and identify potential areas for improvement and collaboration.

3   Case Studies: Comparative Methodology Analysis of Regulations and Media Policies

We will proceed with a comparative law methodology [17], which will give us the possibility to understand whether regulations are present or not, if European directives have been implemented or not, if they are compliant among them, and finally if they have been criticized in a positive or negative way. This will allow us to have a better understanding of the next steps to take to better face the disinformation/Fake News phenomena. We will present a multilingual ontology that when larger and more modular will support the drafting and translation of regulations processes.

A comparative method will be also used to analyze some of the social media policies, and their reporting methods to better understand if, and how to cooperate/regulate them.

3.1 Case Study 1: Comparison of Public Regulations at the International and European Level

A comparative law methodology is employed to analyze public international, European, and national directives, regulations, and proposals regarding disinformation [6, 18–20]. The first European Union attempt to combat disinformation involved the Commission Communication in 2017"Tackling Disinformation: A European approach," which was a non-binding instrument aimed at fostering cooperation with social media platforms. This led to the autoregulation soft law "**Code of Conduct**", which has the goal to stop ads of accounts that disseminate disinformation, improve transparency, cooperate with Social Media in deleting fake accounts/bots, improve official sources with trusted news, and allow academics to access data. However, this approach was criticized for potentially granting excessive power to social media platforms and risking private censorship. Germany and France attempted to implement the code of conduct but faced opposition for potentially infringing upon freedom of speech. The German implentation (**NetzDG**) of 2017, after defining the meaning of "lawful" and "manifestly lawful", makes the social media with more than 200 reports to share a public report on how they faced the issues; added an administrative authority of federal justice; and poses penal sanctions up to 20 millions of euros.

The French implementation (**Loi relative à la lutte contre la manipulation de l'information**) of 2018, was created specifically for electoral periods, and added the obligation of transparency. Indeed, platforms had to create reporting systems, and reported Fake News would have been analyzed within 48 h risking sanctions up to 75.000 euros [20]. These implementations and proposals were rejected because they were seen particularly stringent and raised concerns about the right of freedom of speech. Similarly, Italian proposals (such as **DDL Gambaro** which proposed to add fines; **DDL Zanda-Filippin** that followed the German Model, and **Boldrini's Campaign** which focuses on a campaign of Digital Media Literacy) faced rejection for similar reasons, and the current articles 656-bis/ter still reference disinformation using terminology from 1936 [19]. In 2020, the Code of Conduct was implemented with the **Digital Service Act**, a co-regulation instrument that faced illegal and systemic risks; invite platforms to create conduct codes; empower users with more transparency, and create new independent authorities. However, its adoption remains limited, with only France and Germany having implemented it thus far. Germany, implemented it in 2020 with the "**Mstv**" which regulates algorithms, adds labels of social bots, and improve the findability of public service and journalists' due diligence for social media.

France implemented it in 2020 with the "**Loi visant à lutter contre les contenus haineux sur Internet**", which took down fake accounts about Covid-19 within one hours.

Internationally, regulations and proposals have been tested, except for China, which holds different attitudes toward freedom of expression, and Australia, which focuses more on media literacy. Indeed, China's regulation called "Cybersecurity Law" added fines and detection for those who do not respect the rules. It allows to share only registered news media articles. It also created a platform called Piyao, which broadcasts only real news, allows user to report contents and has AI algorithms that detect rumors.

During the same years, in Australia a Taskfor for Fake News led to different codes and legislations, which focused on: Social Media Literacy; a multi-agency body that could address to disinformation risks (among others); Promotion of self-regulation and a new advertising campaign during elector periods called "Stop and Consider" to encourage users to check the sources.

Also the USA faced this topic during these years, through different measures ("Honest Act", "US National Defence Authorisation act", "Stigler Committee"). These required social media to keep copies of ads, the federal governement to deal with propaganda and disinformation, proposed a Digital Agency, and implement antitrust, data protection laws and media policies. However, most of these regulations have been rejected.

3.2 Case Study 2: Facebook vs Twitter Policies Comparison, and Reporting Methods.

The following is a comparison of the two most important social media related to the news policies: Facebook[1] and Twitter[2].

During the last years, Facebook has continuously implemented its policies. The current one, updated to 2023, is the **"Community Standards - integrity and authenticity"** policy, through which the platform removes: physical harm or violence, harmful health misinformation, miracle cures, manipulated media, and election interferences.

In 2022, Twitter (now "X"), through the **"Crisis Misinformation policy"** punished false or misleading information that could harm crisis-affected populations, and decreased the account visibility, adding a warning note to those accounts that violated this policy. Twitter added then two other policies in 2023 (**"Synthetic and manipulated media policy" and "Civic Integrity policy"**) which punish those manipulated/false contents which are not satire, memes, animations, and opinions. More specifically, the "civic integrity policy" relates to those contents about elections or civic processes. In both cases, accounts risk to be deleted or labeled.

Both Facebook and Twitter trust on users who can report, and on the work of external fact-checkers services. Facebook relies on fact-checkers, certified through the non-partisan International Fact-checking Network (IFCN)[3].

---

[1] https://transparency.fb.com/en-gb/policies/community-standards/misinformation/

[2] https://help.twitter.com/en/rules-and-policies/crisis-misinformation; https://help.twitter.com/ en/rules-and-policies/manipulated-media; https://help.twitter.com/en/rules-and-policies/ele ction-integrity-policy.

[3] https://transparency.fb.com/en-gb/features/how-fact-checking-works/

4   Discussion

4.1   Coping with Issues in Regulations: An Ontological Proposal (Case Study1)

The European situation regarding the regulation of Fake News is fragmented, as member states have not fully adapted to the Digital Service Act [1]. Terminological inconsistencies persist, with some countries, like Italy, having different interpretations of the term "Fake News" or using outdated terminology. Addressing terminological challenges and regulating internet-related events have been explored in previous works using ontologies [9, 12]. Drawing inspiration from Castano's hierarchy and conceptual models [12], We propose to develop a multilingual ontology that can serve as a reference for the definition and creation of regulations and aid in translation processes. Such an ontology could promote harmonization and compliance in international regulations concerning Fake News.

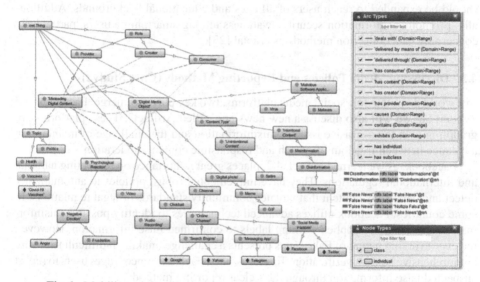

Fig. 1. Multilingual Ontology of terms related to "Misleading Digital Content".

The proposed ontology provides the definitions, translations, and instances of each element. This allows the understanding of a non-unique definition of all elements, in each language, thus eliminating confusion, especially when it comes to the use of "Fake News", "Disinformation", "Misinformation", or neologisms such as "Clickbait". It also provides synonyms for each element in the different languages, modeled as class attributes, and the corresponding semantic relationships, modeled as object properties, showed as arrows in Fig. 1. This ontology can be the starting point for a larger and a modular ontology that would support the regulatory drafting and translation processes.

4.2 Social Campaigns to Regulate Disinformation Through Awareness

At the international level, laws addressing Fake News have faced criticism for potentially infringing upon freedom of speech and granting excessive power to social media platforms. However, positive measures have been observed in the realm of digital literacy and awareness campaigns, which empower individuals to critically evaluate and resist sociocultural manipulation [6, 21]. Educational attainment plays a significant role in enhancing individuals' ability to comprehend and assess information. Younger individuals, especially those with higher levels of education, demonstrate greater awareness and concern for public issues due to their access to reliable information and knowledge [1]. Cybersecurity awareness programs have proven effective in reducing risky online behaviors and fostering a sense of self-responsibility among participants [22]. Consequently, governmental, and private initiatives promoting media literacy, such as "Open the Box[4] "and the "Bad News[5] " game, are promising, although most initiatives primarily target young users. To address this gap, media literacy education through social media should be expanded to reach users of all ages and educational backgrounds. Additionally, promoting information security awareness among smartphone users, particularly concerning data collection methods, is crucial [23].

4.3 Discussing Media Policies and Reporting Methods (Case Study2)

In a world dominated by social media platforms, two key players emerge: Facebook and Twitter, but we will also discuss a new news app called Artifact[6]. These platforms are grappling with the pervasive issue of misinformation and the challenge of maintaining a balance between freedom of speech and responsible content moderation.

Facebook, as we learned in Sect. 4.2, places great importance on respecting national and international regulations. They have implemented robust policies to automatically detect and delete any content that violates community standards or legal regulations. In some countries, Facebook utilizes advanced technologies to identify possible misinformation and promptly applies warning labels to confirmed false information. However, the ever-evolving nature of Fake News presents a challenge, making it difficult to ensure comprehensive content verification. To address this, Facebook encourages users to report suspected false information through their clear reporting method[7].

Meanwhile, Twitter, although relatively new to the policy of addressing misinformation, has made promising strides in combatting false content. They employ a combination of human review, technology, and collaboration with third-party experts to identify and address misleading information. Twitter's unique feature prompts users to reconsider retweeting articles without reading them, reminding them of the potential for misleading headlines. However, the specifics of the technology and expert partnerships remain unclear, and there is a need for more transparency regarding their approach. The reporting system on Twitter offers assorted options, though it can be confusing when it comes to reporting Fake News or false content.

---

[4] https://www.openthebox.io/

[5] https://www.getbadnews.com/books/english/

[6] https://artifact.news/

[7] https://transparency.fb.com/it-it/policies/community-standards/misinformation.

Amidst these challenges, Artifact, a text-based news app driven by AI, presents itself as a possible solution to Filter Bubbles and Clickbaits[8]. Developed by Kevin Systrom, the co-founder of Instagram, Artifact aims to counter the infodemic witnessed during the COVID-19 pandemic. The app asks users to select their topics of interest and official news sources during onboarding, allowing it to tailor the content delivered to individual users. Unlike Facebook, Artifact curates its news sources based on integrity rates analyzed by third-party fact-checking services. Additionally, the app leverages ChatGPT technology to provide easy article summaries for users with time constraints. Nicknamed the "TikTok of news,", because it has the same scrolling feed, but has articles instead of videos, Artifact uses The Transformer machine learning system to offer this scrolling feed of news in text format. This system, initially developed by Google in 2017, considers factors beyond clicks, such as dwell/read time and shares, to avoid serving clickbait and filter bubbles. The app's reinforcement learning algorithm, Epsilon-Greedy, ensures users are exposed to a diverse range of recommendations. According to the founders, Artifact also provides a reporting system to combat clickbait and misleading headlines. However, being new in the social media landscape, little information is available, but it would be useful to continue analyzing and studying it, to see if and how it would differentiate itself from other similar platforms.

4.4 Media Policies, Reporting Methods, and Users: Possible Solutions

In this dynamic landscape of social media and news platforms, Facebook, Twitter, and Artifact are continuously evolving their policies and technologies to tackle misinformation while upholding freedom of speech. Their approaches may differ, but all three platforms are committed to addressing the challenges posed by Fake News and ensuring users have access to reliable information. Media literacy education, and information security awareness should be improved, for users of all ages and education levels [20]. Social media could play a key role. From one hand, if all social media adopted small features like a clear reporting system, the prompt feature on Twitter, or warning/labels on those contents that need to be verified yet, users could be slowly and automatically educated to the use of media. On the other hand, a simple cooperation with social media may not be enough since they may be moved from economic interests, so empowering users could work [20]. Additionally, establishing a national-specific social media platform like Artifact, hosting verified and trusted sources, could provide an official, non-political online channel for news consumption and sharing, akin to traditional newspapers and newscasts. Furthermore, considering the registration requirement for journalists, a similar approach could be implemented for online news platforms, ensuring transparency, and eliminating anonymity. This, in conjunction with algorithmic and AI regulations, forms a comprehensive strategy to address misinformation from a regulatory standpoint. However, it is crucial to complement regulatory efforts with technological advancements, such as the development of automated tools for Fake News detection.

---

[8] https://techcrunch.com/

5   Conclusion and Future Work

Facing the regulatory challenges of Fake News is complex, as evidenced by the analysis of literature and comparative methodologies. Terminological issues, including the definition of terms and the emergence of digital neologisms, pose translation challenges and hinder compliance and harmonization across member states and international borders. This lack of regulation is compounded by the risk of noncompliance with the right to freedom of speech and expression. Furthermore, the European attempt to cooperate with social media platforms has revealed that compliance with agreements depends on economic interests. To address this, enhanced social media regulations are needed, particularly regarding algorithms and the handling of misleading content. By doing so, social media platforms can contribute to the improvement of digital media literacy and educate users effectively. Our future work will focus on the multilingual version of the ontology presented in this paper. We will analyze how compliance among regulations and translation processes can be enhanced. We will also focus on the way the technologies of ChatGPT can support us in translation, Fake News management, and in the analysis of new news app like Artifact.

References

1. Viola, C., Toma, P., Manta, F., Benvenuto, M.: The more you know, the better you act? Institutional communication in Covid-19 crisis management. Technol Forecast Soc Change. **170**, 120929 (2021). https://doi.org/10.1016/j.techfore.2021.120929
2. Berghel, H.: Oh, what a tangled web: russian hacking, fake news, and the 2016 US presidential election. Computer (Long Beach Calif) **50**, 87–91 (2017). https://doi.org/10.1109/MC.2017.3571054
3. Carrieri, V., Madio, L., Principe, F.: Vaccine hesitancy and (fake) news: quasi-experimental evidence from Italy. Health Econ. (United Kingdom). **28**, 1377–1382 (2019). https://doi.org/10.1002/hec.3937
4. Belova, G., Georgieva, G.: Fake News as a Threat to National Security. In: International conference Knowledge-based Organization, vol. 24, pp. 19–22 (2018).https://doi.org/10.1515/kbo-2018-0002
5. Richter, A.: Fake news and freedom of the media. J. Int. Media Entertainmet Law **8**, 1–34 (2020)
6. Goldberg, D.: Responding to "fake news": is there an alternative to law and regulation? Southwest Law Review. **47**, 417–447 (2018)
7. Xu, F., Sheng, V.S., Wang, M.: A unified perspective for disinformation detection and truth discovery in social sensing: a survey. ACM Comput. Surv. **55**, 1–33 (2021). https://doi.org/10.1145/3477138
8. Durach, F., Baragaoanu, A., Nastasiu, C.: Tackling dsinformation: EU regulation of the digital space. Rom. J. Eur. Aff. **20** (2020)
9. Falduti, M., Griffo, C.: Modeling Cybercrime with UFO: an Ontological Analysis of Non-Consensual Pornography Cases. In: Ralyté, J., Chakravarthy, S., Mohania, M., Jeusfeld, M.A., Karlapalem, K. (eds) Conceptual Modeling. ER 2022. LNCS, vol. 13607, pp. 380–394. Springer, Cham (2022). https://doi.org/10.1007/978-3-031-17995-2_27
10. Smal, L., Lösch, A., Van Genabith, J., Giagkou, M., Declerck, T., Busemann, S.: Language Data Sharing in European Public Services-Overcoming Obstacles and Creating Sustainable Data Sharing Infrastructures (2020)

11. Castano, S., Falduti, M., Ferrara, A., Montanelli, S.: The LATO knowledge model for automated knowledge extraction and enrichment from court decisions corpora. Inf. Syst. **106**, 101842 (2022)
12. Castano, S., Falduti, M., Ferrara, A., Montanelli, S.: The CRIKE data-science process for legal knowledge extraction discussion paper. In: Italian Symposium on Advanced Database Systems: 16th–19th June (2019)
13. Castano, S., Falduti, M., Ferrara, A., Montanelli, S.: Law data science and ethics: the CRIKE approach. In: Proceedings of the 1st International Workshop on Processing Information Ethically co-located with 31st International Conference on Advanced Information Systems Engineering, pp. 1–9 (2019)
14. EPRS-ComparativeLaw, europarleuropaeu: La libertà di espressione, una prospettiva di diritto comparato Unione europea. (2019)
15. Longo, L., Goebel, R., Lecue, F., Kieseberg, P., Holzinger, A.: Explainable artificial intelligence: concepts, applications, research challenges and visions. In: Holzinger, A., Kieseberg, P., Tjoa, A.M., Weippl, E. (eds.) CD-MAKE 2020. LNCS, vol. 12279, pp. 1–16. Springer, Cham (2020). https://doi.org/10.1007/978-3-030-57321-8_1
16. Amelio, A., Bonifazi, G., Janusz, A., Briguglio, L., Morpurgo, F., Occhipinti, C.: A multilayer network-based approach for interpreting and compressing convolutional neural networks. A governance and assessment model for ethical artificial intelligence in healthcare. Cognit. Comput. **15**, 1–29 (2022)
17. Eberle, E.J.: Issue 1 Symposium: Methodological Approaches to Comparative Law Article 2 Winter (2011)
18. Zecca, D.: Tutela dell'integrità dell'informazione e della comunicazione in rete: obblighi per le piattaforme digitali fra fonti comunitarie e disciplina degli Stati membri. DPCE Online. 37 (2018)
19. Birritteri Emanuele: Punire la disinformazione. Diritto Penale Contemporaneo - Rivista Trimestrale. 304–334 (2021)
20. Vese, D.: Governing fake news: the regulation of social media and the right to freedom of expression in the era of emergency. Eur. J. Risk Regul. **13**, 477–513 (2022). https://doi.org/10.1017/err.2021.48
21. Kellner, D.: Cultural Studies, Multiculturalism, and Media Culture. SAGE Publications, Thousand Oaks
22. Chaudhary, S., Gkioulos, V., Katsikas, S.: Developing metrics to assess the effectiveness of cybersecurity awareness program, (2022). https://doi.org/10.1093/cybsec/tyac006
23. Bitton, R., Boymgold, K., Puzis, R., Shabtai, A.: Evaluating the information security awareness of smartphone users (2019)

Privacy-Preserving Data Integration for Digital Justice

Lisa Trigiante(✉)⬛, Domenico Beneventano⬛, and Sonia Bergamaschi⬛

University of Modena and Reggio Emilia, Modena, Italy
lisa.trigiante@unimore.it

Abstract. The digital transformation of the Justice domain and the resulting availability of vast amounts of data describing people and their criminal behaviors offer significant promise to feed multiple research areas and enhance the criminal justice system. Achieving this vision requires the integration of different sources to create an accurate and unified representation that enables detailed and extensive data analysis. However, the collection and processing of sensitive legal-related data about individuals imposes consideration of privacy legislation and confidentiality implications. This paper presents the lesson learned from the design and develop of a Privacy-Preserving Data Integration (PPDI) architecture and process to address the challenges and opportunities of integrating personal data belonging to criminal and court sources within the Italian Justice Domain in compliance with GDPR.

Keywords: PPDI · PPRL · Pseudonym · Legal Data · Digitization

1 Introduction and Privacy Context

The digitization of legal and administrative processes, among many others, has resulted in the collection of vast amounts of data describing people and their behavior. The Big Data (BD) revolution offers significant promise to feed multiple research areas and enhance public and private sectors. Within the justice domain, the emergence of Digital Justice in conjunction with advanced Data Analysis techniques presents the opportunity to advance the criminal justice system toward an innovative Data-Driven approach that supports the emerging 5P (Predictive, Preventive, Personalized, Participatory, and Precision) paradigm. The full potential of this vision materializes as a consequence of integrating data from different sources to enable extensive and detailed analysis. The recidiva phenomenon illustrates this concept as it is fundamental in criminal justice to identify the cost-effectiveness of institutional programs and prisons. *Recidivism* is a tendency of an offender to lapse into a previous pattern of criminal behavior,

---

This research was partly funded by the CRUI Foundation (Conferenza dei Rettori delle Universitá Italiane), within the scope of the "Recidivism Data Mart and Criminal Data Warehouse" project.

after he has received sanctions or intervention. Assessing recidivism is a complex measurement problem that necessitate the reconstruction of a subject's criminal history from criminal records, which are usually kept in different autonomous databases. In addition, collecting and processing legal personal data impose the need to consider the privacy legislation. Under the European *General Data Protection Regulation* (GDPR), the classification of data content is based on the twin principles of identifiability and privacy: *Personally Identifiable Information* (PII) refers to attributes that have the ability alone (direct PII) or in combination (*Quasi-IDentifiers* QID) to identify a specific individual e.g. name and address; *Sensitive Personal Information* (SPI) denotes attributes containing confidential data about individuals, e.g. judicial sentences. *Privacy Preserving Data Integration* (PPDI) [1] is a branch of Data Science aimed at providing a unified representation of personal information across multiple heterogeneous data sources while preventing privacy disclosure of the individuals represented in the underlying data. The GDPR leads toward the adoption of specific techniques to prevent both the calculation and the output of the calculation from permitting the possibility of identifying a specific individual, called *re-identification*. *Anonymization* refers to the process of removing PII from the data. *Pseudonymization* [2] refers to the process of replacing PII with a *pseudonym*, to allow further processing while preventing re-identification.

This paper illustrates the design and development of a Proof of Concept for a PPDI process to establish a Data Warehouse across Italian legal data sources and assess the recidivism phenomena. In favor of brevity, this project will henceforth be referred to as "PoC" within the paper. In particular, Sect. 2 outlines the basic concept with respect to the existing related works. Section 3 presents the architecture and PPDI process designed and developed within the PoC. Section 4 summarizes the phases of the PPRL process that incorporates pseudonymization techniques. Section 5 concludes with contributions and future research directions and developments.

2   Related Works

Within the existing body of literature, the categorization of privacy-preserving scenarios is related to different dimensions [3]. On this basis, the PoC envisaged a *multi-party* scenario, in which adversaries were assumed to show a *host-but-curious* behaviour and their *background knowledge* was considered dependent on different aspects, e.g. the external or internal role in the process. The project involved the investigation and comparison of different techniques and approaches [3] and the study of strengths and limitations of existing systems (e.g. EUPID, PRIMAT) and tools (e.g. SPIDER) for gathering and handling sensitive data about patients. A more in-depth discussion of this study was addressed in [4]. The concept that served as the starting point is a *Trusted Third-Party* (TTP) [5] architecture, which represents a reference in the context of decentralized organizations, where legal requirements limit the number of applicable approaches. The PoC is designed and developed to accommodate specific privacy issues and application requirements arising from the Italian Justice Domain.

3   Privacy-Preserving Data Integration Architecture

This Section presents an outline of the architecture designed to support the PPDI process, without delving into the discussion of technical specifics. As we can see in Fig. 1, the TTP serves as PPDI Domain to produce an integrated and consistent representation of the different autonomous data sources within the Justice domain (Source Domain), to provide the Consumer Domain with a unified and privacy-preserving view of the underlying data.

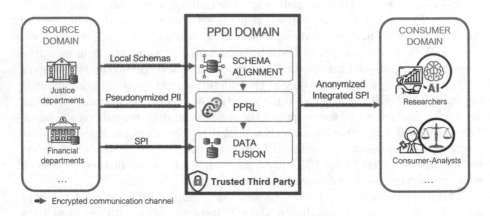

Fig. 1. Schema of the PPDI Architecture

To this end, the TTP carried out the different steps of the PPDI process:

- **Schema Alignment** resolves inconsistencies at schema level by producing an integrated Global Schema from the schemas of the Local Sources. In the PoC, due to the possibility to access to plain-format local schemas, were employed traditional Schema Matching techniques to categorise and establish correspondences between PPI and SPI of the local sources. As a result, a Global Schema across all Italian legal sources was produced and a subset of QID (common to all sources) was selected to carry out the PPRL phase.
- **Privacy-Preserving Record Linkage (PPRL)** resolves inconsistencies at tuple level by identifying and linking records about the same real-world entities (individuals) from different sources while avoiding privacy disclosure. PPRL is the crucial step to achieving the best trade-off between privacy and usability of data resulting from the PPDI process. To this end, the PPRL process represented the core of the project and involved multiple steps (detailed in Subsect. 3.1).
- **Data Fusion** resolves inconsistencies at value level by creating a unique record for each real-world entity. In the PoC basic Data Aggregation techniques were employed. In addition, Data Fusion techniques can be employed to combine duplicate entries with different policies. The result of this step is an anonymized record for each individual that encloses the integrated SPI from the Source Domain.

3.1   Privacy-Preserving Record Linkage (PPRL) Process

Linkage of data about individuals is commonly based on QID since direct PII is more vulnerable to re-identification attacks. However, QID is neither unique nor stable over time and may be subject to recording errors and missing values.

The PPRL process to achieve high linkage quality results in a privacy-preserving setting comprises different steps:

- **Pre-processing** of error-prone QID into a unique and comparable format.
- **Pseudonymization** of normalized QID into a unique pseudonym for each record,

In order to avoid the transmission of non-pseudonymized data from local storage, in the PoC the aforementioned steps are performed at the source site. The pseudonymization techniques employed generate pseudonyms that prevent the TTP from the possibility of re-identify an individual but allow the TTP to conduct the actual linkage, following the subsequent steps of the PPRL process:

- **Blocking** of the pseudonyms that are likely to match into blocks, producing candidate pseudonym pairs to reduce the number of comparisons that need to be conducted.
- **Comparison** of candidate pseudonym pairs in detail using approximate comparison (or similarity) functions.
- **Classification** of candidate pseudonym pairs into a match or not match, using a decision model based on the result of the comparison.

4   Summary of the PoC of the PPDI Process

This section digests the different steps implemented in the PoC, with respect to the key points of the PPRL Process outlined in Subsect. 3.1.

The architecture envisages that the TTP defines a priori the parameters for carrying out the PPRL process: the selection of QIDs, the Pre-processing and Pseudonymization functions and related parameters/keys, and consequently the comparison and classification functions for the actual linkage.

In the PoC, different pseudonymization techniques [6] and PPRL approaches [3] have been employed and compared to comply with GDPR and general IT security practices and to optimize the trade-off between privacy and utility of data, such as Phonetic encryption (PHON), Bloom Filter (BF), Tabulation Min-Hash (TMH) and Cryptographic Long-term Key (CLK).[1]

The selection of appropriate techniques for each PPDI step is an extremely difficult and time-consuming process, which depends on the different privacy and practical requirements and the nature and content of the involved sources. It is worth noticing that the pipeline employed in the PoC was specifically selected to fulfill the scope of the PPDI project within the justice domain. In particular, the

---

[1] The detailed description of these techniques is beyond the scope of this article, please refer to [6].

PPRL process performed in the PoC is illustrated in Fig. 2 (which for simplicity exhibits only the phonetic encryption method) and condensed in the subsequent itemize, with reference to Subsect. 3.1.

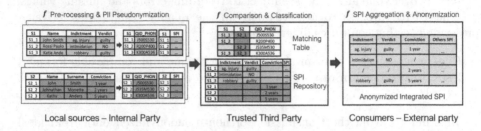

Fig. 2. PPRL process and Data Aggregation

- **Classification and Pre-processing** of QID and SPI;
- **Pseudonymization**: transformation of the QID to pseudonyms by using one or more encoding function, e.g. Phonetic encoding to produce QID_PHON.
- Acquisition by the TTP of record_ID and Pseudonym for each record of the different local sources to be integrated;
- **Blocking** was not implemented in the PoC because there were no strict scalability requirements;
- **Comparison** of pseudonyms conducted using classical functions (e.g. Jaccard similarity) which calculates a similarity value for each candidate pseudonym pairs.
- **Classification**: conducted using a rule-based approach that compares the similarity value for each pair to an appropriate threshold to determine whether it is a match or not.
- **Clustering**: conducted using the connected components technique to obtain groups of record-ID referring to the same real-world entity;
- Storage of the record_ID and Pseudonym for the same real world entity as a Global record into a Matching Table.
- Acquisition by the TTP of SPI and record_ID for each local record and storage into a SPI Repository.
- Data aggregation of SPI for each group of record_ID (representing the same real-world entity) to produce a global record;
- Anonymization of the Integrated SPI by removing pseudonyms from each global record to be transferred to another system.

5   Future Directions and Developments

In conclusion, the Proof Of Concept (PoC) produced the design of a PPDI architecture and the envisaged methodology to address the range of challenges arising from the intersection of GDPR privacy guidelines and the integration of data belonging to the Italian Justice Domain. In particular, we compared different techniques and approaches to identify the optimal PPDI process to address the heterogeneity, issues and practical requirements issues within legal-related datasets. The PoC offers wide contributions as Justice-related PPDI is scarcely covered in the literature while privacy-preserving techniques for legal data analysis are abundant. A potential future development is to exploit Privacy-Preserving Data Publishing techniques to limit the release of SPI (e.g. differential privacy) and the resulting potential for re-identification. In addition, the integrated and anonymized data resulting from the PPDI process enables to leverage advanced data-Centric AI techniques to conduct detailed and extensive analysis of Redivism. Pursuing the potential of this research our ongoing project is aimed at the development of a novel PPDI framework to evaluate and accommodate different real-world application scenarios. For this reason, the project was also presented within the Healthcare Domain [7]. The basic idea is to re-engineer the *MOMIS (Mediator envirOnment for Multiple Information Sources)* [8] Data Integration system toward a microservice architecture with added privacy-preserving functionalities to address distinct domain-specific PPDI challenges.

Acknowledgment. We wish to thank all the members of DBGroup. Lisa Trigiante wishes to mention that her PhD project is founded by MIUR under D.M.351 with the Emilia Romagna region as partner.

References

1. Clifton, C., et al.: Privacy-preserving data integration and sharing. In: ACM 2004, DMKD, pp. 19–26 (2004)
2. Bolognini, L., et al.: Pseudonymization and impacts of big data processing in the transition from the Directive 95/46/EC to the new EU GDPR. Comput. Law Secur. Rev. **33**(2), 171–181 (2017)
3. Vatsala, D., et al.: A taxonomy of privacy-preserving record linkage techniques. Inf. Syst. **38**(6), 946–969 (2013)
4. Trigiante, L.: Analysis and experimentation of State-of-the-Art Privacy-Preserving Record Linkage techniques in Data Integration environments. Master's thesis in Computer Science, Unimore (2022)
5. Schnell, R.: Privacy-preserving Record Linkage. In: Harron, K., Goldstein, H., Dibben, C. (eds.) Methodological Developments in Data Linkage, pp. 201–225. John Wiley & Sons (2015). http://eu.wiley.com/WileyCDA/WileyTitle/productCd-1118745876.html. ISBN 1118745876
6. Vidanage, A., et al.: A vulnerability assessment framework for privacy-preserving record linkage. ACM Trans. Priv. Secur. **26**(3), 2471–2566 (2023)
7. Trigiante, L.: Privacy-preserving data integration for health. In: 31st Symposium on Advanced Database Systems 2023, vol. 3478, pp. 750–756 (2023)
8. Bergamaschi, S., et al.: Data integration. In: Handbook of Conceptual Modeling, pp. 441–476 (2011)

Automated Analysis with Event Log Enrichment of the European Public Procurement Processes

Roberto Nai[1](✉), Emilio Sulis[1] , and Laura Genga[2]

[1] Computer Science Department, University of Turin, Turin, Italy
{roberto.nai,emilio.sulis}@unito.com
[2] Eindhoven University of Technology (TU/e), Eindhoven, The Netherlands
l.genga@tue.nl

Abstract. The length and extension of legal processes are two of the main problems of contemporary justice. Adopting process analysis techniques can serve to understand the course of these processes, to identify bottlenecks, and to propose improvements. This paper showcases an exploration of legal event logs using process mining techniques. First, we discuss the results obtained by applying state-of-the art process discovery techniques to data obtained from public tenders. Then, we show how to use natural language processing to automatically extract events and dates from the texts of the tenders and leverage this information for improving the results of process mining techniques. As a proof-of-concept, we propose a comparison of the French, Spanish, and Italian cases through process discovery based on calls for tenders from the European TED repository.

Keywords: Public procurement · Process mining · Event log enrichment · Natural Language Processing

1 Introduction

Public procurement processes are often lengthy, which has an impact on the legal and political system as a whole. In addition, special attention is paid to the European landscape, where there is difficulty in standardizing the legislative framework because of the different characteristics and practical applications peculiar to each member state. The increasing adoption of legal information technologies opens the way to exploit digital data, both for process analysis and for text analysis (e.g., public competition notices). In the context of public administration and legal information technologies, organizational aspects leading to the delivery and awarding of public tenders can be examined and supported by legal analysis tools. The areas of application are many in the broader field of Business Process Management (BPM) [6]; however, this type of study is not yet common in the analysis of legal processes. We propose the adoption of automated Process Mining (PM) techniques from legal data. First, we present the preparation steps

T. P. Sales et al. (Eds.): ER 2023 Workshops, LNCS 14319, pp. 178–188, 2023.
https://doi.org/10.1007/978-3-031-47112-4_17

for constructing a legal event log to be examined by PM techniques. Second, we show how process discovery can be a useful technique for comparing the procedures (PM models) of public tenders in different member states. Third, we demonstrate how to enrich the event log by extracting features from public tenders by using Natural Language Processing (NLP) techniques. As a case study, we propose to apply this approach to the European public procurement process, starting from the examination of public tenders published only by Tenders Electronic Daily (TED), the online version of the "Supplement to the Official Journal" of the EU, dedicated to European public procurement [15].

We focus on the following research questions (RQs): RQ1) Starting from a legal dataset, can we create a PM model for different countries?; RQ2) Can we compare the legal processes of different countries? RQ3) Starting from a textual legal dataset of a country, can we extract relevant information and events to enrich our PM model?

In the remainder of the article, we describe the background with related work and the case study in Sect. 2, the methodology adopted in Sect. 3. Section 4 introduces the results, and Sect. 5 discusses some concluding remarks.

2 Background

2.1 Related Work

Legal aspects have been typically addressed in a BPM perspective by considering regulatory compliance issues. Recently, the PM discipline gains attention for several application areas, from the typical BPM research area [2], healthcare [8], or education [12]. The conformance checking research seems promising to investigate legal compliance issues [5, 16].

Previous works explored real-world cases of process discovery involving public procurement. For instance, a case study focuses on a heuristic algorithm revealing a concept drift in publication of contract in the Philippines [19], while [17] investigates public procurement procedures in Croatia. A comprehensive application of process discovery in the legal field is [21], where authors applied discovery techniques for the extraction of lawsuit processes from the information system of the Court of Justice of the State of Sao Paulo, Brazil.

Artificial intelligence techniques have been explored in legal domain, including machine learning techniques [13], merging dataset [14], explainability [10]. Finally, a relevant area concerns NLP techniques. Applications are finding increasing interest in BPM [1]. Recent works propose to exploit unstructured information in natural language textual descriptions of processes to enable formal process modeling [9]. Other works investigate the enrichment of an event log based on unstructured text [7].

2.2 Case Study

Legal data of tenders and processes related to publication procedures can be found in the Tenders Electronic Daily (TED) website[1]; TED publishes 735 thousand procurement notices a year, including 258 thousand calls for tenders which are worth approximately €670 billion.

Dataset Overview. In the TED dataset, each tender is identified by an alphanumeric value called *document-number* (the key value) and it has the following relevant features: the *sector* to which it belongs ("services", "works", or "supplies"); the *NUTS* (Nomenclature of Territorial Units for Statistics [4]) of the contracting authority (public administration) that issued the procurement; the *type* of the contracting authority (Ministry, European Institution, Regional or local autority, etc.); and the *amount* of the tender. The complete schema of the CSV is publicly available[2].

The Legal Process and the Activities. We focused on regional authorities of the French (FRA), Italian (ITA), and Spanish (ESP) cases, following the suggestion of domain experts. According to the domain experts, the three member states are similar in the legal review system. There are five main activities in the public procurement process. The initial event is the PUBLICATION of the tender, then the PARTICIPATION of individual entities proposing a bid. Following the adjudication of the tender (AWARD), the contract is issued (CONTRACT-START) with the procedure ending with the deadline provided for in the call (CONTRACT-END). Furthermore, a sixth event can be identified in the notice, i.e. the BID-OPENING which takes place before AWARD.

3 Methodology

3.1 The Process Dataset Construction

The dataset was created from the collection of annual tender and award notices available on the TED portal[3]. For each tender, identified by a document-number, we generated an event log trace. Next, we retrieved the PDF text, and the trace was enriched with the new date found in the text via NLP. Our final collection of French, Italian, and Spanish cases from 2016 to 2022 is publicly available[4].

Event Log of Legal Process. We introduce here the terminology for automated process analysis. The starting point of a PM research is an *event log*, i.e. a set of *traces* where each trace stores a sequence of *events*, each representing the

---

[1] https://ted.europa.eu/TED/browse/browseByMap.do.
[2] https://data.europa.eu/api/hub/store/data/ted-csv-data-information-v3-5.pdf.
[3] https://data.europa.eu/data/datasets/ted-csv?locale=en.
[4] https://github.com/roberto-nai/JUSMOD2023.

execution of an *activity* occurred during a single execution of the process, possibly together with some additional data (e.g., the resource who performed the activity) [2]. Every trace is identified by means of a so-called *case ID*. Different traces in the event log are called *process variants* (*variants*), since they represent alternative ways to execute the same process (cases may perform activities in a different order before the end).

Three attributes are necessary in order to generate an event log from a set of not process-oriented samples, such as the Tender dataset we start from: first, the *case ID* indicating which case of the process is responsible for each event. Second, the *event class* (or *activity name*) specifying which activity the event refers to. Third, the *timestamp* specifying when the event occurred [3]. An event log may carry additional attributes in its payload; these are called *event-specific attributes* (or *event attributes* for short) [2]. In our case study, document-number is the case identifier. The activities include the date of each activity at the level of granularity of the day on which the event occurred. In order to have a consistent event log, we removed cases too short, which are not meaningful according to domain experts. In addition, we add in the log file as attributes the information on the corresponding sector, the NUTS, and the amount. The script in Python to transform TED's CSV files into event log is publicly available[5].

```
Case ID;Activity;Timestamp;Sector;Amount;Nuts;Country
...
2017106814;PUBLICATION;2017-03-17;S;1035000.0;ITC13;IT
2017106814;PARTICIPATION;2017-05-09;S;1035000.0;ITC13;IT
2017106814;AWARD;2017-06-07;S;1035000.0;ITC13;IT
2017106814;CONTRACT-START;2017-09-01;S;1035000.0;ITC13;IT
2017106814;CONTRACT-END;2022-07-31;S;1035000.0;ITC13;IT
2017107959;PUBLICATION;2017-03-20;S;637622.4;ITE19;IT
2017107959;PARTICIPATION;2017-05-03;S;637622.4;ITE19;IT
2017107959;CONTRACT-START;2017-06-01;S;637622.4;ITE19;IT
2017107959;AWARD;2017-06-15;S;637622.4;ITE19;IT
2017107959;CONTRACT-END;2020-05-31;S;637622.4;ITE19;IT
...
```

Fig. 1. Legal event log example in CSV format. Full size image available at https://github.com/roberto-nai/JUSMOD2023

3.2 Process Discovery and Variant Analysis

We imported the event log file in the tool DISCO from Fluxicon[6] to perform an initial analysis of the data involving filtering, exploration with process discovery, and bottleneck search.

Furthermore, we perform variant analysis to discover possible significant differences between subgroups of process executions. Variant analysis is a family of techniques that analyze event logs to identify and explain the differences between two or more processes. In this research, we performed a *manual* comparison of the process models obtained for the identified variants. With this exploratory

---

[5] https://github.com/roberto-nai/JUSMOD2023.
[6] https://fluxicon.com/disco/.

phase, it is possible to gather useful indications as to whether these variants actually show differences and to determine which properties we want to explore further.

3.3 Event Log Enrichment

Information extraction is a challenging task that requires various techniques, including named entity recognition (NER), regular expressions, and text matching, among others [11]. In this research, we dive into date extraction using the Spark NLP `DateMatcher` and `RegexMatcher` *annotators*. Spark NLP[7] is a library built on top of Apache Spark ML [18]. An *annotator* in Spark NLP is a component that performs a specific NLP task on an input text document and produces an output document with additional metadata (e.g., dates or amounts) [20]. To measure the efficiency of these methods, we compared the performance of the pipelines from Spark NLP with the built-in method of Python dedicated to regular expressions (RE). As a proof-of-concept, we extracted dates from the texts of only one state (enriching the ITA log). We used the open source library *PyPDF2*[8] to extract text parts from PDFs to be used as input for NLP tasks[9].

4 Results

4.1 Legal Dataset

As stated in Sect. 3.1, we focused on the period from 2016 to 2022. Following the indications of the domain experts, the time frame is fairly consistent with the validity of the Italian public procurement code, which came into force in April 2016 and was replaced on 31 March 2023. The dataset contains data for 66,382 French tenders, 16,861 Italian tenders and 33,032 Spanish tenders. We started from the dataset to construct the event logs, with which to provide a comparison between the three cases by mean of standard metrics, i.e. the number of traces, instances and variants.

4.2 Legal Event Log

We obtained a main legal event log of 116,275 cases, including 285 variants. We provide an example of the legal event log in Fig. 1; cases from all the countries have a median duration of 18.6 weeks and a mean duration of 35.8 weeks. The initial process discovered from the event log of all tenders is described in Fig. 2. The process model highlights the most common process behaviours; rectangles represent process activities, while edges represent pair-wise ordering relations among the activities. The darker a rectangle is, the more frequently the corresponding activity occurs in the event log. Similarly, the thicker an edge is,

---

[7] https://sparknlp.org/docs/en/quickstart.

[8] https://pypdf2.readthedocs.io/en/3.0.0.

[9] Source code publicly available: https://github.com/roberto-nai/JUSMOD2023.

the more frequently it is observed in the event log that the source activity is eventually followed by the target activity. The exploration of the diagram and the log shows that some data need to be filtered out in order to get a better version of the process. For instance, we notice that some paths (e.g. from PUB-LICATION to CONTRACT-END) are not possible according to domain experts and some cases with extremely high duration also, e.g., the tenders without a stop date (CONTRACT-END). The very high number of variants also includes combinations of activities that were judged to be incorrect (e.g. AWARD to PUBLICATION) or insignificant (e.g. only PUBLICATION and PARTICIPA-TION events). This issue is due to data quality problems, as the tender data were entered manually by the operators of the individual contracting authorities.

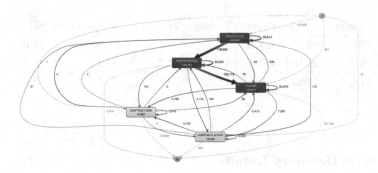

Fig. 2. Process discovery from raw TED data. Full size image available at https://github.com/roberto-nai/JUSMOD2023

Filtering the Event Log. We opted to include only the processes starting after the initial date of this interval (1 January 2016) and concluding before the end (31 December 2022). In addition, to remove incomplete cases, a filter was also inserted to define the endpoints of the process: the first event must be PUBLI-CATION and the final event CONTRACT-END. Filtering the log we obtained a log with 6,502 cases (31,321 events), 5 variants, mean duration of 26.6 months, median case duration 18.3 months and a standard deviation of 16.5. In addition to the number of cases, the number of variants is also reduced compared to the initial number of cases, as the removed cases comprised various combinations of activities (e.g. about 60 per cent of the processes contained the variant of three activities PUBLICATION to PARTICIPATION to AWARD). The corresponding CSV and XES[10] filtered without cases that are not interesting were generated; the event log files are publicly available[11].

Event Log Analysis of Three States. We analyzed process variants by taking into account the performance of every country, expressed in terms of their duration.

---

[10] XES is a standard format for process mining, https://xes-standard.org/.
[11] https://github.com/roberto-nai/JUSMOD2023.

Table 1 summarizes the differences in event logs corresponding to French, Italian, and Spanish processes. Interestingly, we immediately noticed that there are significant differences in the number of procedures and the average duration. France contains a much larger number of cases, not proportionate to the larger population of that state. On the other hand, the median duration is almost 17 months, whereas the Italian case (with a quarter of the procedures) sees a longer duration of twice as much (about 37 months). The Spanish case has the lowest number of cases and a similar median duration to the French case.

Table 1. Comparison of three countries' event logs (Spain, France, Italy) by the number of Cases, Events, Variants, Mean and Median case duration (months), standard deviation (SD)

Country	Cases	Events	Variants	Mean c.d.	Median c.d.	SD
ESP	838	4,063	5	21	17.6	12.6
FRA	22,102	4,601	4	25.7	17.1	16.9
ITA	1,063	5,156	4	34.8	37.1	14.3

4.3   Process Discovery Results

It is possible to clearly visualise both similarities and differences in the three process diagrams for each country in Fig. 3. The diagrams introduce the average duration between two activities in the weight of the arcs, and the size of the arcs easily detect bottlenecks. The average time from the notice of the tender to participation is about 36 days (min. 34, max. 39), the average time to determine the winner of the tender is about 45 days (min. 36, max. 54). The time to decide the winning tender is 19.3 months. In addition to the timing, loops in the Spanish process can be observed in the PUBLICATION and PARTICIPATION activities: according to the domain experts, this is due to the fact that the contracting authorities can re-publish a tender notice, so participation also takes place again on the basis of the new publication. A second observation is that AWARD activity does not always take place; according to domain experts, this is due to 'framework agreements' (multi-year contracts over several tenders, published but automatically awarded).

4.4   Event Log Enrichment

Starting from the 1,063 total cases for the ITA country we found, in the text of tenders, the presence of the bid opening section in 287 cases; in 17 cases, the section contained no text. In the remaining 270 cases, the three NLP techniques described above (Sect. 3.3) were applied and compared. Table 2 shows the results obtained. For this date recognition task, the RE method performed best in our

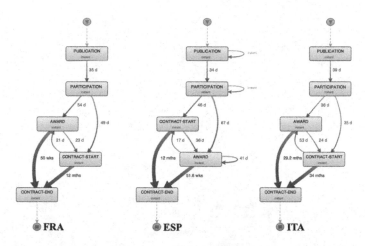

Fig. 3. The European public procurement processes of France, Spain, and Italy extracted from notices published in Tenders Electronic Daily (TED). Full size image available at https://github.com/roberto-nai/JUSMOD2023

experiment. As a demonstration, the dates extracted from the tender notices were associated with the new BID-OPENING event, which was then added to the ITA tender log.

Table 2. Comparison of the three NLP methods used to extract the new BID-OPENING event from PDF texts.

Method	Dates found (tot.)	Timing (min.sec)
RE	270	0.6
DateMatcher	223	1.10
RegexMatcher	209	1.11

To Add a New Event from Dataset. Figure 4 shows the new BID-OPENING event extracted from the tender notices. The new event shows how the opening of received bids takes place in a short time (transition from PARTICIPATION to BID-OPENING) while the main part of the time required for awarding the work to the economic operator is given by the decision-making process of the contracting authorities (transition from BID-OPENING to AWARD).

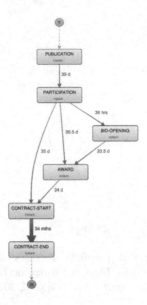

Fig. 4. The public procurement process of ITA, with the new event BID-OPENING extract from the texts. Full size image: https://github.com/roberto-nai/JUSMOD2023

5   Conclusion

In this work, we applied PM techniques to a legal process to understand its execution in reality as well as to identify potential issues and inefficiencies. More precisely, we implemented a combination of process discovery, NLP, and variant analysis techniques. As far as RQ1 is concerned, the transformation of the CSVs has led to the creation of an event log. The consequent discovery techniques allowed to obtain relevant information on the main behaviour of the award process. With regard to RQ2, once the logs were filtered, it was possible to compare the country-based processing in terms of time performance. For RQ3, NLP tasks applied to the notice texts made possible to explore some information extraction techniques as well as to obtain a new event to add to the ITA event log, increasing knowledge about the process. In the future, we intend to further develop our investigation by exploring different techniques for process discovery and variant analysis (e.g. automated variant analysis), also apply it to other features in the log (amount, NUTS, etc.). Furthermore, we intend to further explore other NLP techniques to extract more data from the texts and enrich the process event log even more, also for other countries. Finally, we want to explore the application of prediction algorithms on track prefixes to determine in advance processes that are excessively long or that will not end (e.g. for an appeal to the administrative justice by economic operators).

References

1. van der Aa, H., Carmona, J., Leopold, H., Mendling, J., Padró, L.: Challenges and opportunities of applying natural language processing in business process management. In: Bender, E.M., Derczynski, L., Isabelle, P. (eds.) Proceedings of the 27th COLING, pp. 2791–2801. ACL (2018). https://aclanthology.org/C18-1236/
2. van der Aalst, W.M.P.: Process Mining - Data Science in Action. Springer, Heidelberg (2016). https://doi.org/10.1007/978-3-662-49851-4
3. van der Aalst, W.M.P., Carmona, J. (eds.): Process Mining Handbook. LNBIP, vol. 448. Springer, Heidelberg (2022). https://doi.org/10.1007/978-3-031-08848-3
4. Becker, S.O., Egger, P.H., Von Ehrlich, M.: Going nuts: the effect of EU structural funds on regional performance. J. Public Econ. **94**(9–10), 578–590 (2010)
5. Carmona, J., van Dongen, B.F., Solti, A., Weidlich, M.: Conformance Checking - Relating Processes and Models. Springer, Heidelberg (2018). https://doi.org/10.1007/978-3-319-99414-7
6. Dumas, M., Rosa, M.L., Mendling, J., Reijers, H.A.: Fundamentals of Business Process Management. Springer, Heidelberg (2018). https://doi.org/10.1007/978-3-662-56509-4
7. Geeganage, D.T.K., Wynn, M.T., ter Hofstede, A.H.: Text2EL: exploiting unstructured text for event log enrichment. In: 2022 16th International Conference on SITIS, pp. 1–8 (2022). https://doi.org/10.1109/SITIS57111.2022.00010
8. Munoz-Gama, J., et al.: Process mining for healthcare: characteristics and challenges. J. Biomed. Inform. **127**, 103994 (2022). https://doi.org/10.1016/j.jbi.2022.103994
9. Maqbool, B., et al.: A Comprehensive investigation of BPMN models generation from textual requirements—techniques, tools and trends. In: Kim, K.J., Baek, N. (eds.) ICISA 2018. LNEE, vol. 514, pp. 543–557. Springer, Singapore (2019). https://doi.org/10.1007/978-981-13-1056-0_54
10. Meo, R., Nai, R., Sulis, E.: Explainable, interpretable, trustworthy, responsible, ethical, fair, verifiable AI... what's next? In: Chiusano, S., et al. (eds.) ADBIS 2022. LNCS, vol. 13389, pp. 25–34. Springer, Cham (2022). https://doi.org/10.1007/978-3-031-15740-0_3
11. Mohit, B.: Named entity recognition. In: Zitouni, I. (ed.) Natural Language Processing of Semitic Languages. TANLP, pp. 221–245. Springer, Heidelberg (2014). https://doi.org/10.1007/978-3-642-45358-8_7
12. Nai, R., Sulis, E., Marengo, E., Vinai, M., Capecchi, S.: Process mining on students' web learning traces: a case study with an ethnographic analysis. In: Viberg, O., et al. (eds.) EC-TEL 2023. LNCS, vol. 14200, pp. 599–604. Springer, Cham (2023). https://doi.org/10.1007/978-3-031-42682-7_48
13. Nai, R., Sulis, E., Meo, R.: Public procurement fraud detection and artificial intelligence techniques: a literature review. In: Danai Symeonidou et al. (ed.) 23rd EKAW International Conference on Proceedings CEUR, vol. 3256. CEUR-WS.org (2022). https://ceur-ws.org/Vol-3256/km4law4.pdf
14. Nai, R., Sulis, E., Pasteris, P., Giunta, M., Meo, R.: Exploitation and merge of information sources for public procurement improvement. In: Koprinska, Irena, et al. (eds.) ECML PKDD 2022. CCIS, vol. 1752, pp. 89–102. Springer, Cham (2022). https://doi.org/10.1007/978-3-031-23618-1_6
15. Prier, E., Prysmakova, P., McCue, C.P.: Analysing the European union's tenders electronic daily: possibilities and pitfalls. IJPM **11**(6), 722–747 (2018)

16. Pufahl, L., Rehse, J.: Conformance checking with regulations - a research agenda. In: Koschmider, A., Michael, J. (eds.) 11th Int. Workshop on EMISA, vol. 2867, pp. 24–29. CEUR-WS.org (2021). http://ceur-ws.org/Vol-2867/paper4.pdf
17. Rabuzin, K., Modrušan, N., Križanić, S., Kelemen, R.: Process mining in public procurement in croatia. In: Lalic, B., Gracanin, D., Tasic, N., Simeunović, N. (eds.) IS 2020. Lecture Notes on Multidisciplinary Industrial Engineering, pp. 473–480. Springer, Cham (2022). https://doi.org/10.1007/978-3-030-97947-8_62
18. Salloum, S., Dautov, R., Chen, X., Peng, P.X., Huang, J.Z.: Big data analytics on apache spark. Int. J. Data Sci. Anal. **1**, 145–164 (2016)
19. Sangil, M.J.: Heuristics-based process mining on extracted Philippine public procurement event logs. In: 2020 7th International Conference on BESC, pp. 1–4 (2020). https://doi.org/10.1109/BESC51023.2020.9348306
20. Thomas, A.: Natural Language Processing with Spark NLP: Learning to Understand Text at Scale. O'Reilly Media (2020)
21. Unger, A.J., Neto, J.F.D.S., Fantinato, M., Peres, S.M., Trecenti, J., Hirota, R.: Process mining-enabled jurimetrics: analysis of a Brazilian court's judicial performance in the business law processing. In: Proceedings of 18th ICAIL, pp. 240–244. ACM, New York (2021). https://doi.org/10.1145/3462757.3466137

Formalising Legal Knowledge of Sri Lankan Civil Appellate High Court Domain from Ontological Perspective

Chaminda Liyanage[1], Tharaka Ilayperuma[2]([✉]) [iD], Jeewanie Jayasinghe Archchige[3] [iD], and Faiza A. Bukhsh[3] [iD]

[1] University of Ruhuna, Matara 81000, Sri Lanka
[2] Staffordshire University, College Rd, Stoke-On-Trent ST4 2DE, UK
tharaka.ilayperuma@staffs.ac.uk
[3] University of Twente, Drienerlolaan 5, 7522 NB Enschede, The Netherlands
{j.a.jayasinghearchchige,f.a.bukhsh}@utwente.nl

Abstract. Legal ontologies are used to support structuring concepts and relationships in the legal domain. In this study, we focus on the Sri Lankan Civil Appellate High Court (CAHC) domain and investigate the problem of managing the domain knowledge by building a domain ontology. To answer this problem, in this paper, we propose ontology to represent legal domain knowledge in the Sri Lankan CAHC domain. We use Methontology as the ontology development methodology and the Knowledge Representation (KR) ontology to formalise the CAHC domain concepts and to draw relationships between them. The developed ontology's ability to appropriately model the CAHC domain is evaluated by using a set of anonymized real cases. We believe that the results of this study will benefit knowledge management efforts especially in the Sri Lankan legal context.

Keywords: Legal Ontology · Civil Appellate Domain · Methontology · KR ontology

1 Introduction

A number of research studies have been focused on formalising legal domain knowledge by using ontologies for different purposes, such as developing expert systems to support formalising layout and content in publishing legal documents to support machine processing, developing the generic architecture for building systems that support exchanging legal knowledge between existing legal knowledge systems, and to discuss how to build legal ontologies, etc. [1, 2] and [3]. Further, legal ontologies are used to support information retrieval and Meta-data generation and specification for document annotation in court automation projects [4, 5].

The aim of this research is to build ontology to describe Sri Lankan Civil Appellate High Courts domain. High Court of the Provinces was established into the Sr Lankan Judicial System by a special Act in 1990, which was subsequently amended in 2006

T. P. Sales et al. (Eds.): ER 2023 Workshops, LNCS 14319, pp. 189–194, 2023.
https://doi.org/10.1007/978-3-031-47112-4_18

to grant further powers related to appellate and revisionary jurisdiction in respect of judgments, decrees, and orders delivered and made by any District Court within the respective Province [6].

2 Legal Ontology Development

In this study, we follow a bottom-up approach in building the legal domain ontology. We use the Methontology as the ontology development approach and the top-level ontology, the KR ontology to identify and describe relations between the domain terms.

The METHONTOLOGY proposed by Fernandez et.al. in [7], allows con- structing ontologies from scratch following a bottom-up approach in an iterative and incremental manner meaning that ontology can be modified to meet the domain needs at any time [7]. METHONTOLOGY guides the whole ontology development process through the specification, conceptualization, formalisation, implementation and maintenance of the ontology. METHONTOLOGY conceptualize ontologies using a set of tabular and graphical intermediate representations (IRs). These IRs uses modeling components such as concepts, relations, constants, instances, attributes of instances and concepts, axioms that are used to describe constraints, and rules used to infer knowledge in the ontology.

Knowledge Representation (KR) Ontology in [8] is designed to serve as a foundation for knowledge representation in databases, knowledge bases, and natural language processing. The top-level categories in the KR ontology are derived from a combination of seven primitive categories named as Independent, Relative, Mediating, Physical, Abstract, Continuant and Occurrent. These top-level categories can be used to analyze the concepts, attributes, and relations of a developing ontology by considering the appropriate top-level category of each term of the ontology being developed. Therefore, the KR ontology can be used with the association to an ontology development methodology as a foundation to develop ontology to a particular domain.

3 Methodology

We combine the conceptualization phase of the development activities in the Methontology and the top-level ontology; KR ontology, to develop the legal domain ontology for the civil appellate high court domain. In step 1 we identify important terms by analysing case records, and interviewing court staff. From the terms identified, a glossary of terms is developed in step 2. In this step, the meaning of each term and its synonyms are identified and tabulated. In step 3 the terms are classified into top-level categories of the KR ontology based on the characteristics of each term in the glossary of terms. Considering the relevant categories and rules of the KR ontology, the terms are then classified into concepts, attributes, and relations. In step 4, the ontology was created with the identified components. Finally, the developed ontology was evaluated to identify the accuracy and completeness using protégée tool.

4   Results and Findings

Legal proceedings in the civil appellate high courts in Sri Lanka include a variety of document types a variety of roles. The roles associated with proceedings can be categorized into staff members, lawyers, and people. The people who seek justice from the court are also called parties in a court case or litigants.

Litigants can be categorized again into two categories according to whether they are directly connected to the court case or representing a directly connected person to a court case. There is four directly connected person litigant, they are petitioner/appellant, respondent/appellee, plaintiff, and defendant. Plaintiff and defendant are associated with the court case at the district court. The petitioner/appellant and respondent/appellee are connected with the court case at the civil appellate court. Further, there could be two types of litigants (guardian and next friend) who could represent juvenile.

Lawyers play two roles in a court case as counselors and instructing attorneys. While court staff who works on a court case plays roles such as the clerk, the registrar, and the judge. They play different roles in legal proceedings. All the documents associated with the legal proceedings are kept in a case record. The case record consists of a set of documents sent by the district court and contains the details of the court case at the district court is called Brief. The set of documents added to the case record at Civil Appellate Court is called Docket.

Application of the Methodology
Step 1: Several interviews were done with lawyers and the registrar who work with the Civil Appellate High Court in Matara, Sri Lanka to prepare the questionnaire to obtain domain knowledge. Then a questionnaire was prepared after several refinements to an initial set of questions with the support of the practicing legal experts in the Civil Appellate domain. All the participants surveyed identified Lawyer, Respondent/Appellee, Plaintiff, Clerk, Defendant, Registrar, Petitioner/Appellant, and Judge as important *concepts* in the Civil Appellate Domain. 83% of the participants identified the terms Staff Member, Brief, Document, Docket, Case Record, Civil Appellate Court Case as important concepts in the domain. 66% percent of participants stated that Counselor, Next Friend, Instructing Attorney, Guardian, Litigant were important *concepts* in the domain. All the participants identified Appellate Court name, Nature of District Court case, Category of Appellate Court case, District Court case number, Appellate Court case number, Address of person, Date of Appeal, NIC number of person, District Court name, Name of person as important *attributes* of the domain concepts. 66% of participants identified Juvenile, District Court judgment date, Appellate Court judgment date as important *attributes*.

Step 2: For each identified term, the meaning in the Civil Appellate Domain and synonyms for the term were tabulated as a glossary of terms. A section of the glossary of terms is shown in Table 1 below.

Step 3: In this step, we formalise the domain knowledge according to the top- level categories, roles, and relations as in [8]. Each extracted term in the glossary of terms was categorized into a top-level category, role, or relation, by considering the characteristics of each entity. For example, the concept Court Case connects the respondents and petitioners by acting as a mediating entity between them. Table 2 below summarizes part of the findings.

Table 1. Part of Glossary of Terms

Term	Description
Court Case	A court case which is handled in a Court
Litigant	A person of a party in a court case. Plaintiff, Defendant, Petitioner and Respondent all are litigants. But not witness and Lawyers
Petitioner	The person who files an appeal case using a petition in the Civil Appellate Court (AC)
Guardian	The person who represents a juvenile petitioner or plaintiff in the court
Name	Name of a person
AC case number	The unique number which represents the court case in appellate court
Category	Category or the type of the court case. This can be a Final Appeal, a Revision, a Leave to Appeal, or an Appeal notwithstanding lapse of time
Juvenile	A person whose age is less than 18 years old

Table 2. Results of Categorization

Concept	Type	Top-level Category
Juvenile	Attribute value of Litigant	Phenomenon
Plaintiff	Relation	Role
Defendant	Relation	Role
Respondent/Appellee	Relation	Role
Next Friend	Relation	Role
AC case number	Attribute of Court Case	Characteristic
Date of Appeal	Attribute of Court Case	Characteristic
DC case number	Attribute of Court Case	Characteristic
Litigant	Concept	Physical
Lawyer	Concept	Nexus
Document	Concept	Object
Court Case	Concept	Mediating
Brief	Attribute of Case Record	Attribute
Docket	Attribute of Case Record	Attribute

Step 4: Following the completion of step three, main concepts of the Civil Appellate Domain were identified, and the results are tabulated in table 3. Finally, the relations and attributes of the main concepts also can be extracted. Part of the findings are given in Table 3.

The graphical representation of the ontology created by considering the concepts, relations, and attributes identified above, is shown in Fig. 1.

Table 3. Attributes and Relations for each concept

Concept	Attributes of the concept	Relations from the concept	Relations to the concept
Civil Appellate Court Case	Appellate Court name, Category, Appellate Court case number, Appellate Court judgment date, Date of Appeal, District Court name, Nature, District Court case number, District Court judgment date	hasPlaintiff, hasDefendant, hasPetitioner, hasRespondent, has-Clerk, hasRegistrar, hasJudge, hasCounselor, hasInstructingAttorney, hasProxy, hasPetition, hasAffidavit, hasMotion, hasJournalEntries, Etc	
Litigant	Address,NIC number, Name, Type [juvenile, adult]	hasNextFriend, hasGuardian, hasRepresenter	hasPlaintiff, hasDefendant, hasPetitioner, hasRespondent, hasNextFriend, hasGuardian, hasClient

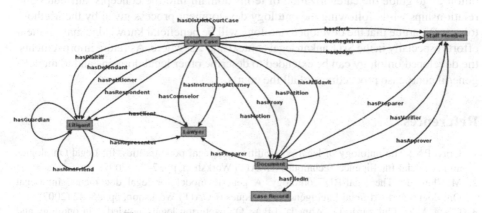

Fig. 1. Civil Appellate Legal Domain Ontology

Step 5: In this step, we use anonymized extractions of five real case details to validate the developed ontology. The resulting model created using protégé is shown in Fig. 2 below. All case documents are not included in this model for the simplicity of the representation.

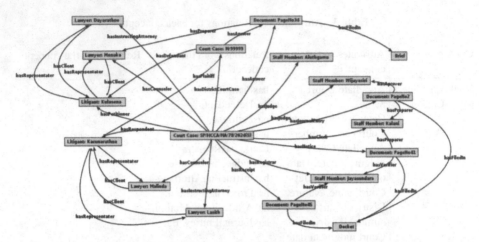

Fig. 2. Civil Appellate Legal Domain Ontology instance.

5   Conclusion

In this research, we focused on developing an ontology to formalise the legal domain knowledge of the Civil Appellate Domain. To achieve this, we used the top-level KR ontology to guide the categorization of terms domain into the concepts, attributes and relationships while following the ontology development process given by the Methontology. We argue that the developed ontology will be beneficial knowledge management efforts especially in the Sri Lankan legal context and elsewhere. As future improvements, the developed ontology can be extended to describe other legal domains and to model a general court case proceedings of all the courts.

References

1. Gray, P.N.: The ontology of legal possibilities and legal potentialities. In: Legal Ontologies and Artificial Intelligence Techniques (LOAIT) Workshop, pp. 7–23 (2007)
2. McClure, J.: The legal-RDF ontology. A generic model for legal documents. In: Legal Ontologies and Artificial Intelligence Techniques (LOAIT) Workshop, pp. 25–42 (2007)
3. Costa, M.Z., Guizzardi, G., Almeda, J.P.A.: On capturing legal knowledge in ontology and process models combined. In: The 35th Annual Conference, Saarbrücken (2022)
4. Reiling, D.: Technology for Justice: How Information Technology can support Judicial Reform (2010). https://doi.org/10.5117/9789087280710
5. Arp, R., Smith, B., Spear, A.D.: Kinds of Ontologies and the Role of Taxonomies. MIT Press, Cambridge (2015)
6. High Court of Provinces. https://www.parliament.lk/uploads/acts/english/5749.pdf. Accessed 16 Sep 2023
7. Fernandez-Lopez, M., Gomez-Pérez, A.: Overview and analysis of methodologies for building ontologies. Knowl. Eng. Rev. **17**(2), 129–156 (2002)
8. Sowa, J.F.: Top-level ontological categories. Int. J. Hum. Comput. Stud. **43**(5–6), 669–685 (1995)

OntoCom

OntoCom – 9th International Workshop on Ontologies and Conceptual Modeling

Sergio de Cesare[1], Frederik Gailly[2], Giancarlo Guizzardi[3], Chris Partridge[1,4],
and Oscar Pastor[5]

[1] Centre for Digital Business Research, Westminster Business School, University of Westminster, UK
s.decesare@westminster.ac.uk
[2] Department of Business Informatics and Operations Management, Ghent University, Belgium
frederik.gailly@ugent.be
[3] Semantics, Cybersecurity and Services (SCS), University of Twente, The Netherlands
g.guizzardi@utwente.nl
[4] BORO Solutions, UK
partridgec@borogroup.co.uk
[5] Valencian Research Institute for Artificial Intelligence (VRAIN), Universidad Politécnica de Valencia, Spain
opastor@dsic.upv.es

The importance of conceptual modeling has grown over the years and it is now common to find examples of conceptual models being developed and used in a range of diverse disciplines not related to computing including, for example, biology, business, construction and engineering. Among the reasons for this disciplinary expansion is the increasing digitalisation of all aspects of modern life as well as the increased complexity that such digitalisation entails in terms of emerging needs and requirements. The natural consequence is a proliferation of conceptual models of multiple real-world domains which sooner or later require data and systems to interoperate and/or integrate. In this emerging scenario ontology-driven conceptual modeling becomes even more fundamental to modern life due to its intrinsic ability to represent reality in a theoretically and semantically consistent manner. Foundational (or upper) ontologies have the potential to resolve the difficult problems that derive from a lack of a consistent and sound ontological theory. The benefits that can derive from the application of a foundational ontology include improved mapping to the real-world domain, an increased level of communication and understanding among stakeholders, model reuse, semantic integration and interoperability, and increased overall efficiency and effectiveness of information systems development and evolution. The application of foundational ontologies can also assist in overcoming the inscrutable nature of most mainstream artificial intelligence methods (i.e. neural networks and machine learning).

The aim of the OntoCom workshop series is to bring together academics, researchers and practitioners in order to develop an agenda of future collaborations that combine research and industrial expertise. The research papers presented at the workshop contribute toward this goal.

Contributions were sought in the form of research and research-in-progress papers. The submissions were Single reviewed by at least two program committee members and evaluated by the organizing committee. We received 13 high-quality submissions. Of these five research papers and one research-in-progress paper were accepted.

Finally, we would like to thank all the authors for their valuable contributions and our entire program committee for their insightful reviews and constructive feedback.

OntoCom Organization

Workshop Chairs

Sergio de Cesare   University of Westminster, UK
Frederik Gailly   Ghent University, Belgium
Giancarlo Guizzardi   University of Twente, The Netherlands
Chris Partridge   University of Westminster and BORO
                                    Solutions, UK
Oscar Pastor   Universidad Politécnica de Valencia, Spain

Program Committee

Mike Bennett   Hypercube, UK
Michael Dzandu   University of Westminster, UK
Pierre Grenon   National Center for Ontological Research,
                                    USA
Paul Johannesson   Royal Institute of Technology, Sweden
Thomas Moser   St. Pölten University of Applied Sciences,
                                    Austria
Yousra Odeh   Philadelphia University, Jordan
Italo Oliveira   Free University of Bozen-Bolzano, Italy
Jeffrey Parsons   Memorial University of Newfoundland,
                                    Canada
Geert Poels   Ghent University, Belgium
Tiago Prince Sales   University of Twente, The Netherlands
Emilio M. Sanfilippo   ISTC-CNR Laboratory for Applied
                                    Ontology, Italy
Pnina Soffer   University of Haifa, Israel
Marzieh Talebpour   University of Westminster, UK
Karsten Tolle   Goethe University Frankfurt, Germany

Towards Semantics for Abstractions in Ontology-Driven Conceptual Modeling

Elena Romanenko[1]($\boxtimes$)(iD), Oliver Kutz[1](iD), Diego Calvanese[1,2](iD),
and Giancarlo Guizzardi[3](iD)

[1] Free University of Bozen-Bolzano, 39100 Bolzano, Italy
{eromanenko,oliver.kutz,diego.calvanese}@unibz.it
[2] Umeå University, 90187 Umeå, Sweden
[3] University of Twente, 7500 Enschede, The Netherlands
g.guizzardi@utwente.nl

Abstract. Ontology-driven conceptual models are precise and semantically transparent domain descriptions that enable the development of information systems. As symbolic artefacts, such models are usually considered to be self-explanatory. However, the complexity of a system significantly correlates with the complexity of the conceptual model that describes it. Abstractions of both conceptual models and ontology-driven conceptual models are thus considered to be a promising way to improve the *understandability* and *comprehensibility* of those models. Although algorithms for providing abstractions of such models already exist, they still lack precisely formulated formal semantics. This paper aims to provide an approach towards the formalization of the abstraction process. We specify in first-order modal logic one of the graph-rewriting rules for ontology-driven conceptual model abstractions, in order to verify the correctness of the corresponding abstraction step. We also assess the entire network of abstractions of ontology-driven conceptual models and discuss existing drawbacks.

Keywords: Semantics for Abstractions · Abstractions of Ontology-Driven Conceptual Models · Networks of Model Abstractions

1 Introduction

Conceptual models (CMs) are high-level abstractions that are used to capture information about a domain of interest or a system that needs to be described. A special class of conceptual models that utilize foundational ontologies to ground modeling elements, modeling languages, and tools, is formed by *ontology-driven conceptual models*, ODCMs [22].

Although CMs are aimed at human comprehension [2] and are usually considered as self-explanatory artefacts, one of the most challenging problems is "to understand, comprehend, and work with very large conceptual schemas" [23]. The main reason for this is the complexity of the described domain or information system, which reveals itself in the complexity of the corresponding model.

© The Author(s), under exclusive license to Springer Nature Switzerland AG 2023
T. P. Sales et al. (Eds.): ER 2023 Workshops, LNCS 14319, pp. 199–209, 2023.
https://doi.org/10.1007/978-3-031-47112-4_19

The problem of making (OD)CMs more comprehensible has been addressed in the literature for quite some time, and different complexity management techniques have been introduced, including clustering methods, relevance methods, and summarization methods [23]. This paper analyses algorithms from the last group, which are aimed to produce a reduced version of the original model (also called summarization or abstraction).

Most of the methods for complexity management of CMs are based on classic modeling notations, such as UML and ER, and rely on syntactic properties of the model [23], such as distance between model elements. However, in the case of ODCMs, there is the possibility to make good use of the built-in ontological semantics. The first version of an abstraction algorithm leveraging foundational ontological semantics was introduced in [12], followed by an enhanced one [18], which is based on 11 graph-rewriting rules.

Although there are some reports on testing the existing algorithm over the FAIR Catalog of OntoUML/UFO models [1] and with users (see [19]), the abstraction mechanism itself still lacks formal semantics. Hence, it is not always clear whether the existing abstraction patterns actually lead to more general models. The main goal of this paper is to propose an *approach for the formalization of this abstraction process*, which is illustrated for one of the existing graph-rewriting rules. This, in turn, enables further investigations of the properties of the process itself and of the resulting abstracted models.

The remainder of the paper is organized as follows. Section 2 reviews existing approaches to formal semantics for abstractions and provides some context for the analysed algorithm. Section 3 translates one of the graph-rewriting rules for producing ODCM's abstractions into first-order modal logic and investigates the relationship between the original and the abstracted model. Section 4 provides a preliminary analysis of networks of ODCM's abstractions that are generated by applying the specified rules. Section 5 elaborates on final considerations and future work.

2   Background and Related Work

2.1   Semantics for Abstractions

Before formulating the desired properties of the abstraction process, we first need to define what we mean by model abstraction. As outlined by Saitta & Zucker [20, p. 49], most existing theories identify abstracting with *a mapping from a ground (original) to an abstracted (intended) space*, and differ in the nature of spaces and the corresponding type of mapping.

The first works in this research field mostly dealt with abstraction at the level of syntax (see [15,17,21]). Giunchiglia & Walsh [10] extended those approaches and proposed a more general theory of the abstraction process. According to these authors, abstracting in problem-solving and theorem proving may be represented as a mapping f between the formal systems Σ_1 (representing the *ground space*) and Σ_2 (representing the *abstract space*). They suggested distinguishing between *Theorem-Decreasing* (TD), *Theorem-Constant* (TC), and *Theorem-*

Increasing (TI) abstractions depending on the changes in theorems of the formal system.

In TC abstractions, the abstract space has exactly the same theorems as the ground space reformulated in another language, so that all well-formed formulas that are theorems of Σ_1 map onto well-formed formulas that are theorems of Σ_2. In a TI abstraction, the abstract space has more theorems than the ground one, while the opposite happens for a TD abstraction. The authors argued that *"certain subclasses of TI abstractions are the appropriate formalization for abstraction"* [10]. Although for the authors, abstracting is a one-step process of mapping from one language to another, they noted that "the process of abstraction can be iterated to generate hierarchies of abstract spaces" [10].

Later Ghidini & Giunchiglia [8,9] proposed a model-theoretic formalization of the abstraction process based on the *Local Models Semantics* and the notion of *compatibility relation*. They claimed that "tuning of the compatibility relation allows for the definition of the many different kinds of abstraction" [8].

Therefore, we can assume that the rules suggested for developing the abstractions for ODCMs [12,18] should generate TI abstractions as defined by Giunchiglia & Walsh [10]. We come back to this discussion in Sect. 3.

2.2 The Unified Foundational Ontology and OntoUML

The *Unified Foundational Ontology* (UFO) is an axiomatic domain-independent formal theory that builds on contributions from analytic metaphysics, cognitive sciences, linguistics, and philosophical logic [14]. UFO addresses fundamental ontological notions via a set of micro-theories that represent types and taxonomic structures, part-whole relations, relations, and events among others. Also, it has been widely used as a foundational ontology in conceptual modeling [22].

The first distinction that UFO makes is highlighting the existence of both *endurants* and *perdurants*. Endurants are object-like individuals that persist in time and are able to qualitatively change while maintaining their identities [13]. Examples include ordinary objects, e.g., 'Car', and existentially dependent entities, e.g., 'Weight'. In contrast, perdurants are entities that unfold in time.

Endurant types in UFO are of different sorts, distinguished by formal meta-properties of *rigidity* and *sortality*. Sortality is defined via the notion of *principle of identity*. A type is *sortal* if all of its instances follow the same identity principle. A *non-sortal* type aggregates properties that are common to different sortals. An example is 'Artwork', which applies to paintings, music compositions, and statues. In a complementary manner, rigidity is a property that describes the dynamics of how the type may be instantiated. Rigid types classify their instances necessarily while anti-rigid types, including *roles* (e.g., 'Wife') and *phases* (e.g., 'Child'), classify their instances contingently. A rigid sortal type providing a uniform principle of identity for its instances is called a *kind* (e.g., 'Person').

One can continue to describe UFO's taxonomy further, but for more examples and formalization, we refer the reader to [5,13]. For the scope of this work, it is essential to mention that the presented taxonomy can be extended by introducing new types and by instantiating at the level of individuals. Both options can be accomplished with the help of existing modeling tools.

OntoUML is a language designed to extend UML with the concepts of UFO. OntoUML defines a set of constructs and semantically-motivated syntactical constraints tailored for ODCMs [3]. In other words, OntoUML shifts the inherent complexity of reality towards the language's definition in such a way that every syntactically valid model represents a sound ontology in terms of UFO [6]. With this in mind, patterns for ODCM's abstractions were initially developed in terms of OntoUML [12, 18].

Although there are several tools that provide support in developing OntoUML models (e.g., OLED [11]), the models that are used as illustrations in this paper were developed using the OntoUML plugin[1] for Visual Paradigm[2].

3   Towards Formalization of Abstraction Rules

Reoccurring ontology modeling situations can be represented by means of *ontology design patterns* [7]. It has been shown, that OntoUML — which was designed to reflect the underlying ontological micro-theories of UFO — is an ontology pattern language [24]. In other words, models expressed in it are constructed by an exemplification of the provided patterns.

For this reason, the graph-rewriting rules for ODCM's abstractions in [12, 18] were designed as patterns for models expressed in OntoUML. Thus, in order to apply them, one needs to substitute the matching model with the replacement in its exemplification. In the following, we illustrate the abstraction process with one of the rules, namely Rule R2 [12].

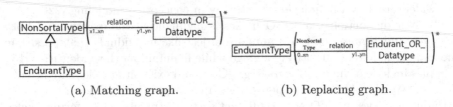

(a) Matching graph.   (b) Replacing graph.

Fig. 1. Rule R2 for abstracting a non-sortal type (from [12]).

Rule R2 should be interpreted as part of the ruleset proposed in [12]. The rationale of that set in a nutshell is: (a) abstract all information in a model by transferring it to the *kinds* in that domain; (b) in order to do that, we need to move all information from lower-level sortals to their upper classes until reaching these kinds and, subsequently, eliminating from the model these lower-level sortal subtypes (Rule R3); (c) before performing Step *(b)*, we need to first move all information from non-sortal types to their sortal subtypes, subsequently, eliminating these non-sortal types. Rule R2 captures exactly Step *(c)* as illustrated in Fig. 1.

---

[1] https://github.com/OntoUML/ontouml-vp-plugin.
[2] https://www.visual-paradigm.com/.

We provide an exemplification of the rule by replacing the placeholders with concrete classes. In Fig. 2b, you can see how this rule is applied to the ODCM given in Fig. 2a. In this example, the pattern in Fig. 1a can be found in the model twice, namely for the cases: (1) "Car as Physical Object has quality Weight", and (2) "Statue as Physical Object has quality Weight".

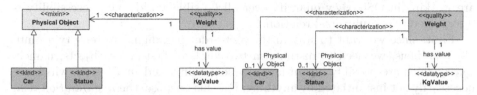

(a) Exemplification of the matching graph. (b) Exemplification of the replacing graph.

Fig. 2. Possible exemplification of Rule R2.

In order to understand the abstraction process, we need to distinguish the following levels (see Fig. 3): (1) the domain-independent level of UFO (on which the patterns are originally described), (2) the level of domain types that exemplify these patterns in a conceptual model, e.g., in OntoUML, and (3) the level of individuals on which the model can be instantiated.

When abstracting, we are changing the pattern exemplification by replacing the sub-model at the level of domain types only. Since the order of pattern applications is not fixed, one may also have another version of ODCM-2 produced by applying the rule first to Statue, instead of Car (for details see Sect. 4).

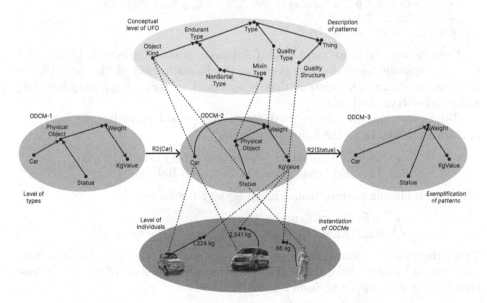

Fig. 3. ODCM abstractions produced by applying Rule R2.

To formalize the given pattern in first-order modal logic we reuse the formalization of UFO [14], which is partially reproduced here in order to ease understanding. Following it, we also use the first-order modal logic QS5 plus the Barcan formula and its converse [4,16]. In other words, we assume a fixed domain of entities for every possible world. The modal operators of necessity ($\Box$) and possibility ($\Diamond$) are used with their usual meaning, and specifically, since we are working in QS5, they quantify over all possible worlds, i.e., accessibility is interpreted over a universal relation.

First, since we want to stay in the first-order paradigm, we reify types into objects. Thus, we need to distinguish two kinds of first-order objects, namely *types*, which are possibly instantiated by something, and *individuals*, which are necessarily not instantiated by anything. For this purpose, the *instantiation relation* (::) is introduced with the possibility to describe first- and second-order types, i.e., respectively objects that are types instantiated by individuals, and objects that are types instantiated by types. In the following axioms, all free variables are implicitly universally quantified.

$$\text{Type}(x) \leftrightarrow \Diamond(\exists y\,(y :: x)) \qquad x :: y \rightarrow (\text{Type}(x) \lor \text{Individual}(x))$$
$$\text{Individual}(x) \leftrightarrow \Box(\neg\exists y\,(y :: x)) \qquad \neg\exists x, y, z\,(\text{Type}(x) \land x :: y \land y :: z)$$

The *specialization relation* between types ($\sqsubseteq$) is defined in terms of necessary extensional inclusion, i.e., the inclusion of the instances. The specialization relation is quasi-reflexive and transitive. Also, whenever two types have a common instance, they must share a super-type or a sub-type for this instance.

$$x \sqsubseteq y \leftrightarrow \text{Type}(x) \land \text{Type}(y) \land \Box(\forall z\,(z :: x \rightarrow z :: y))$$
$$\forall x, y, z\,((z :: x \land z :: y \land \neg(x \sqsubseteq y) \land \neg(y \sqsubseteq x)) \rightarrow$$
$$(\exists v\,(x \sqsubseteq v \land y \sqsubseteq v \land z :: v) \lor \exists v\,(v \sqsubseteq x \land v \sqsubseteq y \land z :: v)))$$
$$x \sqsubseteq y \rightarrow (x \sqsubseteq x \land y \sqsubseteq y)$$
$$x \sqsubseteq y \land y \sqsubseteq z \rightarrow x \sqsubseteq z$$

Considering Fig. 3, we can say that Athena :: Statue, ObjectKind(Car), and Car $\sqsubseteq$ PhysicalObject. Furthermore, in the formalization of the pattern, by *relation* we mean any relation that is specified in UFO, e.g., componentOf, associatedWith, and others [14].

Taking into account all the above mentioned, we can formalize the matching pattern from Fig. 1a as the following schema:

$$\bigwedge \begin{array}{l} \exists x, y\,(\text{NonSortal}(x) \land \text{EndurantType}(y) \land y \sqsubseteq x) \\ \Diamond\exists z\,((\text{EndurantType}(z) \lor \text{Set}(z)) \land \text{Relation}(x, z)) \end{array} \qquad (1)$$

Then the replacing pattern from Fig. 1b is the following:

$$\bigwedge \begin{array}{l} \exists y\,\text{EndurantType}(y) \\ \Diamond\exists z\,((\text{EndurantType}(z) \lor \text{Set}(z)) \land \text{Relation}(y, z)) \end{array} \qquad (2)$$

Thus, the formalization of the exemplification given in Fig. 2 is shown in Formalizations 1 and 2. Also, here KgValue is the non-empty set of possible values that Weight can take, e.g., from Fig. 3:

$$1{,}324\text{kg} \in \text{KgValue} \land 2{,}541\text{kg} \in \text{KgValue} \land 86\text{kg} \in \text{KgValue}$$

Formalization 1. Matching

ObjectKind(Car)
ObjectKind(Statue)
Mixin(PhysicalObject)
Car ⊑ PhysicalObject
Statue ⊑ PhysicalObject
QualityType(Weight)
QualityStructure(KgValue)
characterization(PhysicalObject, Weight)
associatedWith(Weight, KgValue)

Formalization 2. Replacing

ObjectKind(Car)
ObjectKind(Statue)
QualityType(Weight)
QualityStructure(KgValue)
characterization(Car, Weight)
characterization(Statue, Weight)
associatedWith(Weight, KgValue)

Proposition 1 (Rule R2). *Assuming the UFO axiomatisation, every exemplification of the matching graph of Rule R2 entails the corresponding exemplification of the replacing graph.*

Although for this particular case of **Rule R2**, the entailment of the replacing pattern is quite immediate, this is not always the case. In other rules, e.g., in **Rule R3** for abstracting sortal types (again, which proceeds in the opposite direction and moves the relation from lower-level sortal subtypes to kinds [12]), the entailment is possible from the replacing model to the matching model if a specialization of the general sortal type is given.

For example, taking into account the exemplification above we can extend it with a **RentalCar** concept (RentalCar ⊑ Car), the role that **Car** can play when it is rented for a specific **Price**: characterization(RentalCar, Price). Then, the replacing exemplification of the model according to **Rule R3** is: characterization(Car, Price), and if someone provides us with the specialization of the original **Car** type, we can entail the original model from its replacement.

Thus, in the existing rule system [12,18], generated abstractions are not always TI abstractions as suggested by Giunchiglia & Walsh [10], and a full characterisation of how different sequences of rule applications are theory increasing or theory decreasing should be undertaken.

4 Abstractions as Networks of Models

As we have seen, since the rules do not always bring us to TI abstractions, we consider ODCMs generated by them connected by a more general *compatibility relation*. Taking into account that there are different rules, we also consider this mapping process as *an iterative process, where modifications in an ODCM caused by each rule lead to the development of a hicrarchy of abstractions.*

Indeed, even very small models allow several possible abstractions, which are not always fully equivalent. Figure 4 illustrates this case, where the concept name near an arrow specifies the type that has been eliminated at the given step. Namely, we can start by abstracting **Customer**, and then sequentially abstract from **Personal Customer** and **Corporate Customer** (or vice versa, the resulting model will be the same). But we may also start by eliminating **Personal Customer**, and the resulting model will contain at least one isolated class.

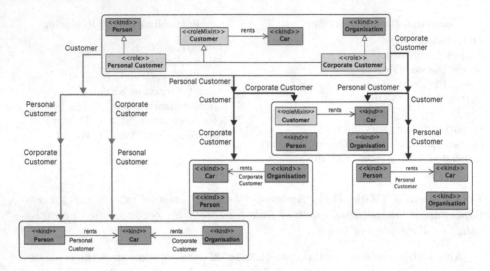

Fig. 4. Hierarchy of abstractions, where a *proper hierarchy* is shown in red (Color figure online).

Although there has been an attempt to specify the order for rule applications (see listings in [18]), those order constraints themselves were part of the methodology for rule application, but not part of the semantics of the rules themselves. As such, formally speaking and without such a constraint, every rule should be used until it is not applicable anymore.

Here, in order to distinguish different types of abstractions, we provide the following definitions.

Definition 1. *An* abstraction of an ODCM *is a model obtained from the given one by applying at least one abstraction rule.*

Definition 2. *A* complete *(or* full*) abstraction is a model to which no abstraction rule can be applied.*

Although in practice it is rare, some OntoUML models are complete abstractions of themselves.[3]

Definition 3. *In an ODCM, a* dependent concept *is either an existentially dependent entity or a concept connected to another one via the parthood or generalization relation.*

Definition 4. *A* directly applicable rule *is a rule that does not eliminate a dependent concept to which another rule could be applied.*

Definition 5. *A* primitive abstraction *is an ODCM obtained from a given one by applying a directly applicable rule.*

---

[3] Examples of such models can be found in the FAIR Catalog of OntoUML/UFO models [1] we mentioned earlier, e.g., `pereira2020ontotrans`.

Definition 6. *A proper hierarchy of abstractions is formed by primitive abstractions only.*

Most of the rules reduce the number of classes during their application, and in a nutshell, we do not want to remove those concepts until there are other rules that are applicable to them. In the original model from Fig. 4, `Personal Customer` and `Corporate Customer` are dependent concepts. Thus, only the left part of the tree (shown in red) is formed by primitive abstractions only.

The next question that arises is whether this process of applying rules is finite and to what kind of ODCMs it leads.

Theorem 1. *A complete abstraction of a domain ODCM is always reachable and not empty.*

The proof of the theorem is out of the scope of the paper. Unfortunately, in the worst case, as a complete abstraction, one can obtain a model with a single class.[4] Although tools for conceptual modeling, including Visual Paradigm, provide opportunities to modularise CMs or extract views, the majority of OntoUML models are developed as connected graphs. However, during the abstraction process, this connectivity can be lost.[5]

5   Conclusions

Ontology-driven conceptual models help in increasing the domain comprehensibility and appropriateness (including explainability) of information systems. However, they also reflect the complexity of the domain in which they are created. Thus, understanding and comprehension of such models can become a challenge without proper tool support. Abstraction — a process of providing a summary of a given CM whilst preserving the gist of conceptualization — is one of the approaches to adopt in this case.

Previously, [18] defined abstraction patterns via graph-rewriting rules. However, by just looking at those rules, and without considering formal semantics, it is not always clear whether they actually lead to more general models. Indeed, we have shown that for some rules the model could be entailed by the original one, and for some others, it cannot. A complete characterisation of how different sequences of rule applications change the theory described by an ODCM is the subject of future work.

The approach that we have specified in this paper is operational, where the abstraction process is a mapping from the original model to a modified, more abstract, version. This mapping defines a meta-theoretical relationship between ODCMs endowed with corresponding formal semantics. We have shown how hierarchies of abstraction spaces can be built and argued that this process is always finite and leads to a complete abstraction.

---

[4] An example of a model whose complete abstraction will include two classes only is `stock-broker2021`.

[5] E.g., for the model `silva2012itarchitecture`.

References

1. Barcelos, P.P., et al.: A FAIR model catalog for ontology-driven conceptual modeling research. In: Ralyté, J., Chakravarthy, S., Mohania, M., Jeusfeld, M.A., Karlapalem, K. (eds.) Conceptual Modeling. ER 2022. LNCS, vol. 13607, pp. 3–17. Springer. Cham (2022). https://doi.org/10.1007/978-3-031-17995-2_1
2. Bork, D.: Conceptual modeling and artificial intelligence: Challenges and opportunities for enterprise engineering. In: Aveiro, D., Proper, H.A., Guerreiro, S., de Vries, M. (eds.) Advances in Enterprise Engineering XV. EEWC 2021. LNCS, vol. 441, pp. 3–9 Springer. Cham (2022). https://doi.org/10.1007/978-3-031-11520-2_1
3. de Carvalho, V.A., Almeida, J.P.A., Guizzardi, G.: Using reference domain ontologies to define the real-world semantics of domain-specific languages. In: Jarke, M., et al. (eds.) CAiSE 2014. LNCS, vol. 8484, pp. 488–502. Springer, Cham (2014). https://doi.org/10.1007/978-3-319-07881-6_33
4. Fitting, M., Mendelsohn, R.L.: First order modal logic. Synthese library 277, Kluwer Acad. Publ, Boston (1998)
5. Fonseca, C.M., Porello, D., Guizzardi, G., Almeida, J.P.A., Guarino, N.: Relations in ontology-driven conceptual modeling. In: Laender, A.H.F., Pernici, B., Lim, E.-P., de Oliveira, J.P.M. (eds.) ER 2019. LNCS, vol. 11788, pp. 28–42. Springer, Cham (2019). https://doi.org/10.1007/978-3-030-33223-5_4
6. Fonseca, C.M. et al.: Ontology-driven conceptual modelling as a service. In: Proceedings of JOWO. CEUR, vol. 2969 (2021). www.ceur-ws.org/Vol-2969/paper29-FOMI.pdf
7. Gangemi, A., Presutti, V.: Ontology design patterns. In: Staab, S., Studer, R. (eds.) Handbook on Ontologies. IHIS, pp. 221–243. Springer, Heidelberg (2009). https://doi.org/10.1007/978-3-540-92673-3_10
8. Ghidini, C., Giunchiglia, F.: A semantics for abstraction. Technical report DIT-03-082, University of Trento (2003)
9. Ghidini, C., Giunchiglia, F.: What is local models semantics? In: Bouquet, P., Serafini, L., Thomason, R.H. (eds.) Perspectives on Contexts. CSLI (2008)
10. Giunchiglia, F., Walsh, T.: A theory of abstraction. Artif. Intell. **57**(2), 323–389 (1992)
11. Guerson, J., Sales, T.P., Guizzardi, G., Almeida, J.P.A.: OntoUML lightweight editor: a model-based environment to build, evaluate and implement reference ontologies. In: Kolb, J. (eds.) 2015 IEEE 19th International Enterprise Distributed Object Computing Workshop, EDOC Workshops 2015, Adelaide, SA, Australia, pp. 144–147. IEEE Computer Society (2015), https://doi.org/10.1109/EDOCW.2015.17
12. Guizzardi, G., Figueiredo, G., Hedblom, M.M., Poels, G.: Ontology-based model abstraction. In: Proceedings of the 13th International Conference on Research Challenges in Information Science (RCIS). pp. 1–13. IEEE (2019) https://doi.org/10.1109/RCIS.2019.8876971
13. Guizzardi, G., Fonseca, C.M., Almeida, J.P.A., Sales, T.P., Benevides, A.B., Porello, D.: Types and taxonomic structures in conceptual modeling: A novel ontological theory and engineering support. Data Knowl. Eng. **134**, 101891 (2021). ISSN 0169-023X. https://doi.org/10.1016/j.datak.2021.101891
14. Guizzardi, G., Botti Benevides, A., Fonseca, C.M., Porello, D., Almeida, J.P.A., Prince Sales, T.: UFO: unified foundational ontology. Appl. Ontol. **17**(1), 167–210 (2022). https://doi.org/10.3233/AO-210256

15. Hobbs, J.R.: Granularity. In: Proceedings of IJCAI, vol. 1. Morgan Kaufmann (1985)

16. Kracht, M., Kutz, O.: Logically possible worlds and counterpart semantics for modal logic. In: Handbook of the Philosophy of Science. Elsevier (2007)

17. Plaisted, D.A.: Theorem proving with abstraction. Artif. Intell. **16**(1), 47–108 (1981)

18. Romanenko, E., et al.: Abstracting ontology-driven conceptual models: Objects, aspects, events, and their parts. In: Guizzardi, R., Ralyté, J., Franch, X. (eds.) Research Challenges in Information Science. RCIS 2022. LNCS, vol. 446, pp. 372–388 Springer. Cham (2022). https://doi.org/10.1007/978-3-031-05760-1_22

19. Romanenlo, E., et al.: What do users think about abstractions of ontology-driven conceptual models? In: Nurcan, S., Opdahl, A.L., Mouratidis, H., Tsohou, A. (eds.) Research Challenges in Information Science: Information Science and the Connected World. RCIS 2023. LNBIP, vol. 476, pp. 53–68. Springer, Cham (2023). https://doi.org/10.1007/978-3-031-33080-3_4

20. Saitta, L., Zucker, J.D.: Abstraction in Artificial Intelligence and Complex Systems. Springer, New York (2013). https://doi.org/10.1007/978-1-4614-7052-6_3

21. Tenenberg, J.D.: Preserving consistency across abstraction mappings. In: IJCAI, pp. 1011–1014 (1987). www.ijcai.org/Proceedings/87-2/Papers/090.pdf

22. Verdonck, M., Gailly, F.: Insights on the use and application of ontology and conceptual modeling languages in ontology-driven conceptual modeling. In: Comyn-Wattiau, I., Tanaka, K., Song, I.-Y., Yamamoto, S., Saeki, M. (eds.) ER 2016. LNCS, vol. 9974, pp. 83–97. Springer, Cham (2016). https://doi.org/10.1007/978-3-319-46397-1_7

23. Villegas Niño, A.: A filtering engine for large conceptual schemas. Ph.D. thesis, Universitat Politècnica de Catalunya (2013)

24. Zambon, E., Guizzardi, G.: Formal definition of a general ontology pattern language using a graph grammar. In: Proceedings of the 2017 Federated Conference on Computer Science and Information Systems. IEEE (2017). https://doi.org/10.15439/2017f001

Misalignments of Social and Numerical Identity—An Ontological Analysis

Birger Andersson, Maria Bergholtz, and Paul Johannesson[✉]

Department of Computer and Systems Sciences, Stockholm University, Stockholm,
Sweden
{maria,pajo}@dsv.su.se

Abstract. Numerical identity is the relation that an individual thing
bears to itself and only itself, which is not dependent on any other
object. However, in the discourse of identity management, identity is
often something else: an identifier that an organization has assigned to
some entity, here called social identity. These two notions of identity are
closely related, as social identities are designed to mirror numerical identities from an organisational point of view. But this mirroring can easily
break down or be misaligned. This paper offers an ontological analysis of
the relationships between numerical and social identity, with a focus on
identifying different forms of their misalignments and potential causes
for these. For the analysis we rely on the Unified Foundational Ontology
(UFO), and for the conceptual modelling we use OntoUML. The result
of the ontological analysis takes the form of a conceptual model. We
envisage that this model can not only clarify theoretical concepts related
to identity, but also have practical applications in addressing issues of
rights and agency in digital identity management.

Keywords: Ontology · Conceptual Model · Social identity ·
Numerical identity

1 Introduction

As digital infrastructures and ecosystems continue to expand, maintaining security, trust, and personalization has become increasingly crucial. A key instrument
for this purpose is the digital identity. It refers to the online representation of
an individual, organization, or entity in the digital world. However, this idea
of identity differs from both our everyday and philosophical understandings of
identity. In this paper, we will analyze and compare these differing notions of
identity, explore the challenges that arise when they overlap, and propose an
ontological analysis as a way of disentangling them.

Two things are identical if and only if they are the same; the notions of
identity and sameness are the same. However, there are different kinds of identity, and a common distinction is the one between qualitative and numerical
identity. Qualitative identity refers to the similarity in properties between two

T. P. Sales et al. (Eds.): ER 2023 Workshops, LNCS 14319, pp. 210–219, 2023.
https://doi.org/10.1007/978-3-031-47112-4_20

entities, which can result in varying degrees of identity. For instance, a Volvo and a Mercedes share the property of being a car, which makes them, to some extent, qualitatively identical. In contrast, "numerical identity requires absolute, or total, qualitative identity, and can only hold between a thing and itself" [12]. A related notion is that of a principle of identity, which has been defined by [8] as "A principle of identity supports the judgment whether two particulars are the same, i.e., in which circumstances the identity relation holds".

Another way of conceptualizing identity, here called "social identity", and common in the field of identity management [2], involves organizations providing identities to individuals or physical entities. Those identities are typically based on some identifier, which is a code or another specific piece of information assigned to uniquely identify an entity within a particular organization or context, e.g. a social security number, an employee number or a vehicle identification number [4]. As expressed by [13], "Most of what we call "identity" isn't. It's identifiers. It's how some organization identifies you: as a citizen, a driver, a member, a student. Those organizations may issue you an "ID" in the form of a passport, license, or membership card, but that isn't your identity. It's their identifier". Provisioning social identities within organisations serves administrative, security as well as compliance purposes, thereby facilitating efficient communication and accountability within an organisational structure.

These two notions of identity, numerical identity and social identity, are closely related, as the purpose of social identities is to mirror numerical identities from an organisational point of view. Nevertheless, the notions are distinct, which in practice becomes apparent when the mirroring breaks down. A number of such breakdowns, or misalignments, are discussed below in the form of vignettes.

The goal of the paper is to offer an ontological analysis of the relationships between numerical and social identity, with a focus on identifying different forms of their misalignments and potential causes for these. The paper is structured as follows. First, Sect. 2 provides some motivating examples through a number of vignettes. Then, Sect. 3 presents the ontological analysis of numerical and social identity, in the form of a UFO-based conceptual model. Finally, Sect. 4 concludes with some open questions and directions for future research.

2   Vignettes of Identity Misalignment

The Dishonest Applicant. A person presents themselves at the citizen registration office with documents from another country, claiming they are sufficient to obtain citizenship. The responsible officer thoroughly examines the documents and deems them satisfactory, ultimately granting citizenship. However, it was later discovered that the documents were fraudulent. In light of this, is the citizenship valid? Can the person be considered a citizen?

The Deceased Citizen. A person who is a citizen passes away, but the citizen registry is not updated to record this fact. Although the person does not exist anymore, various rights related to the citizenship still exist, e.g. allowances and the right to vote. And other persons can actually exercise these rights, even if unlawfully. Does a citizenship and a citizen still exist?

The Corrupt Officer. A record of a new citizenship is fraudulently entered into the citizen registry by an officer at the registration office. This record is then sold to a known criminal who allows multiple individuals to use the citizenship. No one ever discovers these machinations, and several people actually use the citizenship. But is there a new citizenship? Is there a new citizen?

The Involuntary Voter. During a meeting where people are voting, a new-comer enters the room. The chairman requests everyone to vote by raising their hands. However, the newcomer is unaware of the voting process and simply raises his hand to shoo away a fly. The chairman counts the new-comer's raised hand as a vote. Did the newcomer actually cast a vote? (This vignette is in fact not about identity misalignment but serves to generalize the phenomenon).

3   A Conceptual Model of Numerical and Social Identity

3.1   The Unified Foundational Ontology (UFO)

In this paper we rely on UFO [8] as a top-level ontology modelled by means of OntoUML [9]. OntoUML is an extension of UML that incorporates the basic ontological distinctions made in UFO in the form of UML stereotypes. Stereo-types (enclosed between ≪ ≫ symbols) indicate the meta-category (kind of universal) to which a certain UML class belongs, constraining in this way its semantics according to the UFO ontology.

The basic kinds of universals we use in our model are *kind, role* (both sub-classes of sortal universal), *category*, and *rolemixin* (both subclasses of mixin universal). In addition, we shall use the stereotypes *relator, event, mode, quality*, and a few others.

Intuitively, we can see qualities as specific aspects of things we can use to compare them. Qualities inhere in things, where inherence is a special kind of existential dependence relation. Qualities are considered as endurants in UFO. In UFO, qualities belong to the more general class of intrinsic moments which also includes modes such as a thought, a belief, or an intention. Among modes, there are externally dependent modes, such as a particular mental attitude towards another person or object, which are existentially dependent on something else besides their bearer (i.e., the entity they inhere in).

Concerning relations and relationships, while a relation is usually intended as a set of tuples, a relationship should not be considered a tuple (i.e., an ordered set of objects) but rather an object in itself that needs to exist in the world in order for a relation to hold. Relations hold (i.e., relational propositions are true) in virtue of the existence of a relationship; relationships are therefore truthmakers of relations.

A relevant class of relations are extrinsic relations which can not be derived from the intrinsic properties of their relata—'married with' is the prototypical example. Extrinsic relations have corresponding relationships that can be understood as mereological sums of relational qualities: for instance, a marriage can be understood as a sum of mutual commitments and obligations. An advantage of such a position is that, since qualities are assumed to be endurants (i.e., entities that may change in time while maintaining their identity), relationships are endurants as well, whose behavior in time accounts for the way a relation holds in time. In UFO, such relationships are called *relators*.

3.2   The Conceptual Model

The conceptual model of Fig. 1 contrasts the notions of a social identity with the numerical identity of an agent. It intends to capture the mirroring between entities that need to be managed by a social system and those created by that system for identifying them. In particular, it intends to make explicit how rights are related to and exercised by these different kinds of entities. For the sake of brevity, we will focus on entities that are agents.

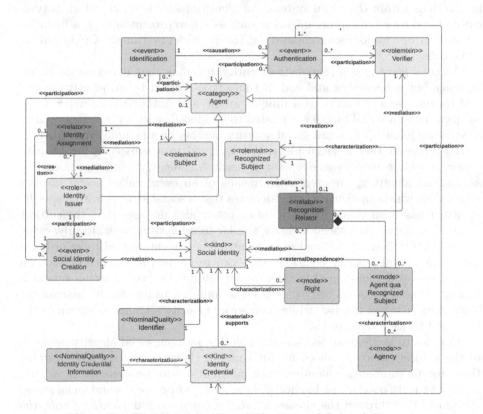

Fig. 1. OntoUML Model of Social Identity and Related Concepts

Agents and Subjects—The Participants of Social Systems. For the notion of a social system, we take a broad view, including any structured society or community, such as nation-states, professional groups, corporations, and communities based on established standards.

An *agent* is an entity that can perceive its environment, process information, make decisions and take actions to achieve specific goals. An agent can be a human or a machine; thus, the class has no uniform principle of identity and is stereotyped as a category.

A *subject* is an agent that is part of a social system and is influenced, affected, or governed by it, e.g. a citizen, an employee or a physician. A subject is dependent on a social system, and the class is therefore stereotyped as a rolemixin.

Social Identities—Identifying Subjects. In order to be recognized by other participants of a social system, a subject needs to be identifiable. It is the responsibility of *identity issuers* to create and maintain such identifiable entities. Identity issuers thereby play a crucial role in establishing trust and enabling secure interactions within the social system. An identity issuer is a role, which is typically played by a trusted organization such as a government agency, a financial institution, or a major tech company, e.g. the Swedish government, Crédit Suisse, or Google.

A *social identity* is a semiabstract entity, meaning that it has no extension in space but a beginning and end in time. It is created by an identity issuer, and its purpose is to serve as a unique, stable and authorized reference for a subject. Thus, a social identity is related to an identifier which is stereotyped as a NominalQuality [14], e.g. a social security number or an employee number.

To represent that a social identity is assigned to a subject by an identity issuer, a relator *identity assignment* is used. An identity assignment, as well as a social identifier, are created by means of an event called *social identity creation*, in which an identity issuer declares that a social identity with a specific identifier has been created. The identity issuer also declares that the subject has been assigned that social identity, similar to a baptism event. Ideally, every social identity should be related to an identity assignment, but mistakes or fraud during the social identity creation event can result in the absence of such an assignment, as demonstrated in the Corrupt Officer vignette. Another reason is that a subject's social identity may become unlinked to its identity assignment if the subject ceases to exist, while their social identity remains, as shown in the Deceased Citizen vignette.

In order to verify a subject's social identity, i.e., the social identity assigned to the subject, there is a need for information that can be used to determine the subject's (numerical) identity. Such information can be a username, a password, a security question or biometric data, such as fingerprints and facial scans. We model this through the classes *Identity Credential* and *Identity Credential Information*.

Social Identities—Serving as a Nexus of Rights. We here base the notion of a *right* on the work by Hohfeld [10], which was further developed by Alexy [1], and applied in UFO-L [7]. Hohfeld distinguishes between different kinds of rights, including privileges, claims, and powers. A person has a privilege to perform an action if they are free to do so in accordance with the rules of a social system. A person has a claim on another person if the latter is required to act in a certain way for the former's benefit. A power is the ability of a person to create or modify claims, privileges or powers. These notions can be extended also to organizations and, more controversially, to machines. However, we choose to model a right as being a mode inherent in a social identity, the reason being that the same rights can be claimed and exercised by different agents. In this way, a social identity can serve as a nexus of rights, where different rights come together in a focal point and can be recognized so that an agent can exercise those rights; more on this will be discussed below.

Authentication and Recognition—From Right to Agency. An *identification* is an event in which an agent claims that it is assigned to a social identity. For example, when an agent provides their name, username, or email address, they are identifying themselves. An identification does not prove that the claimed social identity is assigned to the agent; it is only the agent's assertion that this is the case.

A *verifier* is an agent that verifies whether the claim in an identification event is valid or not. A verifier is stereotyped as a rolemixin.

An *authentication* is an event in which a claimed social identity is to be verified, i.e. when an agent claims that it is assigned to a social identity, a verifier confirms whether this claim is valid or not. An authentication involves presenting evidence that the agent is actually who they claim they are, e.g. a fingerprint, a password or an answer to a security question. The verifier then compares this evidence with the identity credential of the claimed social identity to assess the agent's claim.

If an authentication event verifies an agent's claimed social identity, a *recognition relator* is created. This relator expresses that the social system recognises the agent with their claimed social identity and is prepared to acknowledge their rights accordingly. This relator is between a *recognized subject*, which is an agent, and a social identifier. For example, a check-in clerk at an airport, recognising a person with their claimed social security number, will give access to flights booked by that social identity. Thus, the recognized subject is able to exercise (some of) the rights associated with the claimed social identity. This situation is modelled by means of a qua object, *agent qua recognized subject*, which is a mode that characterizes a recognized subject and is externally dependent on a social identity.

This qua object has a number of agencies that correspond to the rights of that social identity, i.e., it is able to act because the verifier has acknowledged those rights. In the aforementioned example, the person can board the airplane, choose a seat, request assistance from the flight crew, and so on. It is possible that a recognized subject with a social identity is different from the subject assigned to that social identity; in this case, the recognized subject has fraudulently claimed a social identity it did not have. From the perspective of the social system, this is clearly an undesirable situation. However, even when this is the case, the recognized subject will still be able to use their agencies.

3.3 The Vignettes Revisited

In the following we illustrate how the conceptual model mentioned earlier can help elucidate the first three vignettes in Sect. 2. Figure 2 shows a small case study in the form of a domain model that has been tailored specifically to the topic of citizenships.

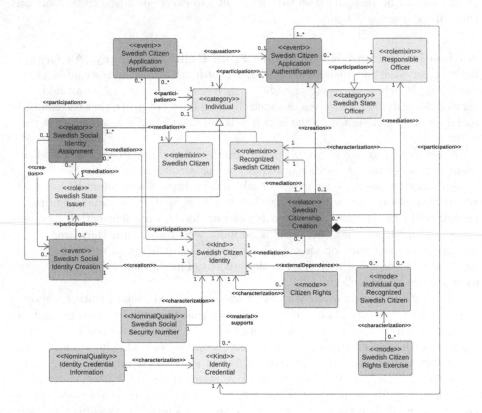

Fig. 2. The Conceptual Model Specialized to Citizenships

The Dishonest Applicant. In the Swedish Social Identity Creation event, a social identity for a Swedish Citizen Identity is created, which will work for recording rights and recognizing a Recognized Swedish Citizen. For these purposes, it doesn't matter that the event was based on faulty evidence.

However, because of this faulty evidence, it can be questioned whether a Swedish Citizen was actually created—the answer to this question depends on the specific circumstance of the case (e.g. whether some error of judgment by the officer was made) and the applicable regulations, as discussed further in Sect. 4.

The Deceased Citizen. In this case, a Swedish Citizen Identity still exists and is available for claiming and exercising the associated rights. But there is no person and, therefore, no Swedish Citizen nor a Swedish Identity Assignment.

The Corrupt Officer. In contrast to the previous case, there has never been any Swedish Identity Assignment. But there is a Swedish Citizen Identity created by the corrupt officer, which is the basis for claiming and exercising the associated rights. And multiple agents can claim and exercise these rights.

4   Concluding Discussion, Open Issues and Future Work

We have proposed a conceptual model, based on UFO, for clarifying the relationships between numerical and social identity. In particular, we have investigated how social identities are assigned to subjects of social systems, how rights are related to social identities, and how authentication events provide subjects with agencies that let them exercise rights. Most of the model's concepts are based on work in the area of identity management, [2], and Identity Management and Access (IAM) systems [11]. There is also work in the information systems area that has acknowledged the need for social (also called institutional) identities that correspond to physical entities, for example [6] and [5]. One distinguishing feature of our work is that we, through the use of a conceptual model, have made explicit the assignment of rights and the verification of social identities, thereby providing a basis for characterizing misalignments between numerical and social identity.

We have only considered rights as being associated with a social identity. Thus, the exercise of these rights requires authentication of that social identity. However, rights can also be related to physical or informational entities, where a subject only must have custody of those entities in order to exercise those rights. For example, a banknote can grant someone the power to purchase items, but they can only make the purchase if they have the banknote in their custody. As pointed out already by [10], if an individual has custody of an entity, they will always have the power that comes with it, but they may not have the privilege to use that power if the entity was acquired illegally—such as if the banknote was stolen. This situation is analogous to a subject claiming a social identity and being recognized for it, even if they were not assigned to it. Thus, these different kinds of violations should be possible to address in a uniform way.

The model does not include physical substrates, or physical carriers, of information entities, such as identifiers, credential information and right records. In particular, it does not include physical credentials, such as passports or access cards, which verifiers would typically use in order to authenticate agents. To address this limitation, the model could be extended by concepts from the Information Artefact Ontology [3].

In the proposed conceptual model, we have focused on persons, but as digital ecosystems become increasingly important, there is a need also to cater for organizations, systems, and things. We envisage that in such an extension of the model, it would be critical to investigate the rights and responsibilities of people, as well as entities such as organizations and systems, in order to ensure fair and ethical practices within the digital ecosystems.

A characteristic of social agency, in contrast to physical agency, is that a subject is only able to exercise its agency when it is recognized by some other agent. Thus, the subject's agency is temporary and only lasts for as long as they are recognized. For example, a person as a bank customer can only carry out transactions for a session, a limited time period. In the model, this fact is partially captured by a recognition relator associating a subject with a verifier and a social identity. However, this way of modelling is not explicitly indicating that the agency of a subject is temporary.

As illustrated in the vignette of the Corrupt Officer, an identity issuer could create a social identity without a corresponding identity assignment—a personal number for a Swedish citizen could be created though there is no corresponding citizen. This situation will come about if some rule of the process for creating a Swedish citizen is violated. And these rule violations may be caused by ordinary mistakes or even fraud. A similar situation may occur when a subject claims a social identity to which it is not assigned, but a verifier still accepts the claim.

One concern is about the consequences of such undesirable situations, i.e. statements about identity not being correct. If they go undetected, numerous social transactions and relationships may be dependent on them without anyone realizing that anything is amiss, and thus they are simply accepted without question. Still, it is (almost) always possible to question those statements, and if they are deemed incorrect and then corrected, the issue arises of what should happen to those transactions and relationships that were dependent on the incorrect statements. There is clearly no general answer to this issue, but in certain cases, they may be partially answered by laws and policies. It might be worthwhile to investigate how such regulations could be modelled.

An additional concern pertains to the recurrent potential for questioning the correctness of identity statements, i.e. if it is always possible to question those statements or if there is some limitation. One peculiar fact of those statements is that there is often some institution with the authority to determine their correctness, e.g. a supreme court. For instance, if someone's Swedish citizenship is in doubt, the case can be taken to various courts, but the ultimate decision of the Swedish supreme court is final—it can no more be questioned. It could be valuable to explore how such regulations could be modelled.

The two concerns above, about the consequences of identity statements and the potential for recurrently questioning them, are not limited to identity statements but apply to all social facts. The vignette of the involuntary voter provides an example of this, where a physical action is interpreted as a social one, even though the person who performed the action did not intend it to be. Thus, the approach suggested in this paper can be expanded and adjusted to address instances of fraud and epistemic confusion in other scenarios.

Acknowledgements. Many thanks to Giancarlo Guizzardi for fruitful discussions on the topics of fraud and epistemic confusion.

References

1. Alexy, R., Rivers, J.: A Theory of Constitutional Rights, 2nd edn. Oxford University Press, Oxford (2002)
2. Bertino, E., Takahashi, K.: Identity Management: Concepts, Technologies, and Systems (Information Security & Privacy). Artech House Publishers (2011)
3. Ceusters, W., Smith, B.: Aboutness: towards foundations for the information artifact ontology. In: Proceedings of the Sixth International Conference on Biomedical Ontology (ICBO). CEUR, vol. 1515. CEUR-WS.org (2015)
4. Clarke, R.: A reconsideration of the foundations of identity management. In: 35th Bled eConference Digital Restructuring and Human (Re) action (2022)
5. Eriksson, O., Ågerfalk, P.J.: Speaking things into existence: ontological foundations of identity representation and management. Inf. Syst. J. (ISJ.12330) (2021)
6. Eriksson, O., Johannesson, P., Bergholtz, M.: Institutional ontology for conceptual modeling. J. Inf. Technol. **33**, 105–123 (2018)
7. Griffo, C., Almeida, J.P.A., Guizzardi, G.: Conceptual modeling of legal relations. In: Trujillo, J.C., et al. (eds.) ER 2018. LNCS, vol. 11157, pp. 169–183. Springer, Cham (2018). https://doi.org/10.1007/978-3-030-00847-5_14
8. Guizzardi, G.: Ontological foundations for structural conceptual models. Ph.D. Thesis, Enschede, The Netherlands (2005)
9. Guizzardi, G., Fonseca, C.M., Benevides, A.B., Almeida, J.P.A., Porello, D., Sales, T.P.: Endurant types in ontology-driven conceptual modeling: towards OntoUML 2.0. In: Trujillo, J.C., et al. (eds.) ER 2018. LNCS, vol. 11157, pp. 136–150. Springer, Cham (2018). https://doi.org/10.1007/978-3-030-00847-5_12
10. Hohfeld, W.N.: Some fundamental legal conceptions as applied in judicial reasoning. Yale Law J. **23**(1), 16–59 (1913)
11. Indu, I., Anand, P.M.R., Bhaskar, V.: Identity and access management in cloud environment: mechanisms and challenges. Eng. Sci. Technol. Int. J. **21**(4), 574–588 (2018)
12. Noonan, H., Curtis, B.: Identity. Metaphysics Research Lab, Stanford University, fall 2022 edn. (2022)
13. Preukschat, A., Reed, D.: Self-sovereign identity. Manning (2021)
14. Suchánek, M.: Quality—OntoUML specification documentation (2018). https://www.ontouml.readthedocs.io/en/latest/classes/aspects/quality/index.html. Accessed 31 July 2023

Using an Ontology for Defining Semantics of Fractal Enterprise Model

Ilia Bider[1,2(✉)] [iD] and Erik Perjons[1] [iD]

[1] DSV, Stockholm University, 7003, 16407 Kista, Sweden
{ilia,perjons}@dsv.su.se
[2] ICS, University of Tartu, Narva Mnt 18, 51009 Tartu, Estonia

Abstract. The paper discusses experience of using a formal ontology for defining semantics of a modeling language. This is done for an enterprise modeling language called Fractal Enterprise Model (FEM). FEM significantly differs from other enterprise modeling languages; it presents the operational activities of an enterprise using a combination of classes and sets. Although a model in FEM does not explicitly represent such concepts as individuals or time, these concepts should be introduced to formally define FEM's semantics. Thus, the ontology underlying FEM needs to include more concepts than FEM itself. The paper presents FEM in an informal fashion, and then suggest the first version of an ontology suitable for defining its formal semantics, which consist of the taxonomy of elements, formally defined relationships between them and a set of axioms.

Keywords: ontology · modeling language · semantics · FEM

1 Motivation

Historically, Fractal Enterprise Model (FEM) [1, 2] was developed as a tool for solving a practical problem–finding all business processes that exist in a given organization. However, to solve this problem, an additional concept–asset–was introduced alongside with the concept of business process, which represented a repetitive behavior. An asset is defined as a set of things that are used in process instances of a given process, playing some role in them. An asset that is used in a process is connected to it by a *used in* relation that has a label specifying the role the asset plays, e.g. *workforce*, *technical & info infrastructure*, etc. Another relation that ties a process to an asset is a *managed by* relation, which means that process instances of the process are aimed at changing the asset, e.g. by adding to or removing elements from the set, or changing them.

With the help of concepts of process, asset and two types of relations between them, it is possible to build a direct graph, starting with a process and going down by adding assets used in it, and then processes that manage these assets. The resulting graph has a recursive structure, which is reflected in the term *fractal* included in the name of the modeling technique. The addition of the asset concept and two relations resulted in the graph built with these concepts has a value on its own, independent of the goal of

T. P. Sales et al. (Eds.): ER 2023 Workshops, LNCS 14319, pp. 220–229, 2023.
https://doi.org/10.1007/978-3-031-47112-4_21

finding all business processes in an organization. It represents operational activities of an organization interconnected via assets, thus becoming a kind of enterprise model.

The basic model with two concepts and two types of relations between them was later extended with new concepts to describe the environment, and new relations that made FEM a relatively rich modeling language, which can be used for various purposes [3]; an example of FEM is presented in Fig. 1. However, the semantics of FEM still remains defined informally, i.e. as text. This make it difficult for users of other enterprise modeling techniques to understand the difference between FEM and more widely spread techniques, such as UML, ArchiMate, BPMN, etc. The metamodel of FEM [4], though quite formal, does not express the difference, as it does not underline an unusual characteristic of FEM of having some concepts representing sets, e.g. an asset, and some - representing classes–a process[1].

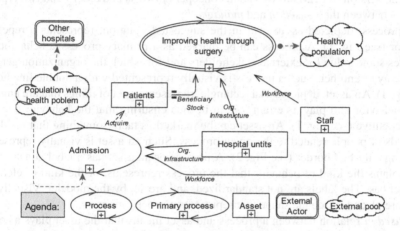

Fig. 1. An example of a FEM

The current paper represents a try to define the semantics of FEM formally, which, may help to get better understanding of its essence among the community of modelers. An approach to defining formal semantics taken is by building a formal ontology underlying FEM that includes axioms expressed in the predicate logic. One approach of building such ontology consists of mapping FEM to one of the upper-level ontologies, such as GFO [5], UFO [6], etc. However, our preliminary investigation has shown that a FEM ontology needs very few concepts, thus we decided that on this stage, it would be better to build an independent ontology that include only concepts that are essential for describing the world that can be expressed by FEM. Though the ontology needs more elements than those that are introduced in FEM to define the formal semantics, they are far less than the number of elements included in upper-level formal ontologies. This decision, however, can be reversed if new reasons of using an upper level ontology arise.

---

[1] The difference between a class and a set as an element of a model is substantial, the class represents the properties and/or behavior of any of its instances, while the set represent a collection of things – active or passive. In FEM, a set is defined by a corresponding label functioning as its intensional definition.

We use the ontology development tool Protégé [7] for defining the taxonomy for FEM ontology, but we refrain from using it for defining relations, due to the limitation built-in in OWL [8] of having only binary relations. Using reification to stay with binary relations would make the definitions more complex and more difficult to understand. We define all relations as predicates that are used in axioms that define the formal semantics of FEM.

The rest of the paper is structured according to the following plan. In Sect. 2, we introduce FEM informally, and in Sect. 3, we present a first version of formal ontology underlying FEM. Section 4 summarizes the results and draws plans for the future.

2 Fractal Enterprise Model–Main Concepts

The basic version of FEM has two main concepts - *process* and *asset* - and two types of relations between them–*used in* and *managed by*.

A process, depicted as an oval in the suggested notation, represents a repetitive behavior (see Fig. 1). A process can be marked as a primary process–a behavior that produces some value for external stakeholders and for which the organization get paid in one way or another. Such a process is visually represented with a double line border (see Fig. 1). An asset, depicted as a rectangle, represents a set of things that are engaged in the behavior and play a certain role in it, thus ensuring that the behavior continue to be repetitive (see Fig. 1). An asset can be marked as tacit–something that resides in the heads of people related to the given process. Such an asset is visually represented as having a dashed border (see Fig. 2). A process or an asset has a label attached to it that explains the kind of behavior that the process represents, or the kind of elements the asset has. The labels are not standardized, and are set by the modeler. Visually, the labels are placed inside the shape that represents a process or asset.

A *used in* relation between a process and asset means that the asset plays a certain role in the process. The relation is visually represented by an arrow with a solid line that goes from the asset to the process (see Fig. 2). A *managed by* relation between an asset and a process means that the process changes the set, i.e. adds or removes elements or changes their properties (see Fig. 2). The relation is visually represented by an arrow with a dashed line that goes from the process to the asset. To identify which role the asset plays in the process, or how the process changes the asset, a label is added to the relation. The set of labels is standardized, more exactly there are 8 labels that can be added to a *used in* relation and three labels that can be added to a *managed by* relation. These labels are listed in Sect. 3.

A straightforward way of building a FEM is to start from a primary process, find all assets engaged in it, find processes that manage these assets, and repeat the search for assets for the management processes. Thus, building of the model can be viewed as alternatively applying two types of archetypes (or patterns), a process-assets archetype and an asset-processes archetype, see visual way of representing archetypes in [3]. In the end, we will get a recursively built graph that represents the operational activities of an organization in question. The idea of recursion is represented by the term fractal in the model name, more on this, see in [1, 3].

Assets and processes can be presented on different levels of granularity. A connection between different levels of granularity, if needed, can be expressed with the help of

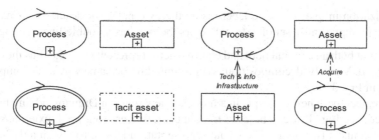

Fig. 2. Basic concepts and relations in FEM

decomposition. As an asset is a set of things, its decomposition means splitting the set into subsets. For the process, which represent behavior, decomposition can be of two types (a) specialization–elements of the decomposition represent concrete versions of generic behavior; (b) part of–elements of decomposition represent parts of the decomposed behavior.

Visually, decomposition is presented via placing elements of the decomposition inside the border of the decomposed element. In this case, the shape of the decomposed element changes to a rectangle with rounded corners; the border also changes from solid line to dashed or dotted line. The dashed border is used for decomposing assets or decomposing processes–behavior–using a *specialization* decomposition. The dotted border is used for decomposing a process using a *part of* decomposition. Details of visual representation of decomposition are presented in Fig. 3.

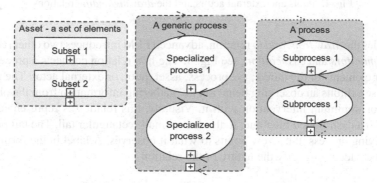

Fig. 3. Decomposition of assets and processes

The two main concepts - process and asset - are used to describe the internal operations of an organization. To be able to represent a context in which an organization operates, an advanced version of FEM is used. It includes two new concepts and three new relations for connecting the concepts. The two new concepts are:

- External pool–a set of the same type of elements as in assets, which is represented by a cloud shape, see Fig. 4. The label inside the external pool describes its content.
- External actor, which is represented by a rectangle with rounded corners. An external actor is an agent, like a company or person, acting outside the boundary of the

organization in question. The label inside the external actor describes its nature. If the shape represents a set of external actors the box has a double line, see Fig. 4.

Note that both an external pool and external actor represent sets, and thus they can be decomposed. Visually, decomposition is represented in the same way as decomposition of assets in Fig. 3.

To connect these new concepts in a model, a new relation Drawing/adding has been introduced, see Fig. 4. It can be used to connect a process to a pool, external actor to a pool, or connect two pools. The visual representation is an arrow with a dashed blue line and rounded tail. If the arrow is pointing to a pool, the arrow tail shows who is adding elements to the pool. In the opposite direction, it shows who is drawing elements from the pool to convert them to own assets. This kind of relations can also connect two pools to show that there is a movement of elements from one pool to another. The labels on these relations are not standardized, a modeler can set any text to explain what it represents.

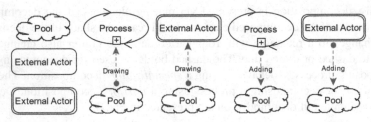

Fig. 4. Pools and external actors and the *drawing/adding* relations

Besides the *draw from/add to* relation, advanced FEM introduces two other relations: *inspects/monitors* and *association*, see Fig. 5. The first relation connects a process with any other elements of the model, i.e. process, asset, pool, or external actor. The process in this case exhibits an observing behavior, i.e. gathering information from the observed with no or minimum intervention with it. Visually, this relation is represented by an arrow with a dash-dotted blue line; it also has a small rectangular tail. The tail points to the observing process, the arrow points to what it observes. A label in the form of free text can be added to specify the nature of the relation.

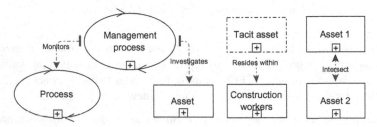

Fig. 5. Relations inspects/monitors–to the left, and *association*–to the right

The relation *association* does not have any definite meaning; it is used whenever there is a need to express something that is not possible to express with other relations. Visually, it is represented by an arrow with a blue dotted line. This relation can be symmetrical–an arrow head on both ends, or asymmetrical–an arrow head only on one end. The meaning of the relation is explained using a free text label.

With this, we are finishing our short introduction to FEM. The readers interested to know more about FEM are referred to [1–3].

3   A Formal Ontology for FEM

3.1   Taxonomy

The taxonomy of elements in the proposed FEM-based ontology are presented in Fig. 6, which is a screen shot from the Protégé tool. Besides the elements that exists in FEM, such as *process, asset, pool* and *external agent*, the taxonomy has additional elements that are needed to provide formal semantics for FEM.

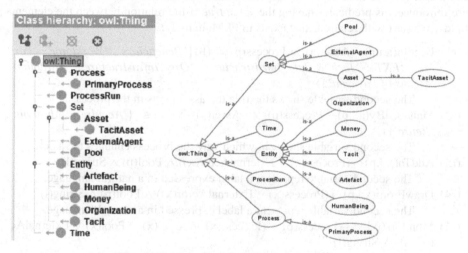

Fig. 6. Taxonomy of the FEM based ontology

In this taxonomy, *Process* represents a type element in terminology of UFO [6]. To be able to instantiate this type, we have introduced a new element called *ProcessRun* that is an individual according to UFO [6]. *Asset, ExternalAgent* and *Pool* are special types of sets. These sets can include members which are represented by *Entity* in our taxonomy. *Entity* is counted as individual according to UFO, it is sub-classed into *Artefact, Money, Organization, HumanBeing* and *Tacit*, which is enough for our purpose. In addition, we introduced the *Time* element that represents time to be able to describe the dynamic world behind a FEM.

For each of the elements in the taxonomy, we introduce a predicate that holds the value *true* if its argument belongs to the respective category of the taxonomy. For example, $Asset(a) = true$, if a is an asset.

3.2 Properties

Right now, there are not many properties defined for the elements of our taxonomy, but this can be changed in the next versions of FEM. The elements of type *Process* and all elements under *Set* have a label which is a string:

(p1) Label(x,l)→(Process(p) ∨ Set(x)) ∧ String(l)

      The second variable represent a label expressed in a natural language

In addition, *ProcessRun* has a Boolean property that expresses whether the run was successful (the goal achieved) or not:

(p2) Result (r,x)→ProcessRun(r) ∧ Boolean(x)

      $x = true$ means the run was successful, $x = false$ means that the run was aborted or has not achieved its goal.

3.3 Examples of Relations

In this section, we introduce relations that connects elements from our taxonomy, they are introduced as predicates having the value *true* if the relation between the elements holds. We start with relations that exists in FEM, namely:

(r1) UsedIn(a,l,p)→Asset(a) ∧ Process(p) ∧ (l∈{*'Beneficiary'*, *'Workforce'*, *'Partner'*, *'EXT'*, *'Tech & Info Infrastructure'*, *'Org. Infrastructure'*, *'Attraction'*, *'Means of Payment'*})

      The second variable shows the role the asset plays in the process.

(r2) ManagedBy(a,l,p)→Process(p) ∧ Asset(a) ∧ (l ∈ {*'Acquire'*, *'Maintain'*, *'Retire'*})

      The second variable shows in which way the process changes the asset.

(r3) AddTo(x,l,p)→(Process(x) ∨ ExternalActor(x)) ∧ Pool(p) ∧ String(l)

      The second variable represent a label expressed in a natural language.

(r4) DrawFrom(x,l,p)→(Process(x) ∨ ExternalActor(x)) ∧ Pool(p) ∧ String(l)

      The second variable represent a label expressed in a natural language.

(r5) Monitors(p,l,x)→Process(p) ∧ (Process(x) ∨ Asset(x) ∨ Pool(x) ∨ ExternalActor(x))) ∧ String(l)

      The second variable represent a label expressed in a natural language.

(r6) PartOf(p1,p2)→Process(p1) ∧ (Process(p2)

      p2 is part of p1.

(r7) KindOf(p1,p2)→Process(p1) ∧ (Process(p2)

      p2 is a kind of p1 (specialization).

(r8) AssociationSym(x, l, y)→(Process(x) ∨ Asset(x) ∨ Pool(x) ∨ ExternalActor(x)) ∧ (Process(y) ∨ Asset(y) ∨ Pool(y) ∨ ExternalActor(y)) ∧ String(l)

      Symmetrical association.

(r9) AssociationAsym(x, l, y)→(Process(x) ∨ Asset(x) ∨ Pool(x) ∨ ExternalActor(x)) ∧ (Process(y) ∨ Asset(y) ∨ Pool(y) ∨ ExternalActor(y)) ∧ String(l)

      Asymmetrical association y is associated with x (but not vice versa).

(r10) SubsetOf (x,y) or y ⊂ x→(Asset(x) ∧ Aset(y)) ∨ (Pool(x) ∧ Pool(y)) ∨ (ExternalActor(x) ∧ ExternalActor(y))

Additional relationships connect entities with types, sets, and time:

(r11) InstanceOf(p,r)→Process(p) ∧ ProcessRun(r)
(r12) ActiveAt(r, t1,t2)→ProcessRun(r) ∧ Time(t1) ∧ Time(t2) ∧ (t2 > t1)
   Time interval at which the process run was active.
(r13) BelongTo(s,e,t) or e ∈ s(t)→Set(s) ∧ Entity(e) ∧ Time(t),
   where: s(t) = {x|BelongTo(s,x,t)}
   The entity e is a member of the set s at the time moment t.

3.4 Examples of Axioms

After we have defined properties of and relations between the elements of the taxonomy, we can define axioms that would form the formal semantics of FEM. At this point, we do not have a goal to define all axioms, but produce enough of them to show the capabilities of ontological design in defining formal semantics of FEM.

(a1) ∀a (Asset(a) ∧ ∃p (Process(p) ∧ UsedIn(p, '*Workforce*', a))→∀e,t (e ∈ a(t)→HumanBeing(e)))
   If an asset fulfills the role of workforce, all its elements at any time should be human beings.
(a2) ∀a (Asset(a) ∧ ∃p (Process(p) ∧ (UsedIn(p, '*Benificiary*', a) ∨ UsedIn(p, '*Partner*', a)))→∀e,t (e ∈ a(t)→(HumanBeing(e) ∨ Organization (e)))
   If an asset fulfills the role of beneficiary or partner, all its elements at any time should be organizations or human beings
(a3) ∀a (Asset(a) ∧ ∃p (Process(p) ∧ (UsedIn(p, '*Technical & Info infrastructure*', a) ∨ UsedIn(p, '*EXT*', a))) →∀e,t (e ∈ a(t)→(Artefact(e) ∨ Tacit (e))))
   If an asset fulfills the role of technical & info infrastructure or EXT, all its elements at any time should be artefacts or tacit.
(a4) ∀a (Asset(a) ∧ ∃p (Process(p) ∧ UsedIn(p, '*Org. Infrastructure*', a))→∀e,t (e ∈ a(t)→Organization(e)))
   If an asset fulfills the role of org. Infrastructure, all its elements at any time should be organizations.
(a5) ∀a (Asset(a) ∧ ∃p (Process(p) ∧ UsedIn(p, '*Means of Payment*', a)) →∀e,t (e ∈ a(t)→Money(e)))
   If an asset fulfills a role of means of payment all its elements at any time should be money.
(a6) ∀a (TacitAsset(a)→∀e,t (e ∈ a(t)→Tacit(e)))
   Tacit asset consists of tacit entities.
(a7) ∀a (Asset(a) ∧¬TacitAsset(a)→∀t∃e (e ∈ a(t) ∧¬Tacit(e))
   Non-tacit asset should have at least one non-tacit entity
(a8) ∀p (PrimaryProces(p)→∃a (Asset(a) ∧ UsedIn(p, '*Benificiary*', a)))
   A primary process should have a beneficiary.
(a9) ∀p,a,r (Process(p) ∧ Asset(a) ∧ ManagedBy(p, '*Acquire*', a) ∧ InstanceOf(p,r) ∧ ActiveAt(r, t1,t2) ∧ Result (r,*true*)→∃x (Entity(x) ∧ x ∉ a(t1) ∧ x ∈ a(t2)))
   Successful run of an acquire process results in adding a new entity to the corresponding asset.

(a10) ∀p,a,r (Process(p) ∧ Asset(a) ∧ ManagedBy(p, '*Retire*', a) ∧ InstanceOf(p,r) ∧ ActiveAt(r, t1,t2) ∧ Result (r,*true*)→ ∃x (Entity(x) ∧ x ∈ a(t1) ∧ x ∉ a(t2)))

    Successful run of an retire process results in removing an entity from the corresponding asset.

(a11) ∀p,a,r (Process(p) ∧ Asset(a) ∧ UsedIn(p, '*Stock*', a) ∧ InstanceOf(p,r) ∧ ActiveAt(r, t1,t2) ∧ Result (r,*true*)→ ∃x (Entity(x) ∧ x ∈ a(t1) ∧ x ∉ a(t2)))

    A stock is depleted after a successful run of the process instance.

(a12) ∀p,x,r (Process(p) ∧ Pool(x) ∧ AddTo(p, l, x) ∧ InstanceOf(p,r) ∧ ActiveAt(r, t1,t2) ∧ Result (r,*true*)→ ∃y (Entity(y) ∧ y ∉ x(t1) ∧ y ∈ x(t2)))

    Successful run of a process that adds to a pool results in adding a new entity to the corresponding pool.

(a13) ∀p1,p2,r   (Process(p1)   ∧   Process(p2)   ∧   KindOf(p1,p2)   ∧ InstanceOf(p2,r)→InstanceOf(p1,r))

    A process run that belongs to a process also belongs to a generalization of this process.

4 Discussion and Plans for the Future

In the previous section, we have presented the first version of an ontology that defines FEM in a more formal way than before. For doing this, we have introduced elements that at the time being are not represented in FEM explicitly, namely individuals in form of process runs and entities. The individuals are not part of FEM, as a model should be, at least relatively, independent from them. Otherwise, any, even insignificant change, e.g. hiring a new employee to substitute the one who has retired, will require to update the model. However, adding individuals has helped to define axioms that can be translated into the rules of correctness of models. For example, an asset that serves as a workforce for some process cannot simultaneously used as EXT for the same or some other process. Some of these restrictions has already been implemented in the FEM toolkit [3], which is a tool to support drawing FEM diagrams.

    The first version is not complete. A straightforward extension would be adding the participant relation between the process run and entities with the same kind of roles as those that connect a process to assets it uses. When it is done, a set of axioms can be added that connect participants of a process run to the assets of the corresponding process.

    In this version of the FEM ontology, we have used the ordinary predicate logic to define relations and axioms. There might be some sense to try a more advanced formalism, e.g. some kind of a modal logic, as was done for UFO [6]. Also, it could be of value, to try to rebuild the FEM ontology using an upper level general ontology, such as GFO [5], UFO [6], or similar. The advantage can be in that these ontologies have already a rich set of axioms that could be applicable for FEM. Both directions are in our plans for the future.

    Another direction that can span from this work is extending FEM to include some individuals. This can be used to represent cases that have process instances with long period of activity, such as software development processes, as was discussed in [9]. This addition may increase the area of usage of FEM to managing specific projects.

Acknowledgement. The first author's work was partly supported by the Estonian Research Council (grant PRG1226). The authors are grateful to the anonymous reviewers whose comments have helped to improve the text.

References

1. Bider, I., Perjons, E., Elias, M., Johannesson, P.: A fractal enterprise model and its application for business development. SoSyM **16**(3), 663–689 (2017)
2. Fractalmodel.org: Fractal Enterprise Model. https://www.fractalmodel.org/. Accessed Feb 2023
3. Bider, I., Perjons, E., Klyukina, V.: Tool Support for Fractal Enterprise Modeling. In: Karagiannis, D., Lee, M., Hinkelmann, K., Utz, W. (eds.) Domain-Specific Conceptual Modeling, pp. 205–229. Springer, Cham (2022). https://doi.org/10.1007/978-3-030-93547-4_10
4. Fractalmodel.org: Meta Model of FEM. https://www.fractalmodel.org/meta-model/. Accessed July 2023
5. Herre, H.: General formal ontology (GFO): a foundational ontology for conceptual modelling. In Poli, R., Healy, M., Kameas, A., (eds.) Theory and Applications of Ontology: Computer Applications, pp. 297–345. Springer, Dordrecht (2010). https://doi.org/10.1007/978-90-481-8847-5_14
6. Guizzardi, G., Botti Benevides, A., Fonseca, C.M., Porello, D., Almeida, J.P.A., Prince Sales, T.: UFO: unified foundational ontology. Appl. Ontol. **17**(1), 167–210 (2022)
7. Musen, M.A.: The protégé project: a look back and a look forward. AI Matters **1**(4), 4–12 (2015)
8. W3C: OWL. https://www.w3.org/2001/sw/wiki/OWL. Accessed Aug 2023.
9. Bider, I., Chalak, A.: Evaluating usefulness of a fractal enterprise model experience report. In: Reinhartz-Berger, I., Zdravkovic, J., Gulden, J., Schmidt, R. (eds.) BPMDS/EMMSAD -2019. LNBIP, vol. 352, pp. 359–373. Springer, Cham (2019). https://doi.org/10.1007/978-3-030-20618-5_24

One Model to Rule Them All

A Demonstration of Ontology-Driven Minimum Viable Product Development for a Local Tourism Platform

Thomas Derave[1]([⊠]) [ID], Lander Maes[2] [ID], Tiago Prince Sales[3] [ID], Frederik Gailly[1,4] [ID], and Geert Poels[1,4] [ID]

[1] Department of Business Informatics and Operations Management, Ghent University, Ghent, Belgium
{thomas.derave,frederik.gailly,geert.poels}@UGent.be
[2] Department of Applied Mathematics, Computer Science and Statistics, Ghent University, Ghent, Belgium
lander.maes@UGent.be
[3] Semantics, Cybersecurity and Services (SCS), University of Twente, Enschede, The Netherlands
t.princesales@utwente.nl
[4] CVAMO FMake@UGent, Ghent, Belgium

Abstract. Developing platform software is a challenging and multidisciplinary task that requires significant time and relies heavily on effective human interaction and teamwork. To enhance communication and expedite the development of a customized and satisfactory Minimum Viable Product (MVP), a method for ontology-driven MVP development in the digital platform domain was introduced in a previous study. In this paper, another Design Science Research (DSR) cycle is executed to demonstrate the method's effectiveness through the development of the 'CreateYourTrip' cooperative tourism platform aimed at promoting sustainable local tourism in developing countries. The key improvement lies in adopting a single model serving multiple purposes throughout the entire development process, modernizing modeling practices in agile MVP development and improving the development speed, internal and external communication, documentation and requirement engineering quality.

Keywords: Ontology · digital platform · modelling · MVP · tourism platform

1 Introduction

The platform economy refers to activities in business, culture and social interaction that are performed on or are intermediated by digital platforms [1]. These digital platforms including Airbnb, Tripadvisor and Uber, intermediate in the interaction between their users. Digital platforms operating within the platform economy can be categorized by platform type including multi-sided platform, digital marketplace, sharing platform, crowdfunding platform and on-demand platform [2]. While these platform types

T. P. Sales et al. (Eds.): ER 2023 Workshops, LNCS 14319, pp. 230–241, 2023.
https://doi.org/10.1007/978-3-031-47112-4_22

share common functionalities, they also exhibit significant differences in the underlying business models they support.

The development of platform software is a multidisciplinary and difficult task, time-consuming and highly sensitive to human interaction and team work [3, 4]. Due to the complexity of the platform economy, developing platform software that offers the right functionality for the intended digital platform is challenging. Nevertheless, the development effort can be controlled using an efficient software development methodology [4]. Therefore, developers adopted agile approaches that offer fast feedback, are more client-focused, capitalize on continuous improvement, and build on cross-functional teams. For rapid deployment of software prototypes and thus early capturing of user feedback, it is advised to launch a Minimum Viable Product [5], or in this case Minimum Viable Platform (MVP) [6]. An MVP is a product with enough features to validate the digital platform idea in an early stage of development.

Existing PaaS tools for developing an MVP, like Sharetribe Go [7], which supports the development of digital marketplaces, and Ever Demand [8], which supports the development of on-demand platforms, have the advantage that a developer doesn't need to start from scratch and the MVP can be developed in little time. Unfortunately, these PAAS tools only focus on one specific digital platform type and do not consider the full diversity within the platform domain. Besides, they do not offer enough flexibility to develop a tailor-made MVP that satisfies all needs of the digital platform initiative. Furthermore, only a limited number of business model choices are configurable using these tools.

A solution to improve the communication between digital platform initiators and software developers and thus fasten the development of a tailor-made and satisfying MVP could be by using a 'platform-specific' ontology, which is an ontology that describes a specific existing or intended digital platform [9]. In [10], we proposed a first version of a method for ontology-driven MVP development in the digital platform domain following the Design Science Research (DSR) guidelines of Peffers et al. [11] and Wieringa [12]. In this paper, we executed another DSR cycle by applying this method in practice for developing an MVP called 'CreateYourTrip', a cooperative tourism platform to boost local, sustainable touristic activities in developing countries. Through this application, we demonstrate, test, and finetune the method, and propose improvements for some issues that popped up. The main improvement is the use of *one single model* during the complete development process serving multiple purposes. This improvement has the potential to modernize modeling in agile MVP development, leading to faster development speed, improved communication between stakeholders, and better documentation and requirement engineering quality.

Our paper will proceed as follows. In Sect. 2, we briefly present our earlier developed method for ontology and model-driven MVP development. In Sect. 3, we describe the case of CreateYourTrip and demonstrate our method on the development of the CreateYourTrip MVP. In Sect. 4, we discuss our results and lessons learned focusing on proposed improvements to the method, particularly the one single model use. We conclude and propose future research in Sect. 5.

2 Background: Method for MVP Development

In the platform domain there was till recently no clear framework or domain ontology that could be reused to avoid developing a platform-specific ontology (i.e., specific to a particular digital platform) from scratch [13]. This gap was filled by (1) the Digital Platform Ontology (DPO), which is a domain ontology that encompasses different digital platforms types [2]; (2) the Business Model (BM) extension to the DPO (i.e., Extended DPO), which makes it easier for developers to analyze the influence of business model decisions on the creation of the platform software [14]; and (3) a method that uses the Extended DPO for developing a platform-specific ontology [9].

In [10], we designed a method for the development of an MVP starting from the platform-specific ontology for that platform. This method includes four main steps which can be iterated: conceptualization, analysis, development and testing.

Conceptualization: The developers and stakeholders need to conceptualize the idea of what they want to accomplish. This includes understanding the domain, goals, and added value of the platform, conducting market research, and choosing a platform type and business model. In the first iteration this will shape a platform-specific ontology, for which the method in [9] is used. The resulting ontology is represented in OntoUML, which is a modeling language that provides a conceptual framework for representing knowledge and information and aims to capture the essence of concepts, relationships, and constraints in a domain. As it is a markup of the Unified Modeling Language (UML) it can support different perspectives and serves as a semantic anchor point for software engineers [15, 16].

Analysis: The developers then analyze the platform-specific ontology and group classes and relationships into user stories, following [17], and create a process model, using [18], to guide the MVP development. User stories are written to illustrate user goals and the desired software functionality. The process model provides a structured representation of the steps, activities, and interactions involved.

Development: The MVP is developed in an agile manner, by implementing the user stories through developing the database, and the platform back-end and front-end software. This includes UI prototyping, ontology-driven database design [19, 20], and implementing the backend and frontend of the MVP using a Model-View-Whatever (MV*) software design pattern [21].

Testing: To ensure continuous development, it is advised to verify the code with tests, validate the MVP with potential users and evaluate it with other stakeholders like investors and employees to ensure it meets their needs and expectations. Based on this feedback, the method iterates the four steps altering the platform-specific ontology to the (changing) needs of the stakeholders.

3 Case Implementation: Developing the CreateYourTrip Platform

In countries like Rwanda and Uganda, the demand for local tourism is rising thanks to its relatively low corruption and safety. Backpackers, hikers and bikers (Word Cup Road Biking 2025 in Rwanda) are finding their way into these regions. Nevertheless, currently, the tourism industry is largely controlled by massive tourism agencies and corporations such as Airbnb, Tripadvisor, and Booking.com. These companies hold significant power and charge large fees. Unfortunately, local initiatives have little knowledge of how to set up their own platforms to easily reach themselves and directly serve tourists, and therefore have difficulties in competing with these established companies.

As address this problem, we are developing the non-profit CreateYourTrip platform to boost local, sustainable tourism in Rwanda and Uganda in cooperation with Ghent University's Global Minds Fund [22]. Through the platform's intermediation, tourists can directly come in contact with local communities and activity providers (e.g., bee-keepers hikes, bike trails, pottery classes, farms, campsites, etc.) in an easy, efficient, and trustworthy manner, and bypassing the large for-profit tourism operators active in those countries.

3.1 Conceptualization

We started with developing the CreateYourTrip ontology. Such platform-specific ontology is the result of reusing and combining classes, relationships and constraints of the extended DPO [14] to describe a specific instance of a digital platform type for platform software development purposes. The modeler has the freedom to provide more concrete names for the ontology classes (e.g., from Listing to Activity Listing), to further constrain cardinalities (e.g., * can become 1..*), and to add classes (e.g., Category) and relationships (e.g., * to * recursive relationships between instances of Category) that align the model with the characteristics and capabilities of the envisioned platform. We would like to note that we have incorporated insights from travel experts from the Victoria Lake area into the development of the ontology model. However, it is important to highlight that an extensive business conceptualization and validation, involving several local partners and experts in the Lake Victoria tourism sector, has not yet been completed. Therefore, date sensitivities related to climate conditions (e.g., rainy season), local languages (e.g., Kinyarwanda) as well as other region-specific business constraints are not yet included Nonetheless, we have plans to undertake this essential step in the near future.

Figure 1 shows the fragment of the CreateYourTrip ontology that describes the creation and search of activity listings and the booking of trips. A **Platform Visitor** can browse through the available activity listings and decide to become a registered platform **User**. Registered users can assume different roles, including **Admin User**, with the right to create an **Activity Listing**, **Service Provider**, who is involved in an **Activity** that is offered on the platform in an **Activity Listing** (e.g., being involved as a guide, accommodation provider or driver), and **Traveler**, meaning a **User** that actually books a **Trip** which includes activities that are listed.

In subsequent MVP development iterations, the conceptual model that was provided by the platform-specific ontology, which is computation and technology independent [20], was further developed into a platform software model reflecting implementation decisions that were taken. We added classes and relationships that align with the implementation of the application software (e.g., create **Subactivity Listing** instead of the recursive *subactivity of* relationship in the ontology, create Customer with objects who can book a trips, and relate travels to the activities he or she plans to participate in),

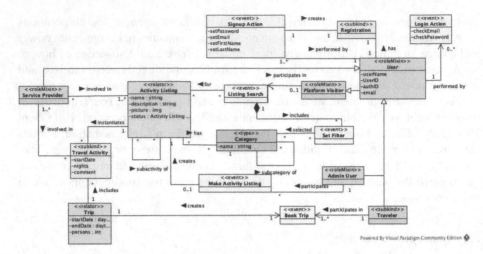

Fig. 1. Fragment from the CreateYourTrip ontology

added enumerations (e.g., *Activity Listing Status*) and added attributes to the classes (e.g., *status* in *Trip*), with their data types (e.g., string, int).

Our latest version of the CreateYourTrip software model is given in Fig. 2. Note, for instance, that we differentiate between an *Activity Listing* created by an *Admin User* and an *Subactivity Listing* created by a *Subactivity Provider*. Subactivity listings are offerings that can be part of an activity. This difference between activities and subactivities is an example of a difference between the initial CreateYourTrip ontology and later developed platform software model.

3.2 Analysis

In Agile software development, requirements documentation is mainly limited to user stories. Nevertheless, the project team may encounter difficulties in maintaining, tracing, and managing user stories, especially in a complex domain where user stories are related to and dependent of each other. Therefore, conceptual models can help to improve communication and understanding of requirements among stakeholders, as well as to validate and verify the quality of user stories [23]. Consequently, we analyze the requirements for the envisioned CreateYourTrip platform, initially starting from the platform-specific ontology and using in later iterations, the platform software model that was developed from the platform-specific ontology.

Requirements analysis entails that classes and relationships of the platform-specific ontology (in the first iteration) and platform software model (later iterations) are grouped into user stories that further guide the MVP development process. User stories are a simple narrative illustrating the user goals that a software function will satisfy [24]. They are usually articulated in the form of 'As a [role], I want [goal], so that [benefit]'. With [role] specifying a type of user, [goal] describing the (inter)actions that the user wants the software to support, and [benefit] motivating the expected functionality from the user's standpoint.

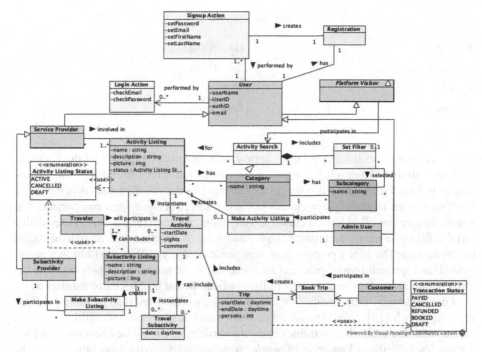

Fig. 2. Fragment from the CreateYourTrip platform model

In [10], we propose not to write user stories in the conventional template-based text format, but to use instead the object-oriented language OntoUML for user stories writing in line with [17]. Our rationale is that user stories can be read from the platform-specific ontology where a certain user role (in red) participates in an event class (in yellow) to create a certain social construct or relator class (in green) between the user role and another user role. Therefore, grouping the classes of a user role, event and relator already grasps the role and goal of a user story (as shown in table 1) while keeping the object-oriented presentation and ontological knowledge within the OntoUML model intact. The benefits part of the user story cannot be fully derived from these figures but could be specified during the validation of the model with potential users. A first validation of the benefits of each user story can be found online[1]. A process model[2] was derived by ordering the event classes of the platform-specific ontology or platform model. This can be used to better understand the dependencies among user stories [10].

3.3 Development

The development of the MVP follows the prioritization of the user stories in an iterative manner and starts with the creation of a UI prototype[3] using Figma [25]. This prototype

---

[1] https://model-a-platform.com/user-stories-of-createyourtrip/.

[2] https://purl.archive.org/domain/processmodelcreateyourtrip.

[3] https://purl.archive.org/domain/uiprototypecreateyourtrip.

Table 1. User Stories

Role (As a..)	Goal (I want to..)
Platform Visitor	Search listings
User	Sign up
Traveler	Book a trip

effectively visualizes the desired flow, appearance, and user experience of the CreateYourTrip platform. We developed a prototype application screen for each event (yellow classes in Figs. 1 and 2) such that the prototype captures the intended actions of the users. Attributes in the relators (green classes in Figs. 1 and 2) become form components that a specific user can fill in (e.g., a create activity listings screen for the admin user with a form to fill in the name, description, status and location together with an upload button for an image). The user's process flow was predominantly captured within this prototype. This approach proved practical as it encompassed not only the process flow but also encompassed the visual aspects of the application and enables early validation.

For the design of the CreateYourTrip database, we followed the one table per hierarchy approach [19], collecting all relationships and attributes of the child and parent classes into one table. The attributes and relationships of the classes *Service Provider*, *Admin User*, *Platform Visitor* and *Traveler* were captured into the *User* table. After, the other object classes (in red in Figs. 1 and 2), relator classes (in green) and type classes (in purple) required for data collection and storage were defined as tables. This was the case for *Activity Listing*, *Activity*, *Category* and *Trip*. We added relationships between tables through the events (e.g., *Traveler* has a one-to-many relationship with *Trip* through the *Book Trip* event class) and converted the OntoUML model into simple Unified Modeling Language (UML) notation, adding primary and foreign keys to specify the relationships while keeping the multiplicity constraints intact. If required, extra tables were included to solve many-to-many relationships. In further iterations, changes to the platform software model influenced the database design in the same way as described above. Therefore, a separate table was created for *Subactivity Listing*, *Subactivity* and *Subcategory*. The last version of the database schema used for the construction of the MySQL database of CreateYourTrip can be found online[4].

The backend and front-end of the MVP were also developed based on our model. We used the Model-View-Controller (MVC) software design pattern to organize code into separate modules that handle specific responsibilities, with the backend including the Model and part of the Controller and the frontend taking care of the View and the other part of the Controller. For the backend we used Node.js, an open-source, cross-platform JavaScript runtime environment that can run on the server-side, and the web framework Express to separate concerns. Following an object-oriented approach in line with our ontology and platform model, we created a Model file for each table in our database during the development of the MVP software. To streamline the interaction with our relational database, we utilized Sequelize, an Object-Relational Mapping (ORM)

---

[4] https://purl.archive.org/domain/databaseschemacreateyourtrip.

library for Node.js. Sequelize serves as an abstraction layer, simplifying the handling of database operations and aligning with the object-oriented principles of our model. The Controller on the other hand triggers events for changing the state of the Model and the View. Within the platform software model this is captured in the relations between the object classes and the event classes. The frontend was developed with Angular [26], a modular framework using the Typescript programming language. Each event (in yellow) in our model is implemented as a separate component, and an overview of the Angular components of CreateYourTrip is given by the Compodoc documentation[5].

3.4 Testing

In the original method of [10], the testing step includes three parallel sub steps: verification, validation and evaluation. Nevertheless, in line with Agile values and principles [27], we verified, validated and evaluated during all steps of the development process. The platform type, business model, ontology and platform software model were validated during conceptualization, and the benefits and priorities of the user stories validated during the analysis. At last, we evaluated the completeness of the UI prototype and MVP software for each user story by interacting with potential users and other stakeholders during the implementation. Involving stakeholders like local travel experts from the start of the modeling process by eliciting feedback, validating assumptions, and co-creating the model has highly improved the end result [23].

4 Lessons Learned and Discussion

In Table 2 we summarize how the platform-specific ontology and platform software model were used in the further steps of our MVP development method, as applied in the case study. This shows that the creation of an MVP is effectively 'driven' by the ontology and platform software model developed from this ontology throughout all development steps including making business model choices, user story writing, UI, database and software development, and validation.

During the development of the CreateYourTrip MVP, we came across multiple issues that were addressed by proposing possible improvements to the MVP development method of [10]. A suggested change to the original method involves excluding the process model development during the analysis phase. It was discovered that this sub-step does not provide additional value as the user flow and dependencies among user stories are already adequately represented in the model. Moreover, due to the user's flexibility to follow events in different orders, the process model contains numerous "or" gates, reducing its readability.

Another change to the method of [10] we propose is to add the sub-step 'Prioritize User Stories' to the analysis step. Prioritization based on the estimated value of the user stories is essential to maximize customer and business value [27]. By adding the most important user stories to the sprint backlog, agile teams can deliver valuable features early, gather feedback, and make necessary adjustments. This approach ensures that the

---

[5] https://purl.archive.org/purl/compodoccreateyourtrip.

Table 2. Influence of ontology/model on MVP development

MVP creation activity	Use of the ontology/model	Purpose/outcome
Create user stories	Complete conceptual model, separated in object (role) – event (goal) – relator (part of benefit).	User stories for agile development
UI prototype	Events	Required UI screens
Database	Objects, relators and relationships	Database tables and relationships
Model (M from MVC)	Objects and relators	Data abstraction and implement business logic
View (V from MVC)	Events	User Interface
Controller (C from MVC)	Relationships between objects and events	Changing the state of the Model and the View
Validation	Complete conceptual model	Demonstrate and iteratively develop the model and MVP in collaboration with stakeholders

product meets the needs of the users and aligns with the business objectives. Additionally, prioritization helps manage risks, optimize resource allocation, and enables incremental and iterative development, leading to faster time-to-market and improved customer satisfaction [28].

At last, to align our method with the Agile values and principles [27] our proposal suggests renaming the fourth step from 'testing' to 'Feedback' and incorporating this step partially into all the preceding steps to independently test each of them. Ultimately, based on these changes the method becomes like the one in Fig. 3.

Reflecting on this DSR cycle of our ontology-driven agile MVP development method, we confirmed that the platform-specific ontology, built from our generic Extended DPO, serves as a tool for early validation and capturing key system features. However, it is important to note that in this process, the ontology becomes a platform software model due to the implementation decisions made during the development process. We learned that the use of ontology and platform software model in agile development provides a shared understanding and visual representation of the software that is being developed. It helps address requirements challenges and facilitates communication among stakeholders. While there is a tension between modeling and Agile values and principles (e.g., "working software over extensive documentation"), our revised approach follows a *one model to rule them all* principle, where the platform-specific ontology is used in the initial iteration and the platform software model is employed in later iterations as a documentation, structuring, and validation tool for all MVP development steps. We argue that for the agile development of an MVP it is more important to have one model that changes over time and can be used for all purposes, than to have multiple models for which the developers must keep all models up to date and in line with each other. This

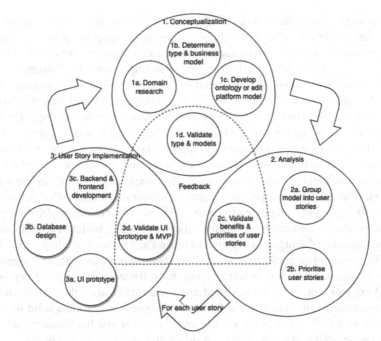

Fig. 3. Revised ontology-driven MVP development method

approach allowed us to prioritize speed while still adhering to good software development practices. Therefore, the model should not be too detailed or rigid as it can hinder the flexibility and adaptability that are essential for agile development. What to include in the platform software model was challenging as various decisions entailed trade-offs between data, process, and domain knowledge.

5 Conclusion and Future Work

Developing a digital platform is considered difficult, due to the implementation of multiple user roles and the required cooperation between different stakeholders of different domains (legal, IT, business, environment, and social). As an ontology describes the common understanding of an envisioned platform, it can provide the requirements needed to integrate complex activities into an application and enables smooth system integration [29]. An ontology can help "detect the structure of the business process", "making it exceptionally easy to integrate different business model variations under the same environment" [13], and therefore "facilitates the design of the database, software, user interface and components of applications" [13].

   In previous research, we have developed a method for Minimum Viable Platform development [10], where the ontology guides the development of the platform software. In this paper, we demonstrate and improve this method with the development of a local tourism platform called 'CreateYourTrip'. One of our main suggested improvements is to use one single model that drives the MVP development, aids in the communication between stakeholders and documents the software for prospective interested parties.

In future research, we plan to evaluate our method by applying it to the development of MVPs for a wide range of platform types operating a variety of business models, hence investigating its flexibility compared to existing PaaS tools. A test case will be set up with aspiring entrepreneurs who plan to develop an MVP of their platform idea originated from a self-constructed, DPO-based platform-specific ontology. During the development process, the version and improvements of each iteration will be monitored using GitHub classrooms, to visualize and analyze how changes in the MVP software can be traced back to ontology modifications. In the end, the efficiency and perceived usefulness of our method will be assessed with a questionnaire towards the software developers. Both single developers and teams are composed for the MVP development, to test the efficiency and communication improvements of our ontology-driven approach.

Additionally, we are currently working on constructing a boilerplate application for MVP development, offering an alternative to using PaaS tools or building an MVP from scratch. This boilerplate application will ensure that individuals without programming proficiency can construct and further expand the software, and therefore can be used as a sandbox that can easily be modified and customized to fit the specific needs of a platform owner. Furthermore, the software is designed to provide a solid foundation for building an MVP and includes a set of basic features and functionality, while also allowing for flexibility and customization depending on the selected properties of the envisioned platform.

References

1. Kenney, M., Zysman, J.: The rise of the platform economy. Issues Sci. Technol. **32**, 61–69 (2016)
2. Derave, T., Prince Sales, T., Gailly, F., Poels, G.: Comparing digital platform types in the platform economy. In: La Rosa, M., Sadiq, S., Teniente, E. (eds.) CAiSE 2021. LNCS, vol. 12751, pp. 417–431. Springer, Cham (2021). https://doi.org/10.1007/978-3-030-79382-1_25
3. Clarke, P., et al.: An investigation of software development process terminology. Commun. Comput. Inf. Sci. **609**, 351–361 (2016)
4. Hasan, S.S., Isaac, R.K.: An integrated approach of MAS-CommonKADS, Model-View-Controller and web application optimization strategies for web-based expert system development. Expert Syst. Appl. **38**, 417–428 (2011)
5. Ries, Er.: The Lean Startup. Currency (2011)
6. Gracia, C.: Your marketplace MVP – How to build a Minimum Viable Platform. https://www.sharetribe.com/academy/how-to-build-a-minimum-viable-platform/
7. Sharetribe: Sharetribe Go (2022). https://github.com/sharetribe/sharetribe
8. Ever Corporation: Ever Demand (2022). https://github.com/ever-co/ever-demand
9. Derave, T., Sales, T.P., Gailly, F., Poels, G.: Sharing platform ontology development: proof-of-concept. Sustainability 1–19 (2022)
10. Derave, T., Prince Sales, T., Gailly, F., Poels, G.: A method for ontology-driven minimum viable platform development. In: Augusto, A., Gill, A., Bork, D., Nurcan, S., Reinhartz-Berger, I., Schmidt, R. (eds.) BPMDS EMMSAD 2022. LNCS, vol. 450, pp. 253–266. Springer, Cham (2022). https://doi.org/10.1007/978-3-031-07475-2_17
11. Peffers, K., Tuunanen, T., Rotherberger, M.A., Chatterjee, S.: A design science research methodology for information systems research. J. Manag. Inf. Syst. **24**, 45–78 (2008)

12. Wieringa, R.J.: Design science methodology: for information systems and software engineering (2014)
13. Mohamad, U.H., Ahmad, M.N., Zakaria, A.M.U.: Ontologies application in the sharing economy domain: a systematic review. Online Inf. Rev. (2021)
14. Derave, T., Sales, T.P., Gailly, F., Poels, G.: Understanding digital marketplace business models: an ontology approach. In: POEM, pp. 1–12 (2021)
15. Gupta, A., Poels, G., Bera, P.: Generating multiple conceptual models from behavior-driven development scenarios. Data Knowl. Eng. **145**, 102141 (2023)
16. Guizzardi, G.: Ontological Foundations for Structural Conceptual Models (2005)
17. Thamrongchote, C., Vatanawood, W.: Business process ontology for defining user story. In: Proceedings of the 2016 IEEE/ACIS 15th International Conference on Computer Information Science, ICIS 2016, pp. 3–6 (2016)
18. Trkman, M., Mendling, J., Krisper, M.: Using business process models to better understand the dependencies among user stories. Inf. Softw. Technol. **71**, 58–76 (2016)
19. Guidoni, G.L., João Paulo, A., Almeida, G.G.: Transformation of ontology-based conceptual models into relational schemas. In: Dobbie, G., Frank, U., Kappel, G., Liddle, S.W., Mayr, H.C. (eds.) ER 2020. LNCS, vol. 12400, pp. 315–330. Springer, Cham (2020). https://doi.org/10.1007/978-3-030-62522-1_23
20. Pergl, R., Sales, T.P., Rybola, Z.: Towards ontoUML for software engineering: from domain ontology to implementation model. In: Cuzzocrea, A., Maabout, S. (eds.) MEDI 2013. LNCS, vol. 8216, pp. 249–263. Springer, Heidelberg (2013). https://doi.org/10.1007/978-3-642-41366-7_21
21. Emmit, A.S.J.: SPA design and architecture: understanding single-page web applications. Manning (2015)
22. VLIR-UOS: Global Mind Fund. https://www.ugent.be/en/research/funding/devcoop/global mindsfund.htm
23. Gupta, A., Poels, G., Bera, P.: Using conceptual models in agile software development: a possible solution to requirements engineering challenges in agile projects. IEEE Access **10**, 119745–119766 (2022)
24. Sh Murtazina, M., Avdeenko, T.V.: The ontology-driven approach to support the requirements engineering process in scrum framework. In: CEUR Workshop Proceedings, vol. 2212, pp. 287–295 (2018)
25. Figma, I.: Figma (2023)
26. Google: Angular (2023)
27. Beck, K., et al.: The Agile Manifesto. Agile manifesto.org
28. Schwaber, K., Sutherland, J.: The Scrum guide. 2, 17 (2011)
29. Dermeval, D., et al.: Applications of ontologies in requirements engineering: a systematic review of the literature. Requir. Eng. **21**, 405–437 (2016)

Enhancing Requirement-Information Mapping for Sustainable Buildings: Introducing the SFIR Ontology

Karim Farghaly[✉] and Kell Jones

Bartlett School of Sustainable Construction, University College London, London, UK
karim.farghaly@ucl.ac.uk

Abstract. Achieving a sustainable and decarbonized built environment is critical to meeting global climate and development goals, yet current approaches face barriers. Top-down policies like standards and certifications aim to drive sustainability, while material/building passports gather data for transparency. However, lack of alignment between these approaches disincentivizes investment and risks progress. This research explores using ontologies to enable resilient information flows between buildings and financers to overcome these barriers. By developing a shared ontology, data and reporting can be streamlined across locations to incentivize and demonstrate sustainable investments. The work proposes a hybrid top-down approach for the development of the Sustainable Finance Information Requirement (SFIR) ontology. Initial design shows promise for improving stakeholder coordination and driving sustainability outcomes. This pioneering research lays the groundwork for optimized sustainability information flows but remains exploratory. With rigorous multi-stakeholder collaboration and iterative ontology refinement, this approach could enable improved policy development, decision-making and reporting to unlock sustainable development of the built environment. While the potential is significant, this research represents only an initial step into applying ontologies to enable sustainability policies and data transparency.

Keywords: Ontology · Sustainable Finance · Material Passport · built environment

1 Introduction

The built environment sector must undergo fundamental changes to meet sustainability goals like those outlined in the Paris Agreement and UN Sustainable Development Goals. However, the sector has been resistant to change due to high fragmentation and misaligned incentives across the entire value chain [1, 2]. This lack of coordination makes shifting towards more sustainable and low-carbon building incredibly difficult, representing a considerable and problematic market failure. To address this failure, many countries are implementing top-down policy interventions like sustainability taxonomies and reporting frameworks that define and incentivize sustainable real estate investments [3]. However, these regulatory measures face barriers in practice due to lack of systemic,

T. P. Sales et al. (Eds.): ER 2023 Workshops, LNCS 14319, pp. 242–248, 2023.
https://doi.org/10.1007/978-3-031-47112-4_23

quality building data and fundamental mismatches between asset-level and investor-level regulations across jurisdictions. In parallel, myriad fragmented bottom-up initiatives like proprietary material and building passports are emerging in pockets to try and compile asset data, but they often use inconsistent structures, schemas, and coding systems. Even when data is compiled, it may not sufficiently meet taxonomy needs, especially for assets located in different countries than the investors and asset managers. Overall, pervasive geographic and reporting silos restrict critical information flows from the built asset level to the financial system and regulators, severely limiting policy efficacy. Bridging this problematic vertical gap between top-down requirements and bottom-up information is therefore critical to accelerate sustainability transitions across the built environment.

Formal ontologies provide a method to align conceptual domains by providing a shared semantic framework and vocabulary. However, most existing ontology efforts focus narrowly on horizontal interoperability between systems and software, not the broader vertical alignment between policies, standards, and data representations needed in a studied context [4, 5]. A top-down ontology holistically mapping regulations, standards, certificates, asset data, and investor requirements could more comprehensively integrate these domains to meet needs across the value chain. Ontology can provide a foundation to close systemic vertical alignment gaps that severely hinder sustainability policies and investment. However, significant engagement and evaluation are needed to refine proposed ontologies across diverse global use cases. This research initiates the methodical development of such ontology using a top-down approach starting from market-leading sustainability assessment schemes, taxonomies, and reporting frameworks. The ontology can then integrate their requirements with other standardized asset-level data structures and schemas. This paper reviews relevant literature on standards, certificates, regulations, and data formats to identify key entities, attributes, and relationships to encode. Initial ontology classes are proposed to connect top-down regulations and bottom-up information sources. Immediate next steps will involve expert reviews and scenario-based testing. Longer-term validation through industry pilot studies across capital and asset owners, financiers, and regulators is also required. While promising, this research is still in the nascent stages and the feasibility of a unified ontology remains largely unproven. While it may take some time to fully tap into its potential, this project lays the groundwork for knowledge integration that may unlock sustainable development of the built environment.

2 Point of Departure

Material Passports aim to offer digitized qualitative and quantitative life cycle data on product attributes to facilitate the implementation of circular principles [6]. Material passports have emerged in recent years as an approach to document building materials and components to enhance transparency, circularity, and sustainability in the built environment. However, a fragmented landscape of passport solutions has developed with inconsistent purposes, maturity levels, scale, and alignment to regulations [7]. For instance, the Cradle to Cradle (C2C) Passport provides detailed chemical ingredient inventories aligned with C2C certification tiers, positioning it as a verification mechanism but with limited adoption beyond the niche standard [8]. BAMB's passport prioritizes circularity data on material recovery, reuse and recycling but lacks regulatory alignment [9]. One Click LCA connects life cycle assessment (LCA) data for greater environmental visibility but mainly serves LCA practitioners [10]. Madaster provides a public materials registry for passports but relies on voluntary self-reporting of data [11]. While varying in focus, common aims emerge across material passports like supply chain transparency, circularity, and life cycle impacts. However, key differences exist in data structure, scope, and quality [12]. From a sustainable finance perspective, passports represent a tool to provide verified ESG data on assets needed for taxonomy compliance and sustainability reporting. However, current passports have restrictions in providing information/outputs that automatically align with regulations. Linkage to taxonomy technical screening criteria, corporate mandatory reporting, and investor-focused metrics remains underdeveloped across most passport solutions. Overall, passports remain fragmented across providers lacking alignment to regulations, standards, and end user needs.

Similarly, sustainability building standards and certifications form a disjointed landscape. Diverse whole-building schemes like LEED, BREEAM, and WELL and standards like EU Taxonomy address energy, water, materials, human health and more [13]. These sustainability building standards and certifications employ varied approaches to break down and assess the components of sustainable construction. Some standards utilize quantitative, measurable indicators while others rely on qualitative, empirical recommendations. This diversity reflects the complex, multi-faceted nature of sustainability, spanning both tangible and intangible factors that influence a building's overall sustainability [14]. However, inconsistent terminology is common, with different terms used for similar indicators or the same terms having distinct meanings across standards. This lack of semantic alignment leads to repetitive, costly efforts to decipher and comply with divergent terminology and requirements. From a sustainable finance lens, most standards lack robust links to taxonomy criteria, mandatory corporate reporting, and investor disclosure needs. Recent efforts to better integrate standards with financial regulations remain limited.

Aligning material passports and certifications could enhance the compliance rigor and transparency of standards while also unlocking verified data for sustainability reporting and financing. Our working hypothesis is that at a granular level (individual data points), these standards and regulations are attempting to capture and assess the same information. Therefore, the need to enable resilient information flows for sustainable finance reporting and investment spurred the inception of this research project to develop the Sustainable Finance Information Requirement (SFIR) ontology. To the best of the authors' understanding, minimal research has been undertaken in this domain. The existing ontologies primarily focus on facilitating horizontal integration across diverse data sources, rather than vertically aligning with extant standards and certifications. Consequently, this study introduces a novel approach, addressing a discernible gap in the field.

3 SFIR Ontology Development

To achieve a top-down ontology, the proposed process development relies a hybrid approach as suggested by [15]. The approach starts with a deductive approach, followed by subsequent iterations utilizing the inspirational, synthetic, collaborative, and inductive approaches consecutively (Fig. 1). The systematic use of each of the five approaches would help to overcome the main challenges in the implementation of the developed ontologies such as missing concepts, expectation gaps, terminology conflicts and correspondences [16]. Missing concepts and expectation gaps can be attributed to the tendency of academic research to propose conceptual frameworks while neglecting stakeholders' data requirements [12]. The issues of terminology disagreements and alignments arise from the absence of a unified scheme in prevailing approaches, which display variance in terminology, content, aggregation depth, technological deployment, and developmental stage [17]. Moreover, while numerous commercial bottom-up solutions present a comprehensive inventory of datapoints, there remains a lack of clarity regarding their methodological foundations and the congruence of these datapoints with established certificates and regulatory guidelines.

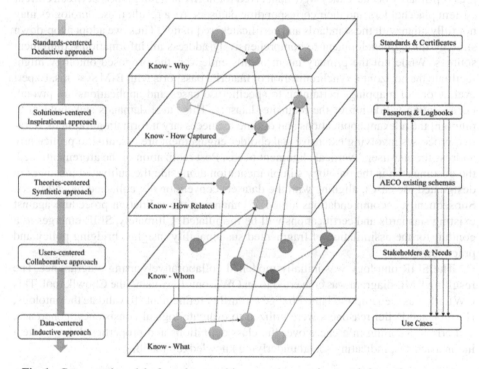

Fig. 1. Conceptual model of ontology multi-approaches requirement-information connect.

The SFIR classes illustrated in Fig. 2 are categorized by color according to the approaches outlined and displayed in Fig. 1. The SFIR ontology aims to vertically map connections between regulations, certifications, passports, existing ontologies, and use cases involving diverse stakeholders. The foundation of this ontology lies in the introduction of novel classes. Subsequently, we plan to align these classes and properties with established top-level ontologies, including OWL, DOLCE, UFO, and BFO. Furthermore, we will also consider mapping to domain-specific ontologies, like those related to digital construction. It originates with key sources like regulations and certifications as subclasses of "Source". These sources contain information requirements like clauses, maturity levels, and passport data categories that comprise the "Information Requirement" class. Each requirement relates to various associated information pieces categorized as "Collection". This includes metadata like location and asset type as well as specific datasets needed for vertical alignment stored as "Datapoint" individuals.

In the realm of diverse naming conventions, the "Semantic Aligner" class serves as a bridge between related requirements and collections. Instances of the semantic aligner forge links between distinct data points that describe identical entities in various sources. Once sources and requirements are interconnected, the subsequent phase involves integrating with prevailing ontologies within the built environment domain, such as BOT, IfcOwl, and Digital Construction. The initial choice to not rely on these existing ontologies is primarily because they were conceived for horizontal integration between current system placeholders and their corresponding datasets. As a result, these ontologies may not fully align with the standards and certificates' requisites. Thus, we adopt a top-down strategy during development to comprehensively address all information requirement sources. While not the primary intent of this endeavor, the proposed ontology might facilitate the horizontal synchronization of material passports with BIM systems. Expert evaluations of mappings pertaining to specific use cases and applications are pivotal, necessitating validation of the requisite datasets. This stage demands stakeholders to pinpoint the relevant applications and outline the necessary information for them. Iterative processes involving extensive stakeholder engagement are essential to proficiently address the end-users' vertical alignment needs. Post realization of the aforementioned, the ensuing step is the ontology's implementation alongside the cultivated instances in designated use cases, aligning with the datasets conventionally collected by end-users. Subsequently, recommendations aimed at enhancing the validation procedure against existing standards and certifications will be formulated. Ultimately, SFIR emerges as a conduit for the assimilation of fragmented sustainability insights, bridging policy and practice.

The SFIR ontology was initially modelled collaboratively using diagram.net. The resulting UML diagram was converted to an OWL ontology using the Chowlk tool. This OWL file was then imported into Protégé for further refinement. To validate the ontology, HermiT and Pellet reasoners were utilized to evaluate logical consistency, coherence, and errors. Running inferences over the class definitions and properties did not detect inconsistencies, indicating sound underlying knowledge modeling.

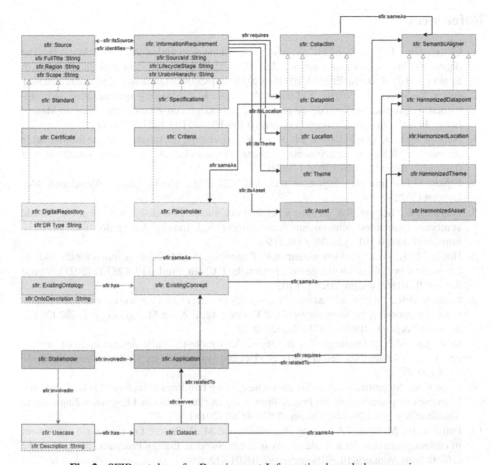

Fig. 2. SFIR ontology for Requirement-Information knowledge mapping.

4 Conclusion and Future Directions

Bridging top-down sustainability requirements with bottom-up building data is paramount for the progression of sustainable built environments. The SFIR ontology, as introduced in this study, stands as a testament to rigorous academic endeavor, bridging the gap between isolated data points and the multifaceted requirements of sustainable design through a particular collaborative modelling approach. The proposed approach for crafting a top-down ontology champions both horizontal and vertical amalgamation of information requirements and data points. This process unfolds through a strategic sequence of approaches: deductive, inspirational, synthetic, collaborative, and inductive. Continuous engagement and collaboration among a broad array of stakeholders are paramount for the ongoing refinement, verification, and evolution of this ontology. It is anticipated that this contribution will prompt further academic exploration in semantic integration to unify the current segmented sustainability framework.

References

1. Gómez-Gil, M., Sesana, M.M., Salvalai, G., Espinosa-Fernández, A., López-Mesa, B.: The digital building logbook as a gateway linked to existing national data sources: the cases of Spain and Italy. J. Build. Eng. **63**, 105461 (2023). https://doi.org/10.1016/j.jobe.2022.105461
2. Jones, K., Mosca, L., Whyte, J., Davies, A., Glass, J.: Addressing specialization and fragmentation: product platform development in construction consultancy firms. Constr. Manag. Econ. **40**, 918–933 (2022). https://doi.org/10.1080/01446193.2021.1983187
3. Nedopil Wang, C., Lund Larsen, M., Wang, Y.: Addressing the missing linkage in sustainable finance: the 'SDG finance taxonomy.' J. Sustain. Financ. Invest. **12**, 630–637 (2022). https://doi.org/10.1080/20430795.2020.1796101
4. Lipkin, J.: Investment management in the UK 2021–2022. The Investment Association (IA), London (2022)
5. Farghaly, K., Soman, R.K., Zhou, S.A.: The evolution of ontology in AEC: a two-decade synthesis, application domains, and future directions. J. Ind. Inf. Integr. **36**, 100519 (2023). https://doi.org/10.1016/j.jii.2023.100519
6. Honic, M., Kovacic, I., Aschenbrenner, P., Ragossnig, A.: Material passports for the end-of-life stage of buildings: challenges and potentials. J. Clean. Prod. **319**, 128702 (2021). https://doi.org/10.1016/j.jclepro.2021.128702
7. Munaro, M.R., Tavares, S.F.: Materials passport's review: challenges and opportunities toward a circular economy building sector. Built Environ. Proj. Asset Manag. **11**, 767–782 (2021). https://doi.org/10.1108/BEPAM-02-2020-0027
8. Braungart, M., McDonough, W., Bollinger, A.: Cradle-to-cradle design: creating healthy emissions–a strategy for eco-effective product and system design. J. Clean. Prod. **15**, 1337–1348 (2007)
9. Rose, C.M., Stegemann, J.A.: Characterising existing buildings as material banks (E-BAMB) to enable component reuse. In: Proceedings of the Institution of Civil Engineers-Engineering Sustainability, pp. 129–140. Thomas Telford Ltd. (2018)
10. Petrovic, B., Myhren, J.A., Zhang, X., Wallhagen, M., Eriksson, O.: Life cycle assessment of building materials for a single-family house in Sweden. Energy Procedia **158**, 3547–3552 (2019). https://doi.org/10.1016/j.egypro.2019.01.913
11. Madaster, U.: Madaster circularity indicator explained. Madaster Services BV Utrecht, The Netherlands (2018)
12. Çetin, S., Raghu, D., Honic, M., Straub, A., Gruis, V.: Data requirements and availabilities for material passports: a digitally enabled framework for improving the circularity of existing buildings. Sustain. Prod. Consumption. **40**, 422–437 (2023). https://doi.org/10.1016/j.spc.2023.07.011
13. Awadh, O.: Sustainability and green building rating systems: LEED, BREEAM, GSAS and Estidama critical analysis. J. Build. Eng. **11**, 25–29 (2017)
14. Massaro, E., Athanassiadis, A., Psyllidis, A., Binder, C.R., Wyss, R.: Ontology-based integration of urban sustainability indicators. Sustain. Assess. Urban Syst. 332–350 (2020)
15. Holsapple, C.W., Joshi, K.D.: A collaborative approach to ontology design. Commun. ACM **45**, 42–47 (2002)
16. Sileryte, R., Wandl, A., van Timmeren, A.: A bottom-up ontology-based approach to monitor circular economy: aligning user expectations, tools, data and theory. J. Ind. Ecol. **27**, 395–407 (2023)
17. Çetin, S., Gruis, V., Straub, A.: Digitalization for a circular economy in the building industry: multiple-case study of Dutch social housing organizations. Resour. Conserv. Recycl. Adv. **15**, 200110 (2022). https://doi.org/10.1016/j.rcradv.2022.200110

On the Use of ChatGPT for Classifying Domain Terms According to Upper Ontologies

Fabrício H. Rodrigues(✉)[iD], Alcides G. Lopes[iD], Nicolau O. dos Santos[iD],
Luan F. Garcia[iD], Joel L. Carbonera[iD], and Mara Abel[iD]

Federal University of Rio Grande do Sul (UFRGS), Porto Alegre, Brazil
{fhrodrigues,aljunior,jlcarbonera,lfgarcia,nosantos,marabel}@inf.ufrgs.br

Abstract. In this paper, we report an experiment to investigate the performance of ChatGPT in the task of classifying domain terms according to the categories of upper-level ontologies. The experiment consisted of (1) starting a conversation in ChatGPT with a contextual prompt listing the categories of an upper-level ontology along with their definitions, (2) submitting a follow-up prompt with a list of terms from a domain along with informal definitions, (3) asking ChatGPT to classify the terms according to the categories of the chosen upper-level ontology and explain its decision, and (4) comparing the answers of ChatGPT with the classification proposed by experts in the chosen ontology. Given the results, we evaluated the success rate of ChatGPT in performing the task and analyzed the cases of misclassification to understand the possible reasons underlying them. Based on that, we made some considerations about the extent to which we can employ ChatGPT as an assistant tool for the task of classifying domain terms into upper-level ontologies. For our experiment, we selected a set of 19 terms from the manufacturing domain that were gathered by the Industrial Ontologies Foundry (IOF) and for which there are informal textual definitions reflecting a community view of them. Also, as a baseline for comparison, we resorted to publicly available classifications of such terms according to DOLCE and BFO upper-level ontologies, which resulted from a thorough ontological analysis of those terms and informal definitions by experts in each of the ontologies.

Keywords: Ontologies · ChatGPT · DOLCE · BFO · Industrial Ontologies Foundry · IOF · Large Language Models · LLMs · Term Classification · Manufacturing

1 Introduction

In Computer Science, an ontology is the specification of a system of categories accounting for a certain view of the world [6], which can provide a standard conceptual framework for effective communication between computers and human

T. P. Sales et al. (Eds.): ER 2023 Workshops, LNCS 14319, pp. 249–258, 2023.
https://doi.org/10.1007/978-3-031-47112-4_24

users and between computers themselves. While much progress has been made in developing ontologies themselves, the task of classifying domain terms into these ontologies, especially upper-level ones, is still challenging [5]. Manual classification is not only time-consuming but also prone to inconsistencies due to the subjective interpretation of the ontology categories by different humans [2]. On the other hand, existing automated methods often struggle with the semantic complexity and variety of domain terms.

In parallel, we have witnessed the recent emergence of Large Language Models (LLMs) – such as GPT-4 [11] – as powerful tools for understanding and generating human language. Based on *deep learning* techniques, LLMs are trained on massive datasets, including books, articles, websites, and other available textual information [4]. Such a large amount of data enables those models to learn a wide range of topics, writing styles, linguistic nuances, and even some common sense reasoning [13]. Also, LLMs have shown the ability to quickly learn new tasks with only a few or even without any specific examples, relying only on general knowledge and context [16], allowing their quick adaptation to new domains and purposes. Their versatility and language abilities have made LLMs invaluable for various applications, including chatbots and virtual assistants [1,14], such as OpenAI's ChatGPT, *i.e.*, a chatbot powered by the GPT-4 model [11,12].

In light of this, we explore the potential of LLMs in the context of ontology-based conceptual modeling – more specifically, to assist with the task of ontological classification of domain terms. In particular, we report an experiment in which we used ChatGPT to classify terms from the industrial manufacturing domain according to two upper-level ontologies, namely, DOLCE (Descriptive Ontology for Linguistic and Cognitive Engineering) [10] and BFO (Basic Formal Ontology) [3]. Then we compared ChatGPT's suggested classification for each term with the classification proposed by experts in their referred ontologies in the context of a proof of concept project proposed by the Industrial Ontology Foundry (IOF)[1]. By evaluating the performance of ChatGPT in classifying domain terms and analyzing the causes of any misclassifications, we aim to give a glimpse of the potential of LLMs as tools for ontology classification.

The remainder of the paper is organized as follows: Sect. 2 details our experiment, Sect. 3 presents and discusses the results of our experiment, and Sect. 4 brings some takeaway reflections.

2   Materials and Methods

2.1   Chosen Terms

Our experiment was based on a list of 19 terms from the manufacturing industry domain provided by the IOF (which we will call *IOF terms* from now on). These terms were originally selected in a proof-of-concept project aiming at, among other things, producing a small initial ontology of the domain [9].

---

[1] https://www.industrialontologies.org/.

Table 1. IOF terms and their respective textual definitions.

Assembly	An assembly is a combination of parts and components that form a functional entity.
Assembly Process	An assembly process is a type of manufacturing process that combines two or more components into a single parent assembly or final product.
Business Process	A business process is a structured set of activities performed to achieve an organizational objective.
Component and Component Part	A component is a part or subassembly that goes into a higher-level assembly or the final product (adapted from APICS). Explanation: A particular artifact can be considered as a component or an assembly depending on the context of a manufacturing process.
Customer	An organization or person that buys or receives a good or service.
Design Specification	A design is a specification that describes a collection of features to be created for a product.
Equipment	Manufacturing equipment is equipment which is operated for directly producing a product, in a manufacturing process (ISO 20140-1:2013).
Feature Specification	A feature description is a description that details some distinctive characteristic about a product.
Manufacturing Machine	A machine is a mechanical system designed expressly to perform a specific task, such as the forming of material or the transference and transformation of motion, force or energy (ISO 22096:2007).
Manufacturing Process	A manufacturing process is a structured set of activities performed to produce a good or service.
Manufacturing Process Plan	A Manufacturing process plan is a specification that prescribes the collection of related activities in a manufacturing process that produces a product with the desired qualities.
Manufacturing Resource	A resource is any person or thing that adds value to a product in its creation, production, or delivery (APICS).
Material Resource	A material resource is any substance from which a product can be made. Explanation: A material substance is often referred to as raw material in the context of a manufacturing process (IOF: Material resource. Possible synonyms: material, input resource. "Former term: material substance").
Plan	A plan is a document that prescribes a collection of related activities that achieve some organizational goal.
Planned Process	(no textual definition was provided).
Product	A product is the tangible outcome of a process (ISO 6707- 3:2017). A product is often produced for sale, barter, or internal use, etc.
Product Quality	A physical or functional characteristic of a product that can be measured or qualitatively evaluated. (Source: Adopted principally from ISO 9000:2016).
Supplier	A provider of goods or services. Source: APICS (Revised: An organization or individual that sells or provides products or services to other organizations and individuals).
Transport Process	A transport process is a process that involves the movement or change in location of some raw material, component or product by some agent or mechanism.

The choice of using IOF's list of terms offered us a couple of benefits. First, such a list was built by gathering suggestions of terms and corresponding informal textual definitions from the manufacturing industry community. With that, we have a list that reflects the community's shared conceptualization of the selected terms in the words of its own members rather than through a set of formal definitions curated by ontologists.

Additionally, this list of terms and textual definitions was subject to the ontological analysis of experts in two of the most relevant top-level ontologies, *i.e.*, DOLCE [7] and BFO [15]. These efforts resulted in the classification and formal definition of the IOF terms according to each of the referred top-level ontologies. Then, we have a high-quality baseline with which to compare the classification suggested by ChatGPT. Finally, we have a setting for our experiment, *i.e.*, we will assign ChatGPT a task already carried out by ontology experts, the ontological classification of domain terms according to a foundational ontology.

Considering that, for our experiment, we selected the IOF terms that were classified both in DOLCE and BFO works so that we could evaluate the correctness of the classification suggested by ChatGPT in both top-level ontologies. It is also important to note that, unfortunately, the textual definitions of IOF terms are not documented in any publicly available resource. For this reason, we use the textual definitions presented in DOLCE work [7] (BFO work does not present textual definitions). The final list of terms and textual definitions we used in our experiments is presented in Table 1.

2.2   Experiment Design

Our experiment was conducted as follows. For each of the chosen ontologies, we first provided a contextual prompt[2] listing the categories of the ontology in an order reflecting the taxonomical skeleton of the ontology, along with their corresponding definitions/descriptions[3]. In this prompt, we also asked ChatGPT to sketch the taxonomy of such categories to assess its understanding of the chosen ontology. We repeated this procedure until ChaptGPT could correctly reproduce the taxonomy of the ontology.

Then, in a follow-up prompt[4], we listed the IOF terms in alphabetical order, along with their respective textual definitions (Table 1), asking ChatGPT to classify them into the ontology categories provided in the previous prompt. In the

---

[2] *"Act as an expert in Formal Ontology and Upper Ontologies. Here are the definitions of categories of the* <DOLCE|BFO> *upper ontology. Make a sketch of the taxonomy of these categories. Present the sketch using bullet points."*.

[3] We used the definitions/descriptions of BFO categories as provided in https://standards.iso.org/iso-iec/21838/-2/ed-1/en/ and definitions of DOLCE categories as provided in https://github.com/gruninger/colore/blob/master/ontologies/dolce/DOLCE-Terms.docx.

[4] *"Now, consider the following list of terms with their respective definitions. Classify each of these terms according to the* <DOLCE|BFO> *taxonomy you sketched and provide a short explanation of why you classified each term in the given* <DOLCE|BFO> *category. Present the classification in a bullet list."*.

same prompt, we also asked ChatGPT for a short explanation of its classification choices. For each ontology, we provided ChatGPT with this follow-up prompt and collected the answer. Then, we asked it to regenerate the answer another 9 times. With that, we had 10 independent classification rounds without any interference from previous answers.

Finally, for each term, we registered the classification proposed by ChatGPT in each of the 10 classification rounds. We considered the most frequent classification (which we called the *major classification*) as the final answer according to ChatGPT. In the cases of equally frequent classifications, the major classification was defined as the first of the tied categories that was suggested on a series of additional classification rounds (which served only as tie-break and were not computed for any other purpose). Then we compared such a major classification with the one proposed by the experts and grouped the results into 5 categories:

(1) **EXACT MATCH**: when ChatGPT and the experts classified the term into the same type;
(2) **PARTIAL MATCH**: when the experts classified the term into a disjunction of types from the ontology, and ChatGPT classified the term into one of the types from such a disjunction;
(3) **SUPERTYPE MATCH**: when ChatGPT classified the term into a supertype of the type chosen by the experts;
(4) **SUBTYPE MATCH**: when ChatGPT classified the term into a type lower than the type chosen by the experts;
(5) **DISTINCT BRANCH**: when ChatGPT and the experts classified the term into distinct types located in distinct branches in the taxonomy of the ontology so that neither is super or subtype of the other.

3   Results and Discussion

Table 2 presents the category suggested by the experts for each term and each ontology along with the various categories suggested by ChatGPT with the number of rounds each category was suggested appearing between parentheses and the major classification in boldface. Category names were abbreviated for space reasons, and the notation "*CAT1/CAT2*" represents the disjunction of categories *CAT1* and *CAT2*. Finally, the table presents the type of match obtained between experts' and ChatGPT's major classifications according to the 5 match categories presented in Sect. 2.2.

Regarding BFO, there were 11 exact matches between ChatGPT's and experts' classifications (*i.e.*, Assembly, Assembly Process, Business Process, Equipment, Manufacturing Machine, Manufacturing Process, Manufacturing Resource, Material Resource, Planned Process, Product Quality, and Transport Process). In addition, there was 1 subtype match (*i.e.*, Product), 5 supertype matches (*i.e.*, Component, Customer, Manufacturing Process Plan, Plan, Supplier), and 2 distinct branch matches (*i.e.*, Design Specification and Feature Specification).

Regarding DOLCE, there were 2 exact matches (*i.e.*, Assembly and Component) and 4 partial ones (*i.e.*, Customer, Material Resource, Product, and Supplier). Also, there were 4 supertype matches (*i.e.*, Design Specification, Feature

Specification, Manufacturing Process Plan, and Plan), 2 subtype matches (*i.e.*, Product Quality and Manufacturing Machine), and 7 distinct branch matches (*i.e.*, Assembly Process, Business Process, Equipment, Manufacturing Process, Manufacturing Resource, Planned Process, and Transport Process).

We can analyze these results in two main ways. First, we can simply evaluate ChatGPT's suggested classifications for each term in comparison to experts' classifications. The other option is evaluating ChatGPT's suggestions in light of the reasons it gave for such choices in order to identify divergent classifications that do not really correspond to errors but rather reveal reasonable alternative accounts for a given term. We cover both approaches in the following sections.

3.1   Overall Performance

Throughout the experiment, ChatGPT carried out 380 classification tasks, *i.e.*, it classified each of the 19 IOF terms 10 times according to DOLCE and 10 times according to BFO. If we consider that ChatGPT succeded in a classification task if its suggested classification exactly matches that of the experts, its overall success rate was 33.68% (128 out of 380 of the classification tasks), being 19.47% for DOLCE (37 of 190) and 47.89% for BFO (91 of 190). From another standpoint, we can consider that ChatGPT only failed a task when it achieved a distinct branch match since, in all other cases (partial, supertype, and subtype matches), the suggested classification covered part of the semantics underlying the classified term. Under this view, the success rates rise to 78.68%, 72.11%, and 85.26%, respectively.

From those 380 classification tasks, we got 2 major classifications (one according to each of the chosen ontologies), ending up with 38 major classifications. If we consider only exact matches as success, the overall success rate was 31.58% (12 of the 38 major classifications), being 10.53% for DOLCE (2 of 19) and 52.63% for BFO (10 of 19). Considering only distinct branch matches as failures, the success rates rise to 76.32%, 63.16%, and 89.47%, respectively.

3.2   Reasonable Divergences

A considerable share of ChatGPT's classification suggestions that disagree with experts' ones is backed up by reasonable justifications. Some cases comprise ordinary issues, such as a divergence in which of the usual senses of a term to adopt. *e.g.*, while DOLCE experts considered *Equipment* as the set of articles used to equip a person or to support an activity, ChatGPT interpreted the term as referring to one of such articles, *i.e.*, a *piece of equipment*. Also, while BFO experts classified *Plan* as a disposition (*i.e.*, the "*intention-to-perform processes on the part of an agent*" [15]), in 3 classification rounds ChatGPT strictly followed the definition of the term, considering a plan as document (which BFO experts called "*plan specification*"). Similarly, while BFO experts classified *Product* as a *Continuant* to account for both material and immaterial products [15], ChatGPT classified the term as a *Material Entity*, doing justice to its definition as

Table 2. List of terms with their respective classifications by experts and by ChatGPT for each top-level ontology.

IOF Term	Upper Onto.	Expert Class.	GPT Classification	Match Type
Assembly	BFO	MAT	**MAT(4)**, OBJ (4), OBJ or AGGR (1), AGGR (1)	=
	DOLCE	POB	**POB (10)**	=
Assembly Process	BFO	PRO	**PRO (10)**	=
	DOLCE	ACC	**PRO (7)**, PD (2), ACC (1)	≠
Business Process	BFO	PRO	**PRO (10)**	=
	DOLCE	ACC	**PRO (7)**, PD (2), ACC (1)	≠
Component	BFO	OBJ	**MAT(4)**, FOP (2), OBJ or FOP (2), OBJ (2)	↑ 1
	DOLCE	POB	**POB (10)**	=
Customer	BFO	MAT	**IND (6)**, SDC (1), MAT (1), OBJ (1), ROLE (1)	↑ 1
	DOLCE	APO/ASO	**ASO (5)**, SAG (5)	≈
Design Specif.	BFO	GDC	**SDC (5)**, GDC (3), OBJ (1), FOP (1)	≠
	DOLCE	NASO	**NPOB (9)**, MO (1)	↑ 2
Equipment	BFO	OBJ	**OBJ (6)**, MAT (4)	=
	DOLCE	COL*	**NAPO (8)**, POB (2)	≠
Feature Specific.	BFO	GDC	**SDC (6)**, GDC (3), FOP (1)	≠
	DOLCE	NASO	**NPOB (8)**, AQ (1), MO (1)	↑ 2
Manufact. Machine	BFO	OBJ	**OBJ (6)**, MAT (4)	=
	DOLCE	POB	**NAPO (8)**, POB (2)	↓ 1
Manufact. Process	BFO	PRO	**PRO (10)**	=
	DOLCE	ACC	**PRO (7)**, PD (2), ACC (1)	≠
Manufact. Proc. Plan	BFO	DISP	**SDC (6)**, GDC (3), FOP (1)	↑ 2
	DOLCE	NASO	**NPOB (9)**, MO (1)	↑ 2
Manufact. Resource	BFO	MAT	**IND(3)**, OBJ(3), MAT(1), MAT/SDC(1), OBJ/ROLE(1), OBJ/GDC(1)	↑ 1
	DOLCE	POB/M	**NPED(3)**, PED/NPED(2), PED(2), POB/NPOB(1), SAG/NAPO(1), POB(1)	≠
Material Resource	BFO	MAT	**MAT (6)**, OBJ (4)	=
	DOLCE	POB/M	**M (10)**	≈
Plan	BFO	DISP	**SDC (6)**, GDC (3), FOP (1)	↑ 1
	DOLCE	NASO	**NPOB (9)**, MO (1)	↑ 2
Planned Process	BFO	PRO	**PRO (8)**, REAL (1), PBD (1)	=
	DOLCE	ACC	**PRO (7)**, PD (2), ACC (1)	≠
Product	BFO	CONT	**MAT (5)**, OBJ (4), OBJ or AGGR (1)	↓ 2
	DOLCE	POB/M	**POB (10)**	≈
Product Quality	BFO	Q	**Q (10)**	=
	DOLCE	Q	**PQ (10)**	↓ 1
Supplier	BFO	MAT	**IND (7)**, MAT (1), OBJ (1), ROLE (1)	↑ 1
	DOLCE	APO/ASO	**ASO (5)**, SAG (5)	≈
Transport Process	BFO	PRO	**PRO (10)**	=
	DOLCE	ACC	**PRO (7)**	≠

BFO Categories: *AGGR* = Object Aggregate; *CONT* = Continuant; *DISP* = Disposition; *FOP* = Fiat Object Part; *GDC* = Generically Dependent Continuant; *IND* = Independent Continuant; *MAT* = Material Entity; *OBJ* = Object; *PBD* = Process Boundary; *PRO* = Process; *Q* = Quality; *REAL* = Realizable Entity; *ROLE* = Role; *SDC* = Specifically Dependent Continuant.

DOLCE Categories: *ACC* = Accomplishment; *APO* = Agentive Physical Object; *AQ* = Abstract Quality; *ASO* = Agentive Social Object; *COL** = Collection (not present in standard DOLCE); *M* = Amount of Matter; *MO* = Mental Object; *NAPO* = Non-Agentive Physical Object; *NASO* = Non-Agentive Social Object; *NPED* = Non-Physical Endurant; *NPOB* = Non-Physical Object; *PD* = Perdurant; *PED* = Physical Endurant; *POB* = Physical Object; *PQ* = Physical Quality; *PRO* = Process; *Q* = Quality; *SAG* = Social Agent.

Match types: (=) Exact, (≈) Partial, (↑n) Supertype *n* levels above, (↓n) Subtype *N* levels below, (≠) Distinct Branch.

the *"tangible* outcome of a process" (in consonance with DOLCE experts, who classified the term as a *Physical Object OR Amount of Matter* [7]).

Other examples are more elaborated, requiring a capacity to analyze the implications of the definitions of the IOF terms and of the ontological categories. For instance, *Product Quality* was classified by DOLCE experts as a *Quality*, while ChatGPT classified the term as a *Physical Quality*. When explaining the choice, ChatGPT described product quality as a characteristic that *"directly inheres to a product, which is a physical endurant"*, which better fits the definition of *Physical Quality* (*i.e.*, a quality that directly inheres to a physical endurant).

Finally, some cases revealed a creative aspect of ChatGPT, such as *Manufacturing Resource* (*i.e.*, a *"thing that adds value to a product in its creation, production, or delivery"*). Experts classified the term as a *Physical Object or Amount of Matter* in DOLCE and as a *Material Entity* in BFO, arguably highlighting the material nature of this type of resource. In contrast, ChatGPT opted for broader categories that also contemplate non-material things, such as *Physical Endurant OR Non-Physical Endurant* in DOLCE, or *Independente Continuant, Material Entity OR Specifically Dependent Continuant*, and *Object OR Generically Dependent Continuant* in BFO. Inspecting the rationale for such choices, we found the reason: for ChatGPT, manufacturing resources include not only physical entities such as *people, machines*, and *raw material* but also things such as *plans, procedures, informational resources, skillsets*, and *intellectual property*.

3.3   Weakenesses and Strengths

ChatGPT did not properly deal with certain finer ontological distinctions. For instance, it could not distinguish existential dependence from other forms of dependence, such as relational or historical. We can observe that in its classification of specifications and plans *Specifically Dependent Continuant*, arguing that a design/feature specification is *"is dependent on a material entity – the thing being designed/described –, and does not exist independently"* and *"specifically dependent on the designed objects."*, and that a plan *"depends on the existence of the entities - processes, materials, goals, etc.– it refers to."*.

It also failed in applying the notion of *agentivity*. While experts classified specifications and plans as *Non-Agentive Social Objects*, ChatGPT classified them as *Non-Physical Objects*. With that, even though ChatGPT refers to plans and specifications as clearly non-agentive entities in its explanations (*e.g.*, as "*a conceptual entity*", "*a unified idea*", or "*a document*"), it implicitly accepts that they may be *Agentive Social Objects, i.e.*, DOLCE's category for social objects to which intentions, beliefs, and desires are ascribed - such as a person.

In addition, ChatGPT was unable to classify the terms referring to industrial processes according to DOLCE, notably due to not understanding the distinction between *eventive* and *stative* perdurants. Here, we may also have observed what we will call the *label effect, i.e.*, the case in which either (1) the label of a term contains one or more words that have a meaning so strongly embedded in the LLM that it takes precedence over the definition provided for the term or for the ontological category, or (2) the label of the term perfectly matches that of a category in the ontology so that the term is classified under such a category despite their conflicting definitions. We intend to investigate this in future work.

Despite those issues, ChatGPT showed impressive capacities, exploring the consequences of definitions when classifying a term (*e.g., Product Quality*). It also resorted to contextual information and general knowledge to infer hidden aspects and relevant related types of entities that were not explicitly mentioned in the information provided (*e.g., Material/Manufacturing Resource*). Notably, ChatGPT was able to make a reasonable analysis of *Planned Process*, even in the absence of any informal definition for the term, *e.g.*, explaining that "*Without a direct definition, it can be inferred that a planned process occurs in time and involves a series of activities based on a plan*".

On top of that, the consistency among the answers of ChatGPT within a given classification round seems to suggest that, at each round, it incorporates, so to speak, an 'expert personality' that commits to a particular worldview. We can observe that the classification of analogous terms in the same categories explanations emphasizing the analogies (*e.g., "Similar to design specification, feature specification also [...]"*, *"A supplier, like a customer, is an [...]"*).

Given that, the results of our experiment indicate the potential of using ChatGPT and similar tools to provide a complementary ontological perspective on domain terms. Irrespective of its accuracy rate, we can certainly entertain the idea of ChatGPT assuming the role of an additional ontologist in an advisory capacity, with its suggested classifications and informative explanations contributing to uncovering overlooked aspects and entities in a domain, challenging modeler's assumptions, and encouraging a reevaluation of her/his modeling choices.

4 Concluding Remarks

In this work, we investigated the use of ChatGPT, powered by GPT-4, in the task of classifying domain terms according to categories of the DOLCE and the BFO upper-level ontologies. Our main contributions lie in the novel use of ChatGPT for classifying domain terms into top-level ontologies categories, the chain-of-thought prompting we used in the experiments, and the critical analysis of the results we achieved. We do not claim that our work proposes a definite way for using LLMs for ontology engineering. Instead, we expect that our work inspires the community to build more experiments based on our examples as well as to propose new approaches for the task we explored here.

Although the results we presented are preliminary yet, they may indicate that in the future ontology engineers may have AI assistants that are capable of properly classifying domain terms with informal textual definitions into top-level ontologies classes, maybe even generating formal definitions into common formats such as OWL. Thus, it is possible that the job of an ontology engineer will be more focused on understanding the domain and creating good informal definitions to be used as inputs for AI models rather than classifying and formalizing terms for a specific ontology.

In future work, we intend to build a larger corpus of terms from different domains to evaluate the performance of ChatGPT classifying terms using DOLCE and BFO, as well as other upper ontologies (*e.g.,* UFO, YAMATO, GFO, SUMO). We also intend to employ the same approach to other tasks, such as ontology alignment and the meta-property analysis proposed in [8].

Acknowledgements. This study was financed by Petwin Project (PeTWIN.org), Coordenação de Aperfeiçoamento de Pessoal de Nível Superior - Brasil (CAPES) - Finance Code 001, CNPq, FINEP, and LIBRA Consortium.

References

1. Abdullah, M., Madain, A., Jararweh, Y.: ChatGPT: fundamentals, applications and social impacts. In: 2022 Ninth International Conference on Social Networks Analysis, Management and Security (SNAMS), pp. 1–8. IEEE (2022)
2. Abel, M., Perrin, M., Carbonera, J.L.: Ontological analysis for information integration in geomodeling. Earth Sci. Inf. **8**, 21–36 (2015)
3. Arp, R., Smith, B., Spear, A.D.: Building Ontologies with Basic Formal Ontology. MIT Press, Cambridge (2015)
4. Brown, T., et al.: Language models are few-shot learners. In: Advances in Neural Information Processing Systems, vol. 33, pp. 1877–1901 (2020)
5. Garcia, L.F., Rodrigues, F.H., Lopes, A., Kuchle, R.d.S.A., Perrin, M., Abel, M.: What geologists talk about: towards a frequency-based ontological analysis of petroleum domain terms. In: ONTOBRAS, pp. 190–203 (2020)
6. Guarino, N.: Formal ontology and information systems. In: International Conference on Formal Ontology and Information Systems (FOIS 1998), pp. 3–15 (1998)
7. Guarino, N., Sanfilippo, E.: Characterizing IOF terms with the DOLCE and UFO ontologies. In: 10th International Workshop on Formal Ontologies Meet Industry (2019)
8. Guarino, N., Welty, C.: Evaluating ontological decisions with ontoclean. Commun. ACM **45**(2), 61–65 (2002). https://doi.org/10.1145/503124.503150
9. Kulvatunyou, B.S., Wallace, E., Kiritsis, D., Smith, B., Will, C.: The industrial ontologies foundry proof-of-concept project. In: Moon, I., Lee, G.M., Park, J., Kiritsis, D., von Cieminski, G. (eds.) APMS 2018. IAICT, vol. 536, pp. 402–409. Springer, Cham (2018). https://doi.org/10.1007/978-3-319-99707-0_50
10. Masolo, C., Borgo, S., Gangemi, A., Guarino, N., Oltramari, A.: WonderWeb deliverable D18: ontology library. Laboratory for Applied Ontology, ISTC-CNR (2003)
11. OpenAI: GPT-4 technical report (2023)
12. Radford, A., Narasimhan, K., Salimans, T., Sutskever, I., et al.: Improving language understanding by generative pre-training (2018)
13. Radford, A., Wu, J., Child, R., Luan, D., Amodei, D., Sutskever, I., et al.: Language models are unsupervised multitask learners. OpenAI Blog **1**(8), 9 (2019)
14. Ray, P.P.: ChatGPT: a comprehensive review on background, applications, key challenges, bias, ethics, limitations and future scope. IoT Cyber-Phys. Syst. (2023)
15. Smith, B., et al.: A first-order logic formalization of the industrial ontologies foundry signature using basic formal ontology. In: JOWO (2019)
16. Wei, J., et al.: Finetuned language models are zero-shot learners. arXiv preprint arXiv:2109.01652 (2021)

QUAMES

QUAMES – 4th International Workshop on Quality and Measurement of Model-Driven Software Development

Beatriz Marín[1], Giovanni Giachetti[1,2], Estefanía Serral[3], and Jose Luis de la Vara[4]

[1] Universitat Politècnica de Valencia, Spain
bmarin@dsic.upv.es
[2] Universidad Andrés Bello, Chile
giovanni.giachetti@unab.cl
[3] KU Leuven, Belgium
estefania.serralasensio@kuleuven.be
[4] Universidad de Castilla-La Mancha, Spain
joseluis.delavara@uclm.es

We are delighted to welcome you to the fourth edition of the International Workshop on Quality and Measurement of Model-Driven Software Development (QUAMES), co-located with the 42nd International Conference on Conceptual Modeling (ER 2023). After the successful editions in Nanjing (2013), Dallas (2014), and Barcelona (2022), this year QUAMES is organized under the leading conference in conceptual modeling in Lisbon.

The success of software development projects depends on the productivity of human resources and the efficiency of development processes to deliver high-quality products. Model-driven development (MDD) is a widely adopted paradigm that automates software generation by means of model transformations and reuse of development knowledge. The advantages of MDD have motivated the emergence of several modeling proposals and MDD tools related to different application domains and stages of the development lifecycle.

In MDD, the quality of conceptual models is critical because it directly impacts the final software systems' quality. Therefore, it is essential to evaluate conceptual models and predict the software products' relevant characteristics. Additionally, MDD project management must be adapted to take into account that programming effort is being replaced by a modelling effort at an earlier stage. Hence, measuring models is crucial to support cost estimation and project management.

To address these challenges, QUAMES aims to attract research on methods, procedures, techniques, and tools for measuring and evaluating the quality of conceptual models that can be used in MDD environments. Its primary goal is to enable the development of high-quality software systems by promoting quality assurance in the modeling process.

This year QUAMES received 5 high-quality paper submissions by authors coming from Europe. Each paper was Single reviewed by at least 3 members of the program committee. The program committee was composed of experts from Europe, South America, and Asia. After discussion with the reviewers, three submissions were accepted for publication.

The program of QUAMES was comprised of a keynote talk and the presentation of the accepted papers. QUAMES started with an encouraging keynote by Isabel Brito entitled *Cyber Physical Systems - Putting Sustainability in the Loop*. After that, the three regular papers were presented. These papers are entitled *Exploring Understandability in Socio-Technical Models for Data Protection Analysis: Results from a Focus Group*, *FlowTGE: Automating Functional Testing of Executable Business Process Models Based on BPMN*, and *An Approach Aligned with Model Driven Development to Evaluate the Quality of Explainable Artificial Intelligence*.

Finally, we want to express our gratitude to all the authors for their valuable contributions, the dedicated program committee members for their reviews and discussion for QUAMES 2023, and the supportive organizers of ER 2023. We enjoyed an outstanding QUAMES 2023 event filled with stimulating presentations and enriching discussions, and are confident that you also thoroughly enjoyed both QUAMES and ER this year in the lovely city of Lisbon!

Acknowledgments. This workshop is partially supported by the European Commission under the ENACTEST project - European innovation alliance for testing education (ERASMUS+ Project number 101055874, 2022-2025), by the"Paradigmas de interacción para la nueva era de resiliencia digital" project (UCLM ref. 2022-GRIN-34436; ERDF A way of making Europe), by the TASOVA PLUS research network (RED2022-134337-T), and by the Ramon y Cajal program (MCIN/AEI/10.13039/501100011033 ref. RYC-2017-22836; ESF Investing in your future).

QUAMES Organization

Workshop Chairs

Beatriz Marín	Universitat Politècnica de Valencia, Spain
Giovanni Giachetti	Universitat Politècnica de Valencia, Spain, and Universidad Andrés Bello, Chile
Estefanía Serral	KU Leuven, Belgium
Jose Luis de la Vara	Universidad de Castilla-La Mancha, Spain

Program Committee

Oscar Pastor	Universidad Politécnica de Valencia, Spain
Monique Snoeck	KU Leuven, Belgium
Clara Ayora	Universidad de Castilla-La Mancha, Spain
Maya Daneva	University of Twente, The Netherlands
Tanja Vos	Universidad Politécnica de Valencia, Spain
Cecilia Bastarrica	Universidad de Chile, Chile
Ignacio Panach	Universidad de Valencia, Spain
Juan Cuadrado-Gallego	University of Alcalá, Spain
Raian Ali	Hamad Bin Khalifa University, Qatar
Yves Wautelet	KU Leuven, Belgium
Mehrdad Saadatmand	RISE Research Institutes of Sweden, Sweden
Martin Solari	Universidad ORT, Uruguay
Juan Carlos Trujillo	University of Alicante, Spain
Dietmar Winkler	Vienna University of Technology, Austria
Jolita Ralyté	University of Geneva, Switzerland
Shaukat Ali	Simula Research Laboratory, Norway

Exploring Understandability in Socio-technical Models for Data Protection Analysis: Results from a Focus Group

Rosa Velasquez[1]([✉]), Claudia Negri-Ribalta[2], Rene Noel[1,3], and Oscar Pastor[1]

[1] Valencian Research Institute for Artificial Intelligence, Universitat Politècnica de València, València, Spain
{rvelasquez,rnoel}@vrain.upv.es
[2] SnT, University of Luxembourg, Luxembourg City, Luxembourg
[3] Escuela de Ingeniería Informática, Universidad de Valparaíso, Valparaíso, Chile

Abstract. The understandability of conceptual models depends not only on the model's inner complexity and representation but also on the personal factors of the model's audience. This is critical when conceptual models are used for achieving common ground during the early stages of requirements engineering for information systems and, moreover, for complex domains such as data protection. In this article, we present the results of an exploratory study consisting of eight focus groups with 21 experts on software development, business analysis and data protection, examining socio-technical models of an information system to identify privacy risks. We surveyed participants on their backgrounds to characterize the personal factors of understandability and performed an initial understandability assessment on a socio-technical model. We compared these values with the outcome of the focus group, i.e., the effectiveness of the participants in identifying privacy risks, annotating whether the risks are identified individually by a participant or collaboratively by two or more participants. The results suggest that most of the privacy risks were identified collaboratively, regardless of the previous understandability scores and personal factors such as experience and background.

Keywords: understandability · requirements engineering · conceptual model quality

1 Introduction

Understandability[1] is a critical quality attribute in conceptual modeling, as stakeholders need to understand the conceptual model to deliver and convey their message effectively [4,8]. Misunderstandings and syntactical errors may occur, affecting the efficiency and usage of artifacts [8]. Previous research has recognized that model and personal factors affect the understandability of artifacts,

---

[1] This article presents some partial results of Negri-Ribalta's PhD thesis.

T. P. Sales et al. (Eds.): ER 2023 Workshops, LNCS 14319, pp. 263–273, 2023.
https://doi.org/10.1007/978-3-031-47112-4_25

such as education, practice, or professional background [4,14]. The differences in how different backgrounds affect the understandability of models could be critical when stakeholders seek to achieve common ground on privacy. This paper focuses on privacy and data protection requirements. On the one hand, privacy experts are specialists in data protection but may not be familiar with modeling, while developers may be familiar with modeling but lack domain expertise.

From a requirements engineering (RE) perspective, data protection requirements should be elicited from the early phases of the software development life cycle [1,6]. However, different studies have shown that data protection requirements are not elicited nor analyzed in early phases [1]. Furthermore, software engineers have difficulties understanding and analyzing these, given their mental model and knowledge, complicating collaborative processes and communication [1,6].

This paper aims to explore the relationship between the stakeholders' personal factors and their effectiveness in eliciting regulatory data protection requirements in a collaborative setting. We conducted eight focus groups with 21 privacy (PRI), business analysis (BUS), and software development (DEV) experts. The participants were asked to elicit privacy risk from a conceptual model based on the Socio-Technical Security modeling language (STS-ml) [2], extended to address GDPR principles by the STAGE language [11]. We previously and individually tested whether the participants could understand a model similar to what was collaboratively discussed in the focus group. We measured the participants' effectiveness in identifying privacy risks during the focus group and their perception of the understandability of the models.

Our results suggest that privacy experts had the highest performance, even though their scores in the understandability pretests were low understandability. However, the privacy experts did not identify most privacy risks individually but collaboratively with business and software development experts. We think the insight of this exploratory study could be relevant for further research on the social factors of understandability of conceptual models in the context of such models being a communication means among team members.

This article is structured as follows: in Sect. 2, we review empirical studies on the understandability of conceptual models. The research method is explained in Sect. 3. The results are detailed and discussed in Sect. 4, and the conclusions are presented in Sect. 5.

2   Related Work

Understandability is the ability of the stakeholder to comprehend, extract information and infer specific elements from a model [4,5,8,15]. To give one precise definition, [4] describes understandability as "... typically associated with the ease of use and the effort required for reading and correctly interpreting a process model". Yet, there are a variety of models on understandability, each evaluating and analyzing different variables [14]. Given this situation, [8] gathered different approaches to understandability available and provided a unified model. Nevertheless, it is still a stretched and flexible concept.

Understandability depends on various factors, including model characteristics and personal variables [4,15]. For instance, [15] distinguishes between "model factors" (related to the model itself, like density and structure) and "personal factors" (related to the reader). This distinction is also discussed by [4], who includes modeling notation, complexity, modularity, approach, visual layout, and coloring as process model factors. Both process model and personal factors influence understandability, though not necessarily perceived understandability [4]. Empirical personal factors include theoretical knowledge, practice, education [15], as well as learning style, motive, cognitive abilities, professional background, and domain familiarity, among others [4].

Previous work has measured if there were significant differences in understanding BPMN and HPN between subjects of health science background and engineers [17]. The study concluded that there seems to be a statistical difference in some aspects of understandability [17]. From a RE perspective, [7] compared the comprehensibility of Tropos versus Use Case (UC), concluding that Tropos was more comprehensible but at the trade-off of higher time consumption. Research on the personal factors of understandability has focused on how individuals process conceptual models, using eye-tracking technologies [13,20] or by exploring the mental models of the audience [9]. These initiatives provide a deep understanding of the context of users reading models individually, which might be valuable when models are used for documentation or code generation purposes.

However, mainly in agile software development contexts, models are used to collaborate among the members of cross-disciplinary teams [16]. In collaborative contexts, models are used to achieve "common ground" between interdisciplinary teams [3]. Common ground can be understood as "establish and achieve shared goals" [18]. Regardless of the agility of the methodology, common ground is relevant in requirements engineering since it "aims to establish a common ground of shared understanding between users and software engineers" [18].

3   Research Method

This article presents partial results of a larger study that analyses how participants interact around conceptual models from an interdisciplinary perspective for regulatory data protection requirements. The models reviewed and the data collected are available online[2].

3.1   Research Questions and Approach

We define the research goal following the recommendation by [21]: Analyze *socio-technical requirement models* for the purpose of *exploring the relationships of personal understandability factors* concerning *the objective and subjective understandability* of the models from the perspective of the *researcher* within the

---

[2] https://zenodo.org/record/7729512.

context of *stakeholders with diverse backgrounds collaboratively performing an understandability task focused on identifying privacy risks in the socio-technical models*. Using Dikici's quality framework [4], we study risk identification's *effectiveness* for objective understandability and participants *perception* for subjective understandability. To address the research goal, we formulate three research questions:

- *RQ1:* What is the relationship between the personal factors of understandability and the understandability task effectiveness of subjects with different backgrounds collaboratively identifying privacy risk in socio-technical requirement models?
- *RQ2:* What is the relationship between the understandability score of a single subject and the understandability task effectiveness of subjects with different backgrounds collaboratively identifying privacy risk in socio-technical requirement models?
- *RQ3:* What is the relationship between the understandability task effectiveness and the perceived understandability of subjects with different backgrounds collaboratively identifying privacy risk in socio-technical requirement models?

The questions are addressed by employing a series of focus groups. In the focus groups, the participants were given tasks through which we measured their understandability and explanation. We qualitatively analyzed the interventions of each participant in order to identify whether they were using their background to identify a privacy risk in a model. We also noted when the identification comes from the collaboration of two or more participants.

3.2   Measurement Design

We define the variables and metrics detailed below to explore the relationship between understandability and the collaborative identification of privacy risks.

Initial Understandability: We measured participants' understandability of a well-formed, simple model as a pretest for the collaborative task. The participants received a handout on STAGE and were asked to complete an online quiz before the focus group. The quiz consisted of six true or false questions about a STAGE model. The questions were designed to assess if the participants were able to extract basic information from the model. The metric for this variable is *Initial Understandability Score (IUS)*, with values from 0 to 100, as a normalization of the six-point score from the quiz.

Personal Understandability Factors: Within the previous day of the focus group, we surveyed the participants on the personal factors which, according to [4], could affect the understandability of the models. They are:

- Education (Ed): The participant received training to work with privacy and data protection. The possible values are university, work training, self-training, and others.

- Experience (Ex): The participant's experience with conceptual modeling. We grouped the participant's experience into the following categories: [0–1] Low, [2–5] Medium, [5–8] High, [8+] Expert.
- Training (Tr): On data protection training (GDPR). Measured in hours (hrs), we categorized the participants into: [0–30] Low, Medium, High, [90+] Expert.
- Familiarity (F): The participant's familiarity with the modeling method, particularly with STAGE. Possible values: yes and no.
- Theory (Th): Concerns the participant's knowledge of GDPR. Possible values: yes and no.

Understandability Task Effectiveness: is if the subjects can understand the "tasks or questions about the process models [4]". This variable is measured by reviewing the transcripts and video from the focus groups and identifying the contributions of each participant to the discussion in the following metrics:

- *Individual Identifications (II):* identifies and discusses a privacy risk based on their understanding of the model without inputs from other participants.
- *Identification and Agreement (IA):* participates in the collaborative identification of privacy risk and discusses it.
- *Agreements (A):* Agrees in privacy risks identified by other participants, without contributing with their perspective on the issue.
- *Percentage of Identifications (% Identification):* Provides an overview of a participant's contribution to the understandability task. Corresponds to the sum of the identifications where the participant contributes with their knowledge (II+IA) with respect to the total seeded privacy risks.

Subjective Understandability: After the focus group, we surveyed the participants' perception of the model. The metrics for his variable are detailed below.

- *Perceived Ease of Use (PEU):* Per [4], it is the perceived easiness of the subject on using this model. It usually "involves a set of questions with answers in Likert scale that aims to capture participants' subjective perception on the ease of use" [4]. Measured using [10].
- *Perceived Usefulness (PU):* how probable does the subject believe it can be of utility to use such model [4], measured using [10] guidelines.
- *Intention to Use (IU):* if the subjects intends to use the model [4], measured using [10] guidelines.

3.3   Focus Group Design

As part of the larger study, we organized focus groups with volunteers we gathered through online surveys. The objective of the larger study is to analyze the interaction between subjects with specific characteristics; thus, we used a purposive sampling approach. Given this, we did focus groups with a triangulation approach; i.e., each focus group ideally would have three participants with different professional backgrounds to perform different roles accordingly: a developer

(DEV), a privacy specialist (PRI), and a business analyst (BUS). We opted for a reduced number of participants to ensure that all of them could participate as much as possible. Before starting the activity, we provided a handout on the STAGE modeling language with examples. The subjects took a quiz on the understandability of a STAGE model, where they answered six questions regarding the model's content, mirroring the approach of [15].

After filling out the understandability quiz, we started the focus group online, presenting the subjects with an unfamiliar scenario and three views of STAGE's models with seeded privacy risks. The participants were tasked with identifying privacy risks through a straightforward reading of the model or by assessing modeling quality within the context. Following the focus group, participants were surveyed to gauge their perception of the usefulness, utility, and intention to use as a subjective assessment of understandability [4].

4   Results and Discussion

4.1   Results

We carried out eight focus groups, with 21 participants - as some subjects did not attend the activity - from October 2022 to February 2023. This made a total of 400 min of recorded focus groups we analyzed, as focus groups lasted a maximum of 50 min. Two participants, a BUS and a DEV, did not speak during the focus group, so they were discarded from the study. Table 1 show the total privacy risks identified per group and their composition (19 subjects in total).

Table 1. Participant roles and total privacy risk identifications per group.

Group	Participant Role	Total ident.
1	PRI, DEV	10
2	PRI, DEV	6
3	PRI, DEV, BUS	11
4	PRI, DEV, BUS	11
5	PRI, DEV	10
6	PRI, DEV, BUS	11
7	DEV, BUS	7
8	PRI, BUS	11

Table 2 shows the percentage of privacy risks identified individually (**II**), collaboratively (IA), or agreed (A) by the participants, grouped by role. As can be seen in column **II**. BUS participants individually identified 5.45% of total privacy risks, while DEVs achieved 9.09%. No role identified more than 10% based solely on their background. On the other hand, most of the identifications were collaborative, as detailed in column **IA** (identification and agreement). PRI participants excelled with 55.80% collaborative identifications, while BUS participants agreed most frequently with' identifications proposed by other participants (as seen in column **A** of Table 2).

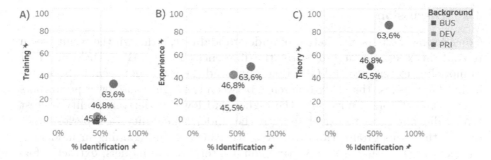

Fig. 1. Personal factors v/s percentage of identification of privacy risks: A) Training, B) Experience, C) Theory.

Table 2. Percentage of privacy risks identified individually, collaboratively or agreed by the participants per role.

Role	II	IA	A	Perc. Identi.
BUS	5.45%	40.00%	32.70%	45.45%
DEV	9.09%	37.70%	29.90%	46.79%
PRI	7.79%	55.80%	15.60%	63.59%

Regarding RQ1 and the relationship of personal factors in the understandability task effectiveness, Fig. 1 depicts the relationship between the training (A), experience (B), and theory (C) personal factors and the percentage of identification of privacy risks, per role. Experimental results for individual understandability assessments show that subjects with higher values for personal factors (PRI) seem to present a higher performance (as seen in Fig. 1). It is worth noting that even subjects with near-to-null training in GDPR (BUS) participated and contributed to understanding the model.

Concerning RQ2, Fig. 2 shows the understandability pretest scores and the percentage of identifications (Perc. Identi.) per role. The understandability pretests do not seem to be positively related to the percentage of identifications since PRIs had the lower pretest scores (71.4%) and the best performance in the understandability task. While DEVs and BUSs scored 81.0% and 86.7% in the pretest and had a lower performance in understandability tasks. Table 3 presents the values for the subjective understandability measurements: PEU, PU and IU, by role, for RQ3. The difference in percentage identification—favoring PRI users—does not seem to be related to a higher or lower subjective perception of the understandability of the model.

Table 3. Perceived understandability measurements: PEU, PU and IU

Role	PEU	PU	IU	Perc. Ident.
BUS	42.5	44.4	32.5	45.50%
DEV	50.6	57.6	53.6	46.80%
PRI	42.9	51.8	50.0	63.60%

4.2   Discussion

Considering the personal factors of understandability, although the results seem to confirm experimental results [15], the differences between BUS, DEV, and PRI in modeling experience and GDPR training and theory do not seem to be significant. There is less than 20% difference between the higher and lower performers in personal factors. Conversely, the higher and lower understandability pretest scores did not seem to correlate with the understandability task performance during the focus group. However, even with little to no training in modeling, subjects in the collaborative setting could interpret the models, extract information, and establish common ground by identifying most of the privacy risks. Even though some design decisions could favor the participants' performance (toy problem, sequential presentation of the views), we think there are two key enablers: the conceptual modeling language and the collaboration between the participants.

Regarding conceptual models, we think the socio-technical approach of STAGE is appropriate for identifying privacy risks for a multi-disciplinary group. Naturally, some misalignments between the domain and the modeling language were identified. Some participants found that some concepts did not clearly represent some domain elements (PRI3: *"Why is that symbol called actor and no Data Controller?"*). Also, most PRIs found the information view overwhelming, while some DEVs were confused with the social view. This is consistent with PRIs and DEVs performing lower on those views, despite their complexity, as seen in Table 4. However, there were just a few comments in this line, and groups identified most of the privacy risks; moreover, two groups identified all of the seeded privacy risks. We think this is evidence of the appropriateness of multi-view, socio-technical models privacy risk identification, since the different problem representations in STAGE models and the problem-solving task (identifying privacy risks), as predicted by Cognitive Fit theory [19].

The second enabler regards collaboration among participants. We think the study format helped participants focus on discussing the model based on their knowledge, experience, and training, as most privacy risks were collaboratively identified. This could be explained by the ontological foundations of collaboration, which is based on coordination, communication, and cooperation [12]. We think the focus group provided a coordination and communication framework, facilitating the participants' cooperation. The focus group defined

Table 4. Percentage of identification (perc. ident.) per view, including Yaqin complexity index [22]

	Social View	Info. View	Auth. View
Yaqin Complexity	54.57	24	19.57
Role	**Perc. Ident.**		
BUS	46.7%	46.7%	44.0%
DEV	33.3%	47.6%	54.3%
PRI	71.4%	57.1%	62.9%

how subjects would interact and the sequence to discuss the model views, helping coordination, while the experimenter supported communication by encouraging the team members to speak. With these two elements facilitated, subjects could focus on cooperation, defined as *"a joint effort in a shared space to achieve some*

Fig. 2. Understandability pretest scores v/s percentage of identifications per role.

goal.", where each participant contributes according to their commitment (i.e., their roles) to the task.

Regarding the study's limitations, the results should be interpreted considering that a single notation (STAGE) was used and that larger groups (more than three participants) could yield different results.

5   Conclusions

Using conceptual modeling artifacts as communication tools between stakeholders with different mental models can be promising for achieving common ground. These artifacts can help interdisciplinary teams discuss regulatory data protection requirements from early phases. However, these conceptual modeling artifacts must be understandable across various disciplines, making the task challenging. Thus, understandability plays a vital role in this aspect.

This article provides evidence of the importance of interdisciplinary approaches when designing conceptual models. Through focus groups, we aimed to explore the relationship between the personal factors of understandability and the task effectiveness of identifying risks within socio-technical requirement models. Using a collaborative methodology, we engaged participants with different privacy backgrounds to analyze and discuss privacy compliance.

Our findings underscore that participants with major training, theoretical knowledge, and experience exhibit enhanced effectiveness in understandability. Although the privacy experts' understandability scores not being the highest, they identified most of the privacy risks and contributed significantly to the discussion. This demonstrates that a collaborative approach effectively mitigates the limitations posed by individual scores. Furthermore, regarding the subjective understandability metrics, the difference between these metrics and the effectiveness of understanding tasks could be because of how participants with different backgrounds interact. This result highlights the intricate and mutually advantageous nature of collaboration, where even a lower perception of ease of use (as observed among privacy experts) does not hinder the ability to achieve the goal of identifying potential risks in the models.

These results offer empirical evidence of the importance of collaboration for leveraging the understandability of models for participants with different personal backgrounds for a complex domain such as data protection and privacy. Moreover, this research encourages us to ponder the concept of interdisciplinarity and its pivotal role in advancing conceptual modeling practices inside collaborative environments. This approach can signal a promising path toward enhanced model understandability.

Acknowledgments. This work was supported by the Generalitat Valenciana through the CoMoDiD project (CIPROM/2021/023) and the Santiago Grisolía fellowship (GRISOLIAP/2020/096), and the Spanish State Research Agency through the DELFOS (PDC2021-121243-I00) and SREC (PID2021-123824OB-I00) projects.

References

1. Breaux, T.D., Norton, T.: Legal accountability as software quality: a US data processing perspective. In: 2022 IEEE 30th International Requirements Engineering Conference (RE). IEEE (2022)
2. Dalpiaz, F., Paja, E., Giorgini, P.: Security Requirements Engineering: Designing Secure Socio-Technical Systems. The MIT Press, Cambridge (2016)
3. Damian, D., Chisan, J.: An empirical study of the complex relationships between requirements engineering processes and other processes that lead to payoffs in productivity, quality, and risk management. IEEE Trans. Softw. Eng. **32** (2006)
4. Dikici, A., Turetken, O., Demirors, O.: Factors influencing the understandability of process models: a systematic literature review. Inf. Softw. Technol. **93**, 112–129 (2018)
5. Engelsman, W., Wieringa, R.: Understandability of goal-oriented requirements engineering concepts for enterprise architects. In: Jarke, M., et al. (eds.) CAiSE 2014. LNCS, vol. 8484, pp. 105–119. Springer, Cham (2014). https://doi.org/10.1007/978-3-319-07881-6_8
6. Hadar, I., et al.: Privacy by designers: software developers' privacy mindset. In: Proceedings of the 40th International Conference on Software Engineering, ICSE 2018. Association for Computing Machinery (2018)
7. Hadar, I., Reinhartz-Berger, I., Kuflik, T., Perini, A., Ricca, F., Susi, A.: Comparing the comprehensibility of requirements models expressed in use case and tropos: results from a family of experiments. Inf. Softw.Technol. **55**, 1823–1843 (2013)
8. Houy, C., Fettke, P., Loos, P.: Understanding understandability of conceptual models – what are we actually talking about? In: Atzeni, P., Cheung, D., Ram, S. (eds.) ER 2012. LNCS, vol. 7532, pp. 64–77. Springer, Heidelberg (2012). https://doi.org/10.1007/978-3-642-34002-4_5
9. Mendling, J., Djurica, D., Malinova, M.: Cognitive effectiveness of representations for process mining. In: Polyvyanyy, A., Wynn, M.T., Van Looy, A., Reichert, M. (eds.) BPM 2021. LNCS, vol. 12875, pp. 17–22. Springer, Cham (2021). https://doi.org/10.1007/978-3-030-85469-0_2
10. Moody, D.L.: The method evaluation model: a theoretical model for validating information systems design methods. In: Proceedings of the European Conference on Information Systems 2003. AIS Electronic Library (2003)

11. Negri-Ribalta, C., Noel, R., Herbaut, N., Pastor, O., Salinesi, C.: Socio-technical modelling for GDPR principles: an extension for the STS-ML. In: 2022 IEEE 30th International Requirements Engineering Conference Workshops (REW) (2022)
12. Oliveira, F.F., Antunes, J.C., Guizzardi, R.S.: Towards a collaboration ontology. In: Proceedings of the Snd Brazilian Workshop on Ontologies and Metamodels for Software and Data Engineering, João Pessoa (2007)
13. Petrusel, R., Mendling, J., Reijers, H.A.: Task-specific visual cues for improving process model understanding. Inf. Softw. Technol. **79**, 63–78 (2016)
14. Reijers, H.A., Mendling, J.: A study into the factors that influence the understandability of business process models. IEEE Trans. Syst. Man Cybern. - Part A: Syst. Hum. **41**, 449–462 (2011)
15. Reijers, H.A., Mendling, J.: A study into the factors that influence the understandability of business process models. IEEE Trans. Syst. Man Cybern. Part A: Syst. Hum. **41**(3), 449–462 (2011)
16. Rumpe, B.: Agile modeling with the UML. In: Wirsing, M., Knapp, A., Balsamo, S. (eds.) RISSEF 2002. LNCS, vol. 2941, pp. 297–309. Springer, Heidelberg (2004). https://doi.org/10.1007/978-3-540-24626-8_21
17. Stitzlein, C., Sanderson, P., Indulska, M.: Understanding healthcare processes. Proc. Hum. Factors Ergon. Soc. **57** (2013)
18. Sutcliffe, A.: User-Centred Requirements Engineering. Springer, Heidelberg (2002)
19. Vessey, I.: Cognitive fit: a theory-based analysis of the graphs versus tables literature. Decis. Sci. **22**(2) (1991)
20. Wang, W., Indulska, M., Sadiq, S., Weber, B.: Effect of linked rules on business process model understanding. In: Carmona, J., Engels, G., Kumar, A. (eds.) BPM 2017. LNCS, vol. 10445, pp. 200–215. Springer, Cham (2017). https://doi.org/10.1007/978-3-319-65000-5_12
21. Wohlin, C., Runeson, P., Höst, M., Ohlsson, M.C., Regnell, B., Wesslén, A.: Experimentation in Software Engineering. Springer, Heidelberg (2012)
22. Yaqin, M.A., Sarno, R., Rochimah, S.: Measuring scalable business process model complexity based on basic control structure. Int. J. Intell. Eng. and Syst. **13**, 52–65 (2020)

FlowTGE: Automating Functional Testing of Executable Business Process Models Based on BPMN

Tomás Lopes[1]([✉])[iD] and Sérgio Guerreiro[1,2,3][iD]

[1] Link Consulting SA, Av. Duque de Ávila 23, 1000-138 Lisbon, Portugal
{tomas.lopes,sergio.guerreiro}@linkconsulting.com
[2] INESC-ID, R. Alves Redol 9, 1000-029 Lisbon, Portugal
[3] Instituto Superior Técnico, Av. Rovisco Pais 1, 1049-001 Lisbon, Portugal
sergio.guerreiro@tecnico.ulisboa.pt

Abstract. Testing business process models is vital for guaranteeing the correct operation of processes and ensuring they comply with requirements and regulatory norms. Often performed manually, automating process model testing expedites testing efforts and reduces human error. This paper presents FlowTGE, a tool for end-to-end black/gray-box testing of executable business process models based on the BPMN language. This tool is built on top of the bPERFECT framework and is designed to tackle the slow and error-prone nature of manual process testing in the context of workflow automation systems. FlowTGE is split into two components—the Generator and the Executor—which, respectively, handle the automated generation of test cases from executable BPMN-based process models and the execution of said test cases through process simulations using workflow automation functionalities. Evaluation shows promising results regarding user perceptions of usefulness and ease of use and demonstrates the tool's suitability to be applied to a significant portion of process-testing-related challenges.

Keywords: BPMN · Business process · Test automation · Business process management · Business process testing · Model-based testing

1 Introduction

Business processes lie at the core of organizations [11]. Defined as "a collection of inter-related events, activities, and decision points that involve a number of actors and objects, which collectively lead to an outcome that is of value to at least one customer" [3], the efficiency and effectiveness of business processes directly influence customer experience and the quality of organizational outputs [11].

Business Process Management (BPM) often relies on using process models that represent people's understanding of how work is done, abstracting away unimportant details. A process model describes all the different ways to execute a business process, differing from the notion of a process instance which consists of one specific execution of a process [3]. Several languages may be used to represent process models, with the most common one being the Business Process Model and Notation (BPMN) language, often regarded as the de facto standard for process modeling.

Technological and organizational growth has led to a massive hike in the complexity of business processes. Maintaining their correctness and compliance with established requirements has become a significant challenge for companies [5]. Business Process Testing (BPT) activities have become essential in guaranteeing that processes continue to bring value as they evolve [5].

Furthermore, the executability of these models has become a critical aspect of BPM, as it determines the extent to which a process can be automated and executed through implementation in BPM systems or other process-aware information systems [3]. BPM systems use process models to guide the flow of work, automatically routing work items to the appropriate people and systems at each step of the process with capabilities for monitoring and process analysis [3].

The main goal of this paper is to discuss the design and implementation of the FlowTGE tool, an automated functional testing solution for assessing BPMN-based executable business process models. Three basic requirements are set: (i) automatically generate test cases from executable process models based on BPMN, (ii) execute process test cases, and (iii) produce easily interpretable test case execution reports.

2   Related Work

A literature review about BPT was recently published, which summarizes and analyzes over 30 studies detailing different approaches and techniques for process testing and verification, focusing on the BPMN language [5]. Many studies reviewed in this paper describe specific testing-related techniques that may be adapted and combined to create a more sophisticated all-encompassing solution:

- [2,6,9,13,14] describe different ways to transform BPMN models into more elementary graph-like structures which facilitate further analysis, namely path discovery.
- [10,12–14] use a Depth-First Search (DFS) based path discovery algorithm to extract the existing execution scenarios from a process.
- [8] implement a recursion delimiter that controls the number of times a given sequence flow may be executed in any given execution scenario.
- [8,14] use a simple tabular format to represent test cases and execution scenarios. These tables contain information such as the corresponding path, pre-conditions, inputs, and post-conditions.

Combining and adapting these techniques shows considerable potential to create new and disruptive BPT solutions. This literature review also proposes a simple classification system for BPT types, as follows [5]:

- Black/gray-box: used to verify basic functional requirements,
- Regression: used to test potentially breaking changes, and
- Integration: used to test specific service-related implementation details

Finally, this review proposes the Business Process Evaluation and Research Framework for Enhancement and Continuous Testing (bPERFECT) framework that aims to guide BPT research, consisting of a series of high-level instructions that all future BPT approaches should follow to facilitate knowledge sharing and boost interoperability and intercompatibility [5]: (i) model the process, (ii) determine constraints, (iii) convert to graph structure, (iv) extract paths, (v) determine test data, (vi) determine testing type, (vii) generate test cases, and (viii) execute test cases.

The classification of testing types, the bPERFECT framework, and the analysis of several existing BPT-related techniques constitute a solid starting point for developing new and innovative solutions that provide added functionality and extend the scope of current BPT approaches.

3 Flow Test Generation and Execution

This section describes the architectural and implementation details of Flow Test Generation and Execution (FlowTGE), an end-to-end black/gray-box business process model testing tool with the capability of automating key parts of the assessment of the correctness of executable process models and providing insights for their improvement.

FlowTGE is designed to automate many steps associated with end-to-end process model testing, including execution scenario determination, test data generation, and test execution. The use of bPERFECT as a reference framework ensures that it is aligned with existing BPT methodologies while also following the best practices in the field.

FlowTGE is split into two components: the Generator and the Executor. Each of these two components is made up of 3–4 core sub-components which are run sequentially to achieve the desired objective.

Table 1 shows the correspondence between bPERFECT steps, specific techniques used in FlowTGE, and the studies from which those techniques were adapted.

3.1 Generator

The Generator is the component of FlowTGE that takes care of everything related to the generation of test cases from process models.

The Generator receives a business process model as input and outputs a set of test cases. Each test case consists of a sequence of detailed steps (such as "Actor A assigns variable X during Task T" or "Actor A executes Task T, meeting constraint C") which, when followed, lead to the successful termination of the process, per the model.

Table 1. Mapping from bPERFECT steps to specific techniques used in FlowTGE.

bPERFECT steps	FlowTGE techniques	References
1. Model the process	Model using BPMN variant	–
2. Determine constraints	Parse conditions from sequence flows	–
3. Convert to graph	Convert to directed graph	[2,6,9,13,14]
4. Extract paths	Apply adapted DFS	[8,10,12–14]
5. Determine test data	Use constraint solver	[8,14]
6. Determine testing type	Perform black/gray-box testing	[1,8,13,14]
7. Generate test cases	Write in tabular format	[8,14]
8. Execute test cases	Simulate using API	–

The Generator comprises four core sub-components – BPMN-to-Graph, Path Extractor, Constraint Extractor, and Test Case Writer. The core sub-components are run sequentially, each taking as input some of the outputs produced by the sub-components run beforehand.

BPMN-to-Graph. The BPMN-to-Graph sub-component, as the name suggests, handles the transformation of the process model into a (directed) graph.

To accomplish this, the model's sequence flows are parsed from the BPMN XML file. Each sequence flow contains a source reference and a target reference which may each be a task, an event, or a gateway. The set of all sequence flows constitutes the edges of the graph, and the union of all source references and all target references constitutes the nodes of the graph. Collapsed subprocess nodes are recursively substituted by the corresponding process graphs.

Path Extractor. The Path Extractor sub-component computes all desired paths from the start node to the end node(s) in the graph outputted by the previous sub-component. In this context, a "path" refers to a single end-to-end execution scenario that is valid per the model. Since these may include parallel activities, one path may actually include several parallel sub-paths.

Beginning with the start event node, a DFS is used to recursively explore each node's successors, updating a `visited` dictionary that keeps track of the number of times each node has been visited in the path currently being explored. A parameter controls how many times any specific node may be visited, allowing the exploration of cycles. Upon reaching an end event, the path is added to a list and the search backtracks, exploring paths via other successors until there are no more alternatives.

After the initial path extraction, a parallel sub-path merging operation between paths that contain the same AND-split node is performed which, under the assumption that there are no race conditions between branches that may execute in parallel, sequentializes parallel tasks.

Constraint Extractor. The Constraint Extractor sub-component determines which constraints must apply for each path to be followed.

For each path, this sub-component of FlowTGE parses the constraints stored as edge data (sequence flow labels) and stores them in a `TestDatum` object containing a path and a collection of constraints. Each constraint is associated with the task where it must be satisfied to proceed with the test case (see Fig. 1).

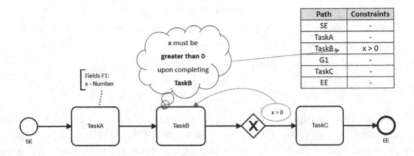

Path	Constraints
SE	-
TaskA	-
TaskB	x > 0
G1	-
TaskC	-
EE	-

Fig. 1. Extracting constraints from a path.

Test Case Writer. Finally, the Test Case Writer generates the test cases based on the acquired information about the possible flows. One test case is generated for each path, representing one possible execution scenario for the process.

Each test case consists of pre-conditions, a sequence of execution steps, and post-conditions. The execution steps are displayed in the test case in tabular form, as done first by [8,14]. Each step may be of one of three different types: (i) *Instantiate process*, (ii) *Assign variable*, or (iii) *Execute task*.

3.2 Executor

The Executor is the component of FlowTGE that handles the execution of test cases generated from business process models. Before execution, test data is also generated for each test case based on the associated constraints.

This component receives a set of test cases as input and outputs relevant information about the execution of the tests. Test execution data includes aspects such as passing and failed tests, reasons for failure, and execution time.

Similarly to the Generator, the Executor's three core sub-components – Pre-Processor, Solution Mapper, and Test Case Simulator – are run sequentially.

Pre-Processor. The Pre-Processor parses the test case specifications and produces additional information about the variables and constraints that may not be explicit in the tests. The tasks performed by this sub-component include variable parsing, constraint parsing, and variable domain estimation.

Solution Mapper. The Solution Mapper sub-component determines possible values for the variables involved in each test case that meet all the constraints specified for said test case.

In order to support arbitrarily complex constraints, a constraint solver was used for this task, similar to the approach presented by [4] and proposed by [12]. More specifically, Microsoft's Z3 Theorem Prover[1], an SMT solver (a generalization of classic Boolean satisfiability/SAT) with bindings for programming languages like Python and C#, was used for this purpose. Each test case's constraints (including implicit ones derived by the previous sub-component) are, thus, converted to Z3 Boolean expressions and added to a solving model which, upon calling the `Solve` method, returns one possible value for each variable (`Status.SATISFIABLE`). Alternatively, if the constraints are incompatible, the solver returns `Status.UNSATISFIABLE`; if that happens, the test case is skipped.

Test Case Simulator. The Test Case Simulator handles the execution (through simulation) of the test cases and collects data concerning this execution.

The implementation of this execution mechanism differs significantly depending on the workflow engine being used to execute the process models. In the current implementation, edoclink, a document management system implemented, commercialized, and maintained by Link Consulting, S.A. designed around workflow automation, is used for that purpose. As such, the execution of test case steps is achieved using the edoclink Public API. This component may be adapted to support other BPM/workflow automation systems.

If any step of a test case fails to execute, that test case is interrupted and marked as failed. Executing every step of a test case with no errors corresponds to a successful test case.

Users have access to information regarding not just the number of failures but also the reasons for failure. This information, combined with other results and metrics such as execution time, enables the extraction of insights that may be used to correct and enhance the process model.

After executing all test cases, an execution report is generated and presented to the user. This execution report contains all execution data collected for all tests executed, such as (i) a statistical overview of the outcomes of all test cases, (ii) total execution time and execution time per test case, (iii) the values assigned to each variable, and (iv) reasons for failure of each test case (if applicable). The insights extracted from this information facilitate the correction and enhancement of the process model.

4   Evaluation

This section contains all the evaluation procedures carried out to assess end user perceptions of FlowTGE.

---

[1] https://www.microsoft.com/en-us/research/project/z3-3/.

4.1 Methodology

To assess the efficiency of the tool, process models with varying amounts of paths were procedurally generated by sequentially chaining cycles. Models were generated using this method for each possible value of n from 0 (1 path) to 16 (65536 paths using the `max` mode) and execution times for both components were measured. This experiment showed exponential time and space growth with respect to the number of cycles (as expected by the exponential growth of the number of paths) and linear time and space growth with respect to the number of execution steps, with the Generator being able to generate over 200 execution steps per second.

The main focus of this evaluation, however, consisted of assessing end user perceptions of the tool regarding usefulness and ease of use. On account of this, eight workers at Link Consulting with prior process testing experience were asked to participate in an experiment.

The experiment consisted of users being asked to detect errors in four process models (with deliberately introduced errors) adapted from previous projects by interacting with the Generator to generate the test cases (with the number of paths for each model varying between 5 and 34), interacting with the Executor to execute them, and analyzing the execution reports.

They were then asked to fill out a survey with eight items, each concerning one of three perception-related constructs of Moody's Method Evaluation Model [7]: (i) Perceived Ease of Use, (ii) Perceived Usefulness, and (iii) Intention to Use. A 5-point Likert scale was used to measure each item, reversing the score of negatively worded items (I3, I4, I5, I7) to enable the calculation of numerical measures. Respondents were able to answer each item for the Generator and the Executor independently. Items were shuffled for each respondent, with each one focusing on one of the three perception-related constructs of the model:

I1 I found the tool easy to learn. (Perceived ease of use)
I2 This tool makes it easier to spot errors in process models. (Perceived usefulness)
I3 I found the tool difficult to utilize. (Perceived ease of use)
I4 Overall, I think this tool does not effectively solve the problem of manually testing large process models being time-consuming and error-prone. (Perceived usefulness)
I5 I would not use this tool to test large process models. (Intention to use)
I6 I believe this tool would reduce the effort needed to test large process models. (Perceived usefulness)
I7 The overall procedure for utilizing this tool is complex. (Perceived ease of use)
I8 I plan on using this tool to test processes over manual testing. (Intention to use)

Both components of FlowTGE were deployed in internal environments for end user validation and logs were used to verify user interaction with both of the tool's components.

4.2 Results

The results of the survey carried out concerning user perceptions can be visualized in Fig. 2 as box-and-whisker diagrams plotted for each tool/model construct pair, where the blue whiskers represent the minimum and maximum scores for that item, the edges of the blue boxes represent the first and third quartiles, and the green horizontal bars represent the second quartiles (the medians). Additionally, mean scores were plotted as black dots with black labels.

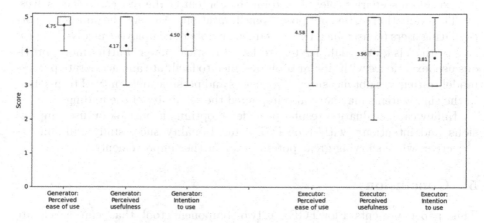

Fig. 2. Box plots of survey scores for each tool/construct pair.

Scores for perceived ease of use were nearly perfect, with both mean scores greater than 4.5 and very little divergence in answers, which likely stems from the lack of effort required from users to use the tool (usage consists of simply pressing one to two buttons and, in the case of the Executor, filling out a simple form, with all of the work being done by the tool in the background).

Scores for items concerning perceived usefulness were still quite positive, albeit taking a noticeable hit when compared to the items concerning perceived case of use. Furthermore, larger interquartile ranges for most items indicate a larger degree of divergence between respondents. It is also worth noting that scores for the Generator are slightly higher than for the Executor. Hence, despite results still being positive, they show mild concerns regarding the tool's practicality and adequacy at solving the problem at hand.

Finally, as for intention to use, although scores for both the Generator and the Executor can still be considered quite positive, similarly to what happens for perceived usefulness, scores for the Generator are noticeably higher than the Executor's. Respondents showed great intent on continuing to use the Generator, which already brings value independently, while showing some skepticism towards using the Executor, which depends on the Generator to function.

Further conversations with the respondents were carried out with the intent of analyzing why scores for the Executor were lower across the board. Two main limitations regarding the Executor were gathered. Firstly, variables with complex data types are unsupported by the Executor's Solution Mapper, leading to the need for manual assignment of field values or process model refactoring efforts (removing variables with those data types) to execute test cases for processes containing variables with such data types. Additionally, flexible processes that allow the execution of activities in an ad hoc fashion cannot be comprehensively tested using FlowTGE. Overall, users seemed keen on using the Generator to derive the valid execution scenarios from a model, showing intent to use the Executor to test critical execution scenarios while opting to validate the remaining execution scenarios solely based on the content of the test cases themselves.

The evaluation carried out poses some limitations. Namely, the number of participating users (constrained by the reduced number of employees involved in process testing) is quite small, leading to low statistical power. Furthermore, previous user experience with similar tools designed to facilitate and accelerate process model testing was not measured. Vagueness and possible ambiguity of terms used in the survey items may have also impacted the reliability of the findings.

In any case, evaluation results provide an optimistic outlook on user impressions and intentions, with FlowTGE's functionality successfully fulfilling its objectives while showing great potential for further improvements.

5   Conclusion

This paper presents FlowTGE, a two-component tool that can accelerate and improve the testing of executable BPMN-based process model testing by automating critical tasks typically done manually. Namely, FlowTGE can automatically generate and execute test cases from business process models, thus helping to ensure the proper operation of the processes and reducing manual process testing efforts.

As of writing this paper, the tool is being used internally at Link Consulting and has been integrated with its process management tool suite (Atlas and edoclink). Both components of FlowTGE are already in use today in the context of process management and business consulting projects.

In future work, the tool will be extended to tackle the limitations pinpointed during evaluation, namely its inability to deal with complex data types. Additionally, the constant context switching involved in a typical process build/test loop leaves an opportunity to improve the cohesion of such activities through, for instance, the creation of a common environment for modeling and testing.

Acknowledgment. This work was supported by a pre-registered project, named as eProcess, which is under national funds with reference C669314338-00003137 (LINK CONSULTING - TECNOLOGIAS DE INFORMAÇÃO S.A.).

References

1. Buchs, D., Lucio, L., Chen, A.: Model checking techniques for test generation from business process models. In: Kordon, F., Kermarrec, Y. (eds.) Ada-Europe 2009. LNCS, vol. 5570, pp. 59–74. Springer, Heidelberg (2009). https://doi.org/10.1007/978-3-642-01924-1_5

2. Dechsupa, C., Vatanawood, W., Thongtak, A.: An automated framework for BPMN model verification achieving branch coverage. Eng. J. Thai. **25**(2), 135–150 (2021). https://doi.org/10.4186/ej.2021.25.2.135

3. Dumas, M., Rosa, M.L., Mendling, J., Reijers, H.A.: Fundamentals of Business Process Management. Springer, Berlin (2018)

4. Jahan, H., Rao, S., Liu, D.: Test case generation for BPEL-based web service composition using colored Petri nets. In: 2016 International Conference on Progress in Informatics and Computing (PIC 2016), pp. 623–628. IEEE, Shanghai, China (2016). https://doi.org/10.1109/PIC.2016.7949575

5. Lopes, T., Guerreiro, S.: Assessing business process models: a literature review on techniques for BPMN testing and formal verification. Bus. Process. Manag. J. **29**(8), 133–162 (2023). https://doi.org/10.1108/BPMJ-11-2022-0557

6. Meghzili, S., Chaoui, A., Strecker, M., Kerkouche, E.: An approach for the transformation and verification of BPMN models to colored Petri nets models. Int. J. Softw. Innovation **8**(1), 17–49 (2020). https://doi.org/10.4018/IJSI.2020010102

7. Moody, D.: The method evaluation model: a theoretical model for validating information systems design methods. In: Proceedings of the 11th European Conference on Information Systems, ECIS 2003, Naples, Italy 16–21 June 2003. Naples, Italy (2003). https://aisel.aisnet.org/ecis2003/79

8. de Moura, J.L., Charão, A.S., Lima, J.C.D., de Oliveira Stein, B.: Test case generation from BPMN models for automated testing of web-based BPM applications. In: 2017 17th International Conference on Computational Science and Its Applications (ICCSA 2017), pp. 1–7. IEEE, Trieste, Italy (2017). https://doi.org/10.1109/ICCSA.2017.7999652

9. Nazaruka, E., Ovchinnikova, V., Alksnis, G., Sukovskis, U.: Verification of BPMN model functional completeness by using the topological functioning model. In: ENASE 2016: Proceedings of the 11th International Conference on Evaluation of Novel Software Approaches to Software Engineering, pp. 349–358. SciTePress, Rome, Italy (2016). https://doi.org/10.5220/0005930903490358

10. Paiva, A.C.R., Flores, N.H., Faria, J.P., Marques, J.M.G.: End-to-end automatic business process validation. Procedia Comput. Sci. **130**, 999–1004 (2018). https://doi.org/10.1016/j.procs.2018.04.104

11. Rosemann, M.: Foreword, 2018. In M. Dumas, M. L. Rosa, J. Mendling, and H. A. Reijers, Fundamentals of Business Process Management. Springer, Berlin, pp. vii-viii (2018)

12. Schneid, K., Stapper, L., Thöne, S., Kuchen, H.: Automated regression tests: a no-code approach for BPMN-based Process-Driven Applications. In: 2021 IEEE 25th International Enterprise Distributed Object Computing Conference (EDOC), pp. 31–40. IEEE, Gold Coast, Australia (2021). https://doi.org/10.1109/EDOC52215.2021.00014

13. Seqerloo, A.Y., Amiri, M.J., Parsa, S., Koupaee, M.: Automatic test cases generation from business process models. Requirements Eng. **24**, 119–132 (2019). https://doi.org/10.1007/s00766-018-0304-3

14. Yotyawilai, P., Suwannasart, T.: Design of a tool for generating test cases from BPMN. In: 2014 International Conference on Data and Software Engineering (ICODSE), pp. 1–6. IEEE, Bandung, Indonesia (2014). https://doi.org/10.1109/ICODSE.2014.7062692

An Approach Aligned with Model Driven Development to Evaluate the Quality of Explainable Artificial Intelligence

Álvaro Navarro[1]([✉])[iD], Javier Sanchis[1,2][iD], Alejandro Maté[1][iD],
and Juan Trujillo[1][iD]

[1] Lucentia Research Group, Department of Software and Computing Systems,
University of Alicante, Carretera San Vicente del Raspeig s/n, San Vicente del
Raspeig 03690, Alicante, Spain
[2] XSB Disseny i Multimedia, S.L., 13 Mercado Street, Onil 03430, Alicante, Spain
{alvaro.nl,amate,jtrujillo}@dlsi.ua.es, javier.sanchis@ua.es

Abstract. Since the fourth industrial revolution began, Artificial Intelligence (AI) systems have become cornerstones in the activities of many organizations. Nonetheless, the application of ML and AI techniques is often limited due to mistrust among users regarding the results obtained by the algorithms. This can lead to decisions not being made correctly. Both facts demonstrate that current algorithms need to be interpretable and transparent. EXplainable Artificial Intelligence (XAI) techniques are paramount to this objective as they translate black-box algorithms into transparent logic for developers and users. However, despite different XAI approaches evaluate dimensions through metrics, there is no consensus on how to evaluate XAI. In an effort to standardize the evaluation of XAI and allow users the comparison of the ideal XAI solution for their problem, we present an approach focused on the evaluation of XAI quality, which is able to be applied in Model Driven Development (MDD) scenarios. The great advantages of our proposal are that it (i) provides a set of quality metrics to evaluate the multiple relevant XAI and AI dimensions, (ii) provides a holistic evaluation thanks to these quality metrics, (iii) emphasizes the relevance of identifying the different target users involved, and (iv) enables the comparison across different XAI alternatives. In order to show the applicability of our proposal, we apply it to a widely-known case study: chess.

Keywords: Artificial Intelligence · EXplainable Artificial Intelligence · Model Driven Development · Quality Metrics

1 Introduction

Nowadays, Artificial Intelligence (AI) systems are playing a crucial role in citizens' everyday life. Even in safety-critical areas, these systems support human-centered processes and decisions, such as in medical diagnosis [1] or floods' forecasting [14]. However, many Machine Learning (ML) and Deep Learning (DL)

T. P. Sales et al. (Eds.): ER 2023 Workshops, LNCS 14319, pp. 284–293, 2023.
https://doi.org/10.1007/978-3-031-47112-4_27

models, which are an integral part of AI systems, are opaque, *i.e.*, the humans are not able to understand their rationale. This is specially true when we consider End-Users, who do not necessarily have technical knowledge about AI systems.

In this context, the eXplainable Artificial Intelligence (XAI) emerged to palliate these problems. Hence, XAI takes into consideration End-Users and includes different techniques to understand the models' outputs, trust in the learned rationale (rules) of the models, and build confidence on the decisions made by them. Furthermore, XAI is being used to help ML developers (henceforth referred to as ML experts) evaluate predictions and validate the results obtained by the models.

Nevertheless, there is no consensus on how to evaluate XAI, which limits confidence in XAI solutions. This lack of consensus is a difficult goal to achieve due to the complexity of (i) detecting the correct dimensions to include, (ii) including the metrics to evaluate them, (iii) determining the relevant actors (the ML Expert who is knowledgeable in ML, the XAI Expert who is knowledgeable in XAI and the Domain Expert who is knowledgeable in the domain of the case study) that should be involved in the evaluation, (iv) objectively verifying that every metric is achieved by defining thresholds for them, and (v) taking into consideration the domain knowledge and correctly adapting the evaluation approach to the specific domain by weighting the different quality metrics defined.

In order to tackle this problem, we present an approach aligned with Model Driven Developments (MDD) scenarios focused on evaluating XAI quality. In order to define it, we have meticulously studied the State-Of-The-Art (SOTA) of XAI. As a result, we have identified several dimensions that should be included in the XAI evaluation approach, providing a holistic view that considers both the AI and XAI aspects of the solution at hand. Consequently, the key aspect of our proposal is that it provides a holistic and objective view, measuring all relevant aspects of the AI solution developed and the XAI techniques for any domain. The main advantages of our proposal are that it (i) provides a set of quality metrics to evaluate the multiple relevant XAI and AI dimensions, (ii) provides a holistic assessment through the application of these quality metrics, (iii) identifies different target users and adapts the evaluation according to their profile, and (iv) enables the comparison across different XAI approaches, identifying which are ideal for each type of user and problem.

The rest of the paper is structured as follows. Section 2 presents an introduction about XAI and MDD. Section 3 presents the most relevant related work in the area. Section 4 describes our proposal, an MDD-aligned approach to evaluate XAI quality. Section 5 presents the case study of applying XAI in chess and presents the XAI evaluations for these applications. Finally, Sect. 6 outlines conclusions and sketches the future work to be done in this field.

2   Background

Once detected that AI systems need to be more transparent and interpretable, the key relevance of XAI in the current AI field has been exposed. In this way, the XAI field presents a clear distinction among different models in XAI. This dichotomy is widely accepted as a classification between (i) transparent models, which present an enough degree of interpretability, and (ii) opaque models, where we should apply post-hoc techniques to understand their decision-making processes. As argued in recent works (*e.g.*, [2]), these post-hoc techniques are categorized as: simplification, text, visual, local, feature relevance and example explanation techniques.

Moreover, the MDD field can provide crucial advantages to achieve goals in different areas, such as explanations in XAI. In Fig. 1, the main MDD elements are included, where the Computer Independent Model (CIM) is -manually, semi-automatically or automatically- derived into the Platform Independent Model (PIM) by applying Query/View/Transformation (QVT) rules [13]. The same process is applied to derive the PIM into the Platform Specific Model (PSM) and the PSM into the CODE layer. In this context, the MDD environment is useful to derive different abstraction levels in XAI, which can help to achieve explanations in AI scenarios.

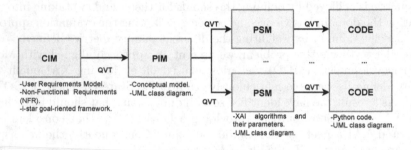

Fig. 1. Model Driven Development (MDD) framework.

Consequently, as presented in Fig. 1, it is possible to bridge the XAI and MDD fields. Specifically, we can achieve an MDD implementation for XAI through (i) applying the i-star goal-oriented framework [5] in the CIM layer to capture the User Requirements Model and Non-Functional Requirements (NFR), (ii) defining a Unified Modeling Language (UML) [11] class diagram that capture the key concepts in XAI, (iii) selecting the XAI algorithms and their parameters by modeling another UML class diagram, (iv) implementing them thanks to capture the python code by modeling a final UML class diagram, and (v) applying QVT rules to derive different layers from each other. However, although we present an MDD-aligned proposal, defining these models is out of our paper scope, which will be studied in future work (Sect. 6).

3 Related Work

In this section, we will discuss existing explainable techniques applied. Moreover, we will also discuss the problems, guidelines, opportunities and challenges of

XAI. Finally, once summarized these works, we motivate our proposal, which is aligned with the MDD field.

In [6], the authors developed a text explainable approximation for chess, called Rationale-Generating Algorithm (RGA), and RGA+, which includes domain knowledge. Finally, the authors make an essential question about the possibility that an AI system and its explanations associated can help humans to improve. Consequently, they were focused on improving human task-performance. In [14], Kadiyala *et al.* analysed different ML models by means of applying different XAI techniques for weather forecasting, that were focused on extracting feature relevance learnt. Furthermore, different works (*e.g.*, [2,8,17]) present XAI motivation, guidelines, opportunities and challenges. In these works, the paramount importance of creating a consensus on how to evaluate XAI is outlined as one of the most important research goal in the XAI and AI fields. Moreover, the authors emphasize that (i) it is possible by means of including different metrics and involving different actors, and (ii) this consensus should not depend on the specific case study, *i.e.*, the specific domain or ML model to explain. However, this consensus has not been achieved yet.

Summarizing, XAI presents a large amount of techniques and methods that are being exploited in their current state. Researchers are working on making XAI techniques to help ML Experts to evaluate models. Moreover, there are works where XAI is being evaluated in different ways. However, there is no consensus on how to evaluate XAI, which is one of the most important current AI and XAI goals. This consensus should be defined and should be able to be applied with independence of the specific case study. Furthermore, to completely and objectively evaluate XAI, this consensus should involve different actors, define different quality metrics, achieve a threshold for each of these metrics, and weight each metric depending on the specific case study.

Thus, we tackle this lack of consensus and present an MDD-aligned approach focused on evaluating XAI quality. This approach has been defined taking into account all above-presented key elements in a quantitative and objective manner. The great advantages of our proposal are that it (i) provides a set of quality metrics to evaluate the multiple relevant XAI and AI dimensions detected, (ii) provides a holistic evaluation thanks to these quality metrics, (iii) identifies different target users and adapts the evaluation according to their profile, and (iv) enables the comparison across XAI alternatives, identifying which are ideal for each type of user and problem.

4   An MDD-Aligned Approach to Evaluate XAI Quality

This section presents our proposal, a novel approach to evaluate the XAI quality, whose different steps will be also presented in this section. In the MDD context, our proposal has been designed to be aligned with: (i) the previously-described CIM, PIM, PSM and CODE layers, and (ii) the QVT rules that allow the different layers to be derived to each other. Moreover, thanks to the full implementation of our proposal, a continuous value between 0 (the worst quality of the explanation) and 1 (the best quality of the explanation) is finally obtained. Detailing the proposed approach, there are four main concepts that will be presented, as follows.

First, the different actors involved in the XAI and AI processes should be defined, who play different roles in the these processes. These four actors are: the Machine Learning (ML) Expert, who is specialized in the development of ML systems. Furthermore, the ML Expert is responsible for validating the results obtained by the ML model; the Domain Expert, who has expertise in the field where the AI system and its associated XAI solutions are applied. This contribution in very important due the huge knowledge in the application Domain; the eXplainable AI (XAI) Expert, responsible for the development and application of XAI methods, to make more transparent ML models; and End-Users, who consume ML models and can be ML Experts, XAI Experts, Domain Experts or none of them (e.g., in medicine, the patients and the doctors are stakeholders and possible recipients from XAI approaches applied).

Second, motivated by evaluating the detected XAI and AI dimensions, our proposal provides four different quality metrics. Specifically, the included metrics are: M_1 = Prediction Quality (PQ), which aims to verify whether the model on which explanations are based is reliable (we should continue the evaluation process) or not (the model should be improved before basing an explanation on it); with the aim of verifying the coherence between the model decisions-explanations relationship, it is necessary to achieve M_2 = Consistency (C). By evaluating this metric, we can conclude whether the explanations are correctly linked to the decisions made by the model; as the domain knowledge is a key factor in both AI and XAI processes, there is included M_3 = Domain Review (DR) to verify whether the Domain knowledge of the case study has been correctly included in the explainable approach; in order to asses the result of applying the explainable approach on End-Users, our approach takes into consideration the M_4 = Social Impact (SI), which is essential to verify whether the model and the explanations based on it are playing their role correctly in society. Due to space constraints, we can not detail how each M_i is calculated. However, it is important to clarify that (i) M_1 represents the ML traditional metrics results, (ii) M_2 aims to represent the three axioms for explanation Consistency [3,10] (Identity, Separability and Stability), (iii) M_3 evaluates the three Cs of interpretability [16] (Completeness, Compactness and Correctness) and can take advantage of the Cohen's kappa coefficient [4] and the Fleiss's kappa coefficient [7] depending of the number of Domain Experts involved, and (iv) M_4 can be evaluated by means of defining and achieving specific goals (1 if the goals have been achieved and 0 if the goals have not been achieved) or evaluating specific metrics (e.g., percentile) in values between 0 and 1.

Once defined the metrics and the actors, we can easily observe a visual approach of how both concepts are linked, which are presented in Fig. 2. In this Fig. 2, we can observe how our approach works. First, the ML Expert evaluates M_1 = PQ. Second, the XAI Expert, who has implemented the XAI technique, evaluates M_2 = C. Third, the Domain Expert evaluates the M_3 = DR. Fourth, the XAI Expert evaluates the M_4 = SI by applying XAI experiments on End-

Users. Furthermore, we can also observe how the MDD layers are included in our proposal. Specifically, the different user roles -and their goals- will be captured in the CIM layer, the PIM layer will capture the ML model, the most adequate XAI algorithms and their parameters will modeled in the PSM layer, and the final python code that implements them will be captured in the CODE layer.

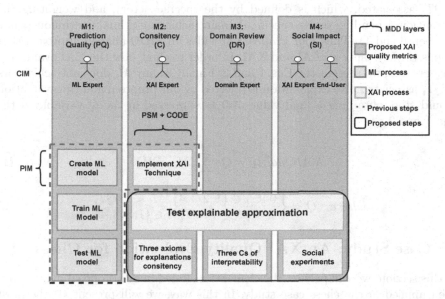

Fig. 2. Proposed workflow where the actors and metrics of our proposal are linked.

Third, the MQT are defined by the different actors included. Hence, the proposed quality metrics should achieve these MQT, whose values depend on the knowledge of the actors and their criterion in the specific case study, to correctly evaluate explainable approximations before reaching End-Users. Otherwise, our proposal will reject the explanation as the explanation and/or the model should be improved and re-evaluated. In this context, (i) the MQT_1 for the M_1 (Prediction Quality, PQ) is defined by the ML Expert, (ii) the MQT_2 for the M_2 = Consistency (C) is defined by the XAI Expert, (iii) the (ii) the MQT_2 for the M_2 = Domain Review (DR) is defined by Domain Expert, and (iv) the MQT_4 for the M_4 = Social Impact (SI) is defined by the XAI Expert. Finally, by following all above-described points, the different actors involved in the proposed approach, who make use of the proposed metrics, will be able to evaluate explainable approximations with independence of the specific case study.

Fourth, the final XAI value is calculated by the weighted average of the four proposed metrics. As some metrics may be more important than others in obtaining the XAI evaluation, these weights are defined depending on the domain. Thus, for each Metric M_i, it is necessary to define a DW_i. Then, the XAI Expert should completely understand the relevance of each metric M_i for the specific case study. Next, this expert should weight each metric M_i, where $\sum_{i=1}^{4} DW_i = 1$. Consequently, thanks to define each DW_i, the four proposed quality metrics will be correctly weighted and adapted in any domain.

Summarizing, our proposed approach, which evaluates the quality of XAI, is computed by involving its actors, computing its quality metrics (where each $M_i \in [0, 1]$), applying thresholds (where each M_i is filtered depending on its MQT_i associated, which is defined by the specific actor), and weighting the evaluation depending on its specific domain (achieving a final continuous value between 0 and 1). Thus, our proposal provides an MDD-aligned approach that scopes the key aspects of AI and XAI. In order to show how the XAI evaluation is calculated, we present the Eqs. 1 and 2. Finally, if any M_i does not achieve its MQT_i associated, the explanation is rejected by our proposal and the evaluation should stop, obtaining a final value of 0 as expressed in the Q variable in the Eqs. 1 and 2.

$$XAIQuality = Q \cdot \sum_{i=1}^{4} M_i \cdot DW_i, \tag{1}$$

$$where \quad Q = \begin{cases} 0 \ \text{if} \exists i \in \{1,2,3,4\}/M_i < MQT_i \\ 1 \ \text{if} M_i \geq MQT_i \forall i \in \{1,2,3,4\} \end{cases} \tag{2}$$

5   Case Study: An XAI Quality Evaluation for Chess

In this section, we will present the elements included in our proposal, which have been applied in the chess case study. In this way, we will present (i) the most relevant SOTA XAI approaches and (ii) our implemented experiments. However, due to space constraints, we will not detail every step applied.

In the chess domain, there are different actors involved. First, the ML Expert role will be developed by us, who have a lot of experience in the ML and AI fields. Second, as the first author of this paper has a huge knowledge in chess, this author can develop the Domain Expert role. Furthermore, as there is only one rater, it is not necessary to take into account Cohen's kappa coefficient or Fleiss' kappa coefficient, that were presented in the proposal (Sect. 4). Third, as we also have experience in XAI, we have developed the XAI Expert role. Fourth, different chess players have been developed the End-Users role.

Moreover, it is necessary to define the MQT to achieve each quality metric. First, the ML Expert has defined an MQT_1 of $\frac{11}{20} = 0.55$ for M_1. Second, the XAI Expert has defined an MQT_2 of $\frac{2}{3} \approx 0.67$ for M_2. Third, the Domain Expert has has defined an MQT_3 of $\frac{1}{2} = 0.5$ for M_3. Fourth, the XAI Expert has defined an MQT_4 of $\frac{1}{2} = 0.5$ for M_4.

Finally, playing the XAI role, we define the DW for the chess domain. Due to the chess domain complexity, we have emphasized the $M_2 = DR$ quality metric. Thus, we have defined a $DW_1 = 0.2$ for M_1, a $DW_2 = 0.2$ for M_2, a $DW_3 = 0.4$ for M_3 and a $DW_4 = 0.2$ for M_4.

In the following, we will present the different XAI approaches implemented in XAI and the evaluations calculated for each of them.

First, Devleena Das and Sonia Chernova presented a Rationale-Generating Algorithm (RGA) and a more complex version that includes domain knowledge (RGA+) [6]. In this work, the authors implemented visual and text explanations to help to understand why the chess AI model makes its decisions, aiming to improve the chess knowledge on End-Users thanks to these explanations. In more detail, they implemented the explanations based on interpreting an User Chess Interface (UCI) that uses one of the best chess engines. Second, in [9], the authors implemented a visual technique for different strategy games also focused on interpreting an UCI that applies a powerful chess engine. Their explanation aimed to help chess players, who played the End-Users role in their work, to improve their chess knowledge. This implemented explanation was called Specific and Relevant Feature Attribution (SARFA) and is achieved by means of applying saliency maps on chessboards, as is explained in their work. Third, we have implemented an explanation simplification technique to explain chess. In this way, we have created a less complex model. Next, we have applied SHAP [12] and LIME [15] based on this less complex model. Consequently, there are two techniques to support this simplification technique to explain chess. More specifically, there are two techniques applied in this simplification technique: 1) SHAP that is feature relevance explanation, and 2) LIME that is a local explanation.

In order to easily show the XAI evaluations obtained for these explanations, we should apply the above-defined DW for the results presented in Table 1. Thanks to this Table 1 and the Eqs. 1 and 2, we can calculate the final results for each explainable approximation, as follows.

Table 1. Evaluation of each explainable approximation.

XAI evaluation				
Explainable approximation	Prediction Quality	Consistency	Domain Review	Social Impact
Simp. SHAP	0	-	-	-
Simp. LIME	0	-	-	-
SARFA	0.6	0.67	0.5	0.72
RGA	0.6	0.67	0.5	0.5
RGA+	0.6	0.67	1	0.75
MQT	0.55	0.67	0.5	0.5
DW	0.2	0.2	0.4	0.2

On the one hand, the simplification techniques for both SHAP and LIME, which are based on a less complex model implemented in this work, have not achieved the threshold to PQ ($M_1 < MQT_1$). Then, these explanations have obtained a Q value of 0. Consequently, our approach rejects these explainable approximations and it is not necessary to calculate the followed proposed quality metrics for them (as is expressed as "-" in Table 1). On the other hand, SARFA,

RGA and RGA+ have achieved all proposed quality metrics, obtaining a Q value of 1. Consequently, regarding the values and the DW presented in Table 1, we can obtain the values for each M_i. More specifically, the XAI values that have been obtained are (i) $1 \cdot (0.6 \cdot 0.2 + 0.67 \cdot 0.2 + 0.5 \cdot 0.4 + 0.72 \cdot 0.2) = \frac{299}{500} = 0.598$ for SARFA; (ii) $1 \cdot (0.6 \cdot 0.2 + 0.67 \cdot 0.2 + 0.5 \cdot 0.4 + 0.5 \cdot 0.2) = \frac{277}{500} = 0.554$ for RGA; and (iii) $1 \cdot (0.6 \cdot 0.2 + 0.67 \cdot 0.2 + 1 \cdot 0.4 + 0.75 \cdot 0.2) = \frac{201}{250} = 0.804$ for RGA+. Summarizing, RGA+ has obtained the best value: $\frac{201}{250} = 0.804$.

6 Conclusions and Future Work

In this paper, we have identified the lack of consensus on how to evaluate XAI as one of the most important current research goals. Therefore, we have presented an approach focused on evaluating XAI quality, which can be implemented in an MDD environment. The main advantages of our proposal are that it (i) provides a set of quality metrics to evaluate the XAI and AI dimensions, (ii) provides a holistic evaluation thanks to these quality metrics, (iii) identifies the different target users in order to adapt the evaluation to them, and (iv) enables the comparison across XAI implementations.

In order to show the applicability of our proposal, we applied it to a widely-known case study: chess. In this complex case study, it was clearly presented that the explainable approximations based on making a less complex model worked worse than the explanations based on interpreting a chess engine by means of using an UCI. Specifically, the RGA+ presented in [6] obtained the best value.

Our plans for the immediate future are to provide a complete XAI framework and detail the different abstraction levels included in the MDD field for XAI. In more detail, we plan to develop the different models for the different abstraction layers (CIM, PIM, PSM and CODE) and the QVT rules that allow the different layers to be derived to each other.

Acknowledgments. This work has been co-funded by (i) the AETHER-UA project (PID2020-112540 RB-C43), (ii) the BALLADEER (PROMETEO/ 2021/ 088) project, and (iii) the Program for the Promotion of R+D+I (UAIND20-03B).

References

1. Amoroso, N., et al.: A roadmap towards breast cancer therapies supported by explainable artificial intelligence. Appl. Sci. **11**(11), 4881 (2021). https://doi.org/10.3390/app11114881
2. Arrieta, A.B., et al.: Explainable artificial intelligence (XAI): concepts, taxonomies, opportunities and challenges toward responsible AI. Inf. Fusion **58**, 82–115 (2020). https://doi.org/10.1016/j.inffus.2019.12.012
3. Carvalho, D.V., Pereira, E.M., Cardoso, J.S.: Machine learning interpretability: a survey on methods and metrics. Electronics **8**(8), 832 (2019). https://doi.org/10.3390/electronics8080832
4. Cohen, J.: A coefficient of agreement for nominal scales. Educ. Psychol. Meas. **20**(1), 37–46 (1960)

5. Dalpiaz, F., Franch, X., Horkoff, J.: istar 2.0 language guide. arXiv preprint arXiv:1605.07767 (2016)
6. Das, D., Chernova, S.: Leveraging rationales to improve human task performance. In: Proceedings of the 25th International Conference on Intelligent User Interfaces, pp. 510–518 (2020). https://doi.org/10.1145/3377325.3377512
7. Fleiss, J.L.: Measuring nominal scale agreement among many raters. Psychol. Bull. **76**(5), 378 (1971). https://psycnet.apa.org/doi/10.1037/h0031619
8. Guidotti, R., Monreale, A., Ruggieri, S., Turini, F., Giannotti, F., Pedreschi, D.: A survey of methods for explaining black box models. ACM Comput. Surv. **51**(5), 1–42 (2018). https://doi.org/10.1145/3236009
9. Gupta, P., et al.: Explain your move: understanding agent actions using specific and relevant feature attribution (2020). https://par.nsf.gov/biblio/10166401
10. Honegger, M.: Shedding light on black box machine learning algorithms: Development of an axiomatic framework to assess the quality of methods that explain individual predictions. arXiv preprint arXiv:1808.05054 (2018). https://doi.org/10.48550/arXiv.1808.05054
11. Jacobson, I.: The Unified Software Development Process. Pearson Education India (1999)
12. Lundberg, S.M., Lee, S.I.: A unified approach to interpreting model predictions, vol. 30 (2017)
13. (OMG), O.M.G.: 2nd revised submission: Mof 2.0 query/views/transformations. https://www.omg.org/cgi-bin/doc?ad/05-03-02. Accessed 08 Sept 2023
14. Prasanth Kadiyala, S., Woo, W.L.: Flood prediction and analysis on the relevance of features using explainable artificial intelligence. In: 2021 2nd Artificial Intelligence and Complex Systems Conference, pp. 1–6 (2021). https://doi.org/10.1145/3516529.3516530
15. Ribeiro, M.T., Singh, S., Guestrin, C.: "why should i trust you?" explaining the predictions of any classifier. In: Proceedings of the 22nd ACM SIGKDD International Conference on Knowledge Discovery and Data Mining, pp. 1135–1144 (2016). https://doi.org/10.1145/2939672.2939778
16. Silva, W., Fernandes, K., Cardoso, M.J., Cardoso, J.S.: Towards complementary explanations using deep neural networks. In: Stoyanov, D., et al. (eds.) MLCN/DLF/IMIMIC -2018. LNCS, vol. 11038, pp. 133–140. Springer, Cham (2018). https://doi.org/10.1007/978-3-030-02628-8_15
17. Zhang, Y., Weng, Y., Lund, J.: Applications of explainable artificial intelligence in diagnosis and surgery. Diagnostics **12**(2), 237 (2022). https://doi.org/10.3390/diagnostics12020237

SmartFood

SmartFood – 1st Workshop on Controlled Vocabularies and Data Platforms for Smart Food Systems

Renata Guizzardi[1], Catherine Faron[2], Filipi Miranda Soares[1,3],
and Gayane Sedrakyan[1]

1 University of Twente, The Netherlands
r.guizzardi@utwente.nl, g.sedrakyan@utwente.nl
2 Université Côte d'Azur, France
faron@i3s.unice.fr
3 University of São Paulo, Brazil
filipisoares@usp.br

The trajectory of future food production and consumption is currently being sculpted by an amalgamation of diverse technological advancements. These include, but are not limited to, big data, mobile technologies, robotics, remote-sensing services, virtual and augmented reality, distributed computing, the Internet of Things (IoT), adaptive systems, and Semantic Web technologies. This burgeoning field is variously termed as Smart Food Systems, Digital Agriculture, e-Agriculture, Agriculture 4.0, and Smart Agriculture. Despite its pivotal role in global sustenance, agriculture remains one of the world's least digitized sectors. However, it stands to gain immensely from digitization, underscoring the importance of research in this domain.

Semantic Web applications have found their niche in various facets of agriculture and smart food systems. They are instrumental in ensuring data interoperability, sharing, and reuse. Controlled vocabularies, which are systematic arrangements of concepts curated by specific communities, cater to their unique data description needs. By formalizing these vocabularies in standard Semantic Web languages like RDF and OWL, we not only facilitate their reuse by other communities but also enable machines to conduct more precise data analyses.

The "SmartFood" theme resonated with a diverse audience, including researchers, industry professionals, farmers, and consumer advocacy groups. These stakeholders share a common vision: the belief that Semantic technologies, epitomized by controlled vocabularies, ontologies, and data platforms, are central to devising solutions for this sector. Additionally, sustainable business models in the realm of data-driven agri-food were a focal point of discussion in this forum.

For this workshop, the paper review process was executed in a single-blind fashion. Each submission underwent a stringent review by at least three and at most five experts in the field. Out of the five papers submitted, three met the rigorous standards set by the committee and were subsequently approved for presentation.

Acknowledgments. FMS would like to thank São Paulo Research Foundation (FAPESP) for the research grants (process numbers 21/15125-0 and 22/08385-8).

SmartFood Organization

Workshop Chairs

Renata Guizzardi — University of Twente, The Netherlands
Catherine Faron — Université Côte d'Azur, France
Filipi Miranda Soares — University of São Paulo, Brazil;
University of Twente, The Netherlands
Gayane Sedrakyan — University of Twente, The Netherlands

Program Committee

Anand Gavai — University of Twente; Wageningen
University & Research, The Netherlands
Alexandre Delbem — University of São Paulo, Brazil
Antonio M. Saraiva — University of São Paulo, Brazil
Bruno Albertini — University of São Paulo, Brazil
Clément Jonquet — French National Research Institute for
Agriculture, Food and Environment,
France
Cynthia Parr — United States Department of Agriculture,
USA
Debora Drucker — Embrapa Digital Agriculture, Brazil
Dilvan Moreira — University of São Paulo, Brazil
Fernando Corrêa — University of São Paulo, Brazil
Kelly Rosa Braghetto — University of São Paulo, Brazil
Nadia Yacoubi Ayadi — National Institute for Research in Digital
Science and Technology, France
Patrice Buche — French National Institute for Agriculture,
Food and Environment, France
Sérgio M. S. da Cruz — Federal Rural University of Rio de Janeiro;
Federal University of Rio de Janeiro,
Brazil
Valeria Pesce — Food and Agriculture Organization, Italy

Unveiling Knowledge Organization Systems' Artifacts for Digital Agriculture with Lexical Network Analysis

Filipi Miranda Soares[1,2]([✉]) [iD], Ivan Bergier[3] [iD], Maria Carolina Coradini[4] [iD],
Ana Paula Lüdtke Ferreira[5] [iD], Milena Ambrosio Telles[3] [iD],
Benildes Coura Moreira dos Santos Maculan[6] [iD],
Maria de Cléofas Faggion Alencar[3] [iD], Victor Paulo Marques Simão[3] [iD],
Bibiana Teixeira de Almeida[3] [iD], Debora Pignatari Drucker[3] [iD],
Marcia dos Santos Machado Vieira[7] [iD], and Sérgio Manuel Serra da Cruz[7,8] [iD]

[1] University of São Paulo, São Paulo, SP, Brazil
`filipisoares@usp.br`
[2] University of Twente, Enschede, Netherlands
`f.mirandasoares@utwente.nl`
[3] Brazilian Agricultural Research Corporation, Brasilia, Brazil
{`ivan.bergier,milena.telles,cleofas.alencar,victor.simao,`
`bibiana.almeida,debora.drucker`}`@embrapa.br`
[4] Brazilian Institute of Information in Science and Technology, Brasilia, Brazil
[5] Federal University of Pampa, Bagé, RS, Brazil
`anaferreira@unipampa.edu.br`
[6] Federal University of Minas Gerais, Belo Horizonte, MG, Brazil
[7] Federal University of Rio de Janeiro, Cidade Universitária, RJ, Brazil
`marcia@letras.ufrj.br`
[8] Federal Rural University of Rio de Janeiro, Seropédica, RJ, Brazil
`serra@ufrrj.br`

Abstract. This article presents a bibliometric and terminological study of a corpus composed of abstracts and titles of 278 articles retrieved by a review protocol planned for surveying initiatives on building artifacts for modeling knowledge related to agricultural production systems. The original corpus comprised a 53,379-word linguistic extract filtered to 111 interconnected major terminologies by combining AntConc and VOSViewer tools. The reduced data were imported into the Gephi tool for analysis of lexical network graphs. Emergent clusters and their central terms underscore the thematic areas that prominently shape the landscape of agricultural Knowledge Organization Systems (KOS) and highlight the interplay between technological advancements, semantic enrichment, and domain-specific challenges. Our analysis of term occurrences and clusters contributes to a broader understanding of these concepts, inferring their significance, roles, and interconnections within the agricultural landscape. It also sheds light on the roles played by KOS in Digital Agriculture.

Keywords: Corpus Linguistics · KOS · Semantic artifacts ·
Ontologies · Thesaurus · Metadata · Knowledge graph

© The Author(s), under exclusive license to Springer Nature Switzerland AG 2023
T. P. Sales et al. (Eds.): ER 2023 Workshops, LNCS 14319, pp. 299–311, 2023.
https://doi.org/10.1007/978-3-031-47112-4_28

1   Introduction

Agricultural production sustains life on Earth and, as human population increases and climate changes, there is a growing effort to make productive systems more resilient and sustainable, in order to assure food security and nutrition to all. At the same time, agriculture production systems, agricultural sciences, and digital agriculture technologies tightly bound with computer science are generating a growing amount of valuable data that can drive decision-making toward a sustainable future.

However, to make sense of the data deluge in agriculture, it is imperative to adequately model agricultural knowledge using Knowledge Organization Systems (KOS) [20,22]. KOS encompass several types of artifacts with different vocabulary control (terminology) levels that seek to minimize the ambiguities of natural language. KOS represent the knowledge of a domain (scientific, educational, professional) based on an agreement accepted among peers. In general, KOS are composed of a set of terms representing concepts, delimited by their meanings and linked by semantic relationships. Each type of KOS support a specific function, although there may be some overlap of functions between them. Computationally, KOS can support the indexing of document resources, search (as an extension of semantic search), and information retrieval (as a browsing mechanism).

There are several efforts to represent agricultural knowledge, such as AGROVOC Multilingual Thesaurus (https://agrovoc.fao.org/browse/agrovoc/en/), NAL Agricultural Thesaurus (NALT) (https://data.nal.usda.gov/datas et/nal-agricultural-thesaurus-and-glossary), Chinese Agricultural Thesaurus (CAT) (https://bartoc.org/en/node/18606), CAB Thesaurus (Centre for Agricultural Bioscience International (https://www.cabi.org/cabithesaurus/) and the ones aggregated by Agroportal (https://agroportal.lirmm.fr/). Finding and mapping existing semantic artifacts is essential to understand their structure and representation models to better apply such resources to agricultural systems and products. Furthermore, within the scope of the GO FAIR initiative (https://www.go-fair.org/), the regional office in Brazil supports the implementation of the FAIR principles (Findable, Accessible, Interoperable and Reusable) [39]. In this context, the GO FAIR Agro Brazil network convened experts in the Ontology Working Group (WG) [15].

The WG aims to explore and propose methods and standards for constructing, adapting, or incorporating existing ontologies into data-centric agronomic systems with a tropical agriculture focus. The primary emphasis is on achieving improved semantic data descriptions and enhancing software interoperability.

In this paper, we describe the WG's efforts to map existing semantic artifacts for the Agriculture field, in whichever form. Particularly, we describe a bibliometric and terminological analysis of the articles' abstracts found with a literature scoping review protocol.

The next sections are structured as follows: Sect. 2 analyses related literature review initiatives concerning the use of technology on Digital Agriculture, focusing on Semantic Web models for Knowledge management. Section 3 provides information on the literature review protocol and bibliometric analysis strategies. Section 4 discusses the main findings. Final remarks are presented in Sect. 5.

2 Related Work

The use of Information and Communication Technologies (ICTs) in agricultural and livestock systems has been the subject of several reviews in the literature. For example, [16] provides a comprehensive review of semantic web technologies for agriculture, finding 13 resources by the end of 2018. It is the only article directly related to our work focusing on the type of artifacts.

The literature reviews highlight that reliable and secure data repositories, metadata, semantic analysis, and systems interoperability are essential for the objectives of Digital Agriculture. Another article concerned with the FAIR principles in Agriculture is [5], which discusses precision livestock farming technologies in relation to their use of public data, open standards, interoperability, and tools. The authors did not follow any literature review protocol. Several initiatives towards standardization of metadata are presented, including ISO standards for agriculture machinery. Ontologies appear as a means to give semantics for gathered data, without discussing specific research/initiatives in that direction. It is mainly concerned with decision support tools aiming at smart livestock farming.

The other papers we are aware of do not focus on semantic web technologies *per se*. However, such papers lay on how technology is used in agricultural settings, and on what are the obstacles to achieving full. Particularly, [12] uses a systematic literature review protocol to uncover term definitions, technologies, barriers, advantages, and disadvantages of Digital Agriculture. [38] uses a systematic literature review protocol to survey the main features and obstacles to the use of Farm Management Information Systems. [1] reviews the literature related to crop farming regarding the extent of digital technology adoption in the context of service type, technology readiness level, and farm type. [30] surveys the technologies used in smart livestock farming, focusing on biometric and biological sensors, big data analytics, data science, machine learning, and blockchain.

The need for compatibility between technological artifacts is considered a barrier to Digital Agriculture to develop. However, no discussion about conceptual models or data integration was issued in [12]. [4] focus their review on research activities related to smart farming within the EU, with sensing techniques, robotics, IoT, and decision support systems being the most frequent themes. [34] uses a literature review protocol to analyze 378 articles concerning technological advances in smart farming. It states that the most agriculture work is still performed manually by the farmers, mainly due to the high maintenance and deployment cost. Finally, [31] surveys articles related to data-driven decision support systems in livestock farming.

Despite ongoing efforts and existing results to achieving the objectives of Digital Agriculture, there remains the need in improving standardized data/metadata formats [1,30,34,38], data and system integration [4,31,38], data security [31,34], language regionalisms [38], reusable models [31], availability [31,38], semantic analyzes [4,30], lack of provenance metadata [10] and system interoperability [1,4,30,34,38].

3   Material and Methods

We used the literature scope review described in [2] to identify and analyze different types of artifacts produced to structure KOS and terminologies in Agriculture. Scoping methods aim to identify the nature and extent of research evidence, including ongoing research. The working method was collaborative, supported by the cloud-based software Parsifal (https://parsif.al/).

We are mainly interested in knowing how these artifacts were produced, that is, the methods, their use, their main technological characteristics and for what purpose they were developed. These are the reasons why we conducted our review rather than searching for artifact repositories, which do not directly provide this information.

The literature review protocol was collaboratively developed and refined from February to July 2023 and establishes the following research questions:

1. What are the artifacts that structure knowledge and terminology in the field of agriculture and are available for use?
2. What are the main characteristics of the artifacts found and analyzed?

The search strings were derived from a set of terms considered relevant. Table 1 presents the search terms and related terms used to search the articles.

Table 1. Search terms and related terms

Keyword	Related terms
Agriculture	Agricultural, Agronomy, Livestock
Literature review	Literature mapping, Scope review, Survey
Ontology	Conceptual model, Controlled vocabulary, Data description, Data integration, Dictionary, Glossary, Metadata, Thesaurus

The search strings development aimed to organize the authors' work. Each author was responsible for applying one of the strings to the chosen search sources. Several strings had their format modified to meet the selected bibliographic databases requirements regarding syntax or the number of operators' restrictions. The list of generic search strings is as follows:

- ((Ontolog* OR Taxonom*) AND (Agricult* OR Agronom* OR Livestock))
- ((Ontolog*) AND (Agricult* OR Agronom* OR Livestock))
- ((Metadata OR Data descriptor*) AND (Agricult* OR Agronom* OR Livestock))

- ((Thesaurus OR Data descriptor* OR controlled vocabular*) AND (Agricult* OR Agronom* OR Livestock))
- ((Glossar* OR vocabular* OR dictionar*) AND (Agricult* OR Agronom* OR Livestock))
- ((Conceptual map*) AND (Agricult* OR Agronom* OR Livestock))
- ((Subject headings list*) AND (Agricult* OR Agronom* OR Livestock))
- ((Data model*) AND (Agricult* OR Agronom* OR Livestock))
- ((Conceptual scheme for agro thesaurus integration) AND (Agricult* OR Agronom* OR Livestock))

The digital bibliographic databases used as search sources were: Dimensions (https://www.dimensions.ai/), ISI Web of Science (http://www.isiknowledge.com) and Scopus (http://www.scopus.com). We chose these bibliographic databases because of their interdisciplinary character, mirroring the diverse nature of the domain we examined.

The selection criteria consider both inclusion and exclusion sub-criteria. The inclusion sub-criteria define what are the obligatory subjects of an article to be included in the analysis, while the exclusion sub-criteria disqualify prospective subjects for an article not to be included, provided that it meets at least one of them. Table 2 shows the selection criteria defined within the protocol.

Table 2. Selection (inclusion and exclusion) criteria

Selection Criterion	
Inclusion Criterion	Exclusion Criterion
. Articles written in PT, EN, ES, or FR	.Strictly theoretical review articles
.Articles published between 2002 and 2022	.Articles that mention polysemous terms in other domains of knowledge
.Articles that answer at least one of the research questions	.Articles without full-text digital access
.Articles that present artifacts that structure knowledge and terminology of some domain of agriculture	.Duplicated texts

The data extraction form organizes the articles' reading objectives, focusing on the analysis' main aspects. We have defined a set of information to be extracted, including artifact type, building methods, modeling language, serialization/exchange formats, used license, specific application domain within Agriculture, building tools, used objectives, semantic formalization degree, as well as a qualitative analysis to capture aspects not presented in the form. This data will be used in future analyses since the myriad of answer possibilities showed us that a prior, automatized analysis would be beneficial, so we conducted first a bibliometric and terminological analysis, as described below.

The lexical network metrics for graph analysis were computed using a linguistic corpus [33] composed of abstracts and titles of the selected articles, totaling 53,379 words. In order to extract the lexical-conceptual knowledge [13], i.e., the terms and their semantic meaning and relations, to have a first analysis of the distribution of how the searched artifacts appear in the retrieved texts, the corpus was processed using AntConc (https://www.laurenceanthony.net/software/antconc/) with the minimum frequency of occurrence, i.e., all terms that occur at least one time were collected, and a preliminary inspection of full terms' frequencies and occurrences was obtained. Further, VOSviewer (https://www.vosviewer.com/) was used for the visualization and analysis of the occurrence and co-occurrence of concepts.

In Antconc, to optimize the results and reduce the wordlists, a stoplist was used and a list of terms composed of 1 to 5 words was generated to verify the main terms mentioned in the literature. Each term was searched in its singular and plural forms. The frequencies of the singular and plural forms of the same term were added, considering each pair as a synset, reaching the absolute frequency of each term.

Following the inspection of the file, the default settings of the VOSViewer application were then employed and 6,605 terms were first identified and further arrived in 111 network nodes. The data was then exported to the latest version of Gephi (https://gephi.org/) via a .gml file, which enabled calculating order degree distributions, betweenness centrality [8], and clustering nodes according to modularity classes [25].

Graphs were produced using the ForceAtlas2 algorithm [23]. Node sizes are related to order degree, whereas link sizes are proportional to the strength between nodes' connections. Colors are associated with classes created by means of modularity, while node label size is associated with betweenness centrality, which is responsible for shortening the path between nodes.

4   Results and Discussion

This first approach to analyzing the corpus using the absolute frequency of the searched artifacts highlighted ontological and thesauri artifacts, while the analysis of terms relations by graphs analysis provided a comprehensive portrait of semantic relations around the possible concepts. The absolute frequencies of the terms of interest are: *ontology* = 956; *agriculture* = 326; *thesaurus* = 156; *controlled vocabulary* = 28; *livestock* = 24; *taxonomy* = 21; *data model* = 17; *agronomy* = 12; *dictionary* = 9; *subject headings* = 7; *conceptual map* = 2; *glossary* = 1; *conceptual scheme* = 0; and *data descriptor* = 0.

This first approach to analyzing the corpus using the absolute frequency of the searched artifacts highlighted ontological and thesauri artifacts, while the analysis of terms relations by graphs analysis provided a comprehensive portrait of semantic relations around the possible concepts.

The graph with the 111 terms selected by the method is displayed in Fig. 1. It evidences a greater degree of centrality of terms *vocabulary, analysis, farmer,* and

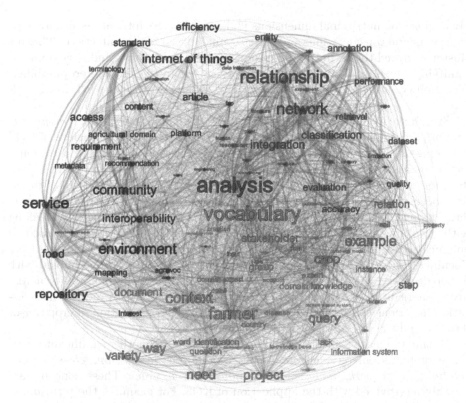

Fig. 1. Network graph of selected terms colored by modularity, nodes sized by order degree and labels sized by betweenness centrality. Link thickness is related to the strength between two adjacent nodes.

relationship, while important terms responsible for shortening the path between nodes are *network, context, service*, and *environment*.

Node distributions segregated by clusters are illustrated in Fig. 2. The four networks were filtered according with the attribute modularity [25]. Within the blue cluster highlighted in Fig. 2, in the first level, notable recurring terms include *environment* and *service*. That observation likely reflects the increasing utilization of these terms within the context of environmentally-oriented KOS in Agriculture – e.g., [29] –, and the extensive application of KOS in agricultural web services, such as in [19]. The convergence of these terms signaled a thematic intersection that resonates with the imperative of sustainable agricultural practices in the face of environmental challenges.

Within the second tier, there is a notable resurgence of key terms, which encompass *food, interoperability, metadata, repository, community, content, access*, and *Internet of Things (IoT)*. The term *food* emerges with heightened frequency as a pivotal domain of application within KOS, extending its influence across a spectrum of domains. This encompasses its relevance to food supply chains [26], nutritional dimensions [32], as well as its intricate interplay with environmental considerations and the sustainability of food production [29]. This cluster mirrored the technical dimensions of KOS application and development, signifying the pivotal role that KOS play in enhancing data interoperability, accessibility, and the utilization of emerging technology like the IoT in agriculture.

Still in the second tier, the *interoperability, metadata, repository, community, content, access,* and *IoT* concepts pivot towards the technical dimensions of KOS development and application. The strength of those terms might be a response to the increasing environmental demand for interoperable access to standardized data for sustainable compliance in supply chains [24]. Besides, the use of metadata becomes a recurrent strategy for achieving seamless interoperability among datasets and repositories, as demonstrated in [17]. Moreover, metadata, along with ontologies and other forms of KOS, serve as frequent tools harnessed to enhance accessibility to data content, as illustrated by the insights gleaned from Bechini's study [6]. IoT has demonstrated extensive utility when coupled with ontologies, serving as a catalyst for bolstering semantic interoperability across diverse agricultural domains. This symbiotic integration finds practical manifestation in various contexts, including the descriptive encapsulation of crop sensor data [3] and the dynamics of pest management [9].

Within the scarlet-hued cluster illustrated in Fig. 2, another significant recurrence of terms emerges, encompassing *annotation, integration, classification, relationship, network, analysis, performance,* and *accuracy*. These concepts are notably intertwined with the application of KOS. For example, the term *annotation* predominantly pertains to the semantic annotation of data, involving the utilization of KOS like the crop ontology and AGROVOC for annotating data resources. This practice is exemplified in studies such as [7,18,35]. Data annotation is a commonly utilized technique aimed at facilitating the harmonization of data from diverse sources, with the ultimate goal of achieving data *integration*, such as in [14,27,37]. The study illuminated their roles in semantic annotation, data harmonization, and systematic categorization, illustrating the diverse ways KOS support information organization and retrieval in agriculture.

KOS are also extensively employed in the systematic categorization of objects based on established canonical knowledge within a specific domain. The utilization of ontologies and taxonomies is instrumental in formulating rules, often in the form of axioms, for precise object classification. Noteworthy examples include the utilization of an ontology for soil classification [21], as well as the application of ontologies for the classification of agricultural products [28].

The term *relationship* frequently emerges to delineate the modeling of interconnections between concepts within KOS. Networks find diverse applications, ranging from ontology networks (which also refers to the relationship between concepts in ontologies), exemplified by [9], to sensor networks leveraging ontologies, as demonstrated by [36].

The remaining detached terms within the cluster, namely *analysis, performance,* and *accuracy*, possess notably comprehensive and wide-ranging connota-

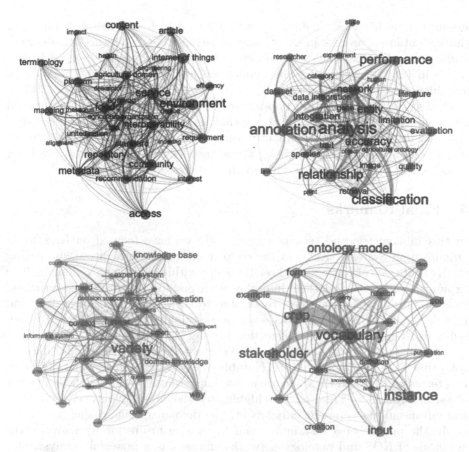

Fig. 2. Network graphs of selected terms clusterized by modularity. Measurements of centrality are recalculated for each cluster. Link thickness is related to the strength between two adjacent nodes. (Color figure online)

tions. These terms are prominently utilized as processes to scrutinize the application of KOS in specific instances or to assess the developmental trajectory of the KOS itself.

The yellowish cluster in Fig. 2 is about the application to farm data and *farmers'* needs and the related challenges of *variety* and *context*-specificity (context-specific *domain knowledge*) for supporting farmers' *decision making* (*decision support systems, decision making, expert system*). This cluster reveals needs and challenges for farmers, and therefore it does not even include technical concepts related to KOS.

The pink cluster, the last one in our analysis, draws attention to specific terms such as *vocabulary, crop, stakeholder, instance,* and *ontology model*. Notably, the strong connections between the first three terms hint at the significance of establishing controlled vocabularies for crops, especially concerning stakeholders within the agricultural domain. This interpretation is further supported by satellite nodes like *class*, *definition*, and *relation*, which are intricately linked to the instantiation process in ontological modeling [11]. In essence, this cluster underscores the crucial role of standardized terminology and ontological frameworks in facilitating effective communication and knowledge dissemination in agriculture.

Moreover, the pink cluster appears to reflect the landscape of international research, where ontologies and conceptual models enjoy substantial recognition and utilization. It predominantly centers around crops and soil as primary domains of application, highlighting the acknowledged importance of involving stakeholders and gaining their support in these research endeavors.

5   Final Remarks

In this bibliometric and terminological study we have focused on Knowledge Organization Systems (KOS) in the context of Digital Agriculture, unraveling the intricate web of concepts and relationships within this dynamic domain. The exploration of KOS and their interplay with ontology networks has provided insights into how these systems contribute to the advancement of agricultural knowledge, decision-making processes, and sustainable practices. Also, it would guide us to improve our data extraction form in the literature review ahead.

Our analysis of term occurrences and clusters shed light on the multifaceted roles that KOS play in agriculture. Notably, the emergent clusters and their central terms underscored the thematic areas that prominently shape the landscape of agricultural KOS. These clusters highlighted the interplay between technological advancements, semantic enrichment, and domain-specific challenges.

In the pursuit of a sustainable and innovative future for Agriculture, the synthesis of KOS and ontology networks emerges as a powerful catalyst. Our study contributes to a broader understanding of these concepts, elucidating their significance, roles, and interconnections within the agricultural landscape.

As a prospect for future research, the continuation of a systematic review of the literature is currently underway to delve deeply into diverse applications and developmental aspects of KOS within the digital agricultural domain, and their level of compliance with FAIR principles. Thus, the final results of the study should support the recommendation of semantic artifacts (KOS) by the GO FAIR Agro Brazil network to Brazilian institutions that maintain agricultural information systems to enhance the dissemination and use of the knowledge generated and managed by them. We also expect the insights gleaned from the future research can guide researchers, practitioners, and policymakers toward digital data-driven decision making based on KOS for a more transparent, efficient, sustainable and resilient agricultural ecosystem.

Acknowledgments. This study was financed in part by the Coordenação de Aperfeiçoamento de Pessoal de Nível Superior - Brasil (CAPES) - Finance Code 001. FMS would like to thank São Paulo Research Foundation (FAPESP) for the research grants (process numbers 21/15125-0 and 22/08385-8). SMSC would like to thank Brazilian National Council for Scientific and Technological Development (CNPq) for the research grants (process numbers 400044/2023-4 and 306115/2021-2). MSMV thanks CNPq and

FAPERJ (Fundação Carlos Chagas Filho de Amparo à Pesquisa do Estado do Rio de Janeiro/Carlos Chagas Filho Foundation for Research Support in the State of Rio de Janeiro) for the research support (409043/2021-4, 312423/2022-5, E-26/201.209/2022 (273339)).

References

1. Abbasi, R., Martinez, P., Ahmad, R.: The digitization of agricultural industry - a systematic literature review on agriculture 4.0. Smart Agric. Technol. **2**, 100042 (2022). https://doi.org/10.1016/j.atech.2022.100042, https://www.sciencedirect.com/science/article/pii/S2772375522000090
2. Arksey, H., O'Malley, L.: Scoping studies: towards a methodological framework. Int. J. Social Res. Methodol. **8**(1), 19–32 (2005)
3. Aydin, S., Aydin, M.N.: Semantic and syntactic interoperability for agricultural open-data platforms in the context of IoT using crop-specific trait ontologies. Appl. Sci. **10**(13) (2020). https://doi.org/10.3390/app10134460,https://www.mdpi.com/2076-3417/10/13/4460
4. Bacco, M., Barsocchi, P., Ferro, E., Gotta, A., Ruggeri, M.: The digitisation of agriculture: a survey of research activities on smart farming. Array **3–4**, 100009 (2019). https://doi.org/10.1016/j.array.2019.100009, https://www.sciencedirect.com/science/article/pii/S2590005619300098
5. Bahlo, C., Dahlhaus, P., Thompson, H., Trotter, M.: The role of interoperable data standards in precision livestock farming in extensive livestock systems: a review. Comput. Electron. Agric. **156**, 459–466 (2019). https://doi.org/10.1016/j.compag.2018.12.007, https://www.sciencedirect.com/science/article/pii/S0168169918312699
6. Bechini, L., et al.; Improving access to research outcomes for innovation in agriculture and forestry: the VALERIE project. Italian J. Agron. **12**(2) (2017). https://doi.org/10.4081/ija.2016.756, https://www.agronomy.it/index.php/agro/article/view/756
7. Beneventano, D., Bergamaschi, S., Sorrentino, S., Vincini, M., Benedetti, F.: Semantic annotation of the CEREALAB database by the AGROVOC linked dataset. Ecol. Inf. **26**, 119–126 (2015). https://doi.org/10.1016/j.ecoinf.2014.07.002, https://www.sciencedirect.com/science/article/pii/S1574954114000843. Information and Decision Support Systems for Agriculture and Environment
8. Brandes, U.: A faster algorithm for betweenness centrality. J. Math. Sociol. **25**(2), 163–177 (2001)
9. Chougule, A., Jha, V.K., Mukhopadhyay, D.: Using IoT for integrated pest management. In: 2016 International Conference on Internet of Things and Applications (IOTA), pp. 17–22. IEEE (2016). https://doi.org/10.1109/IOTA.2016.7562688, http://ieeexplore.ieee.org/document/7562688/
10. da Cruz, S.M.S., et al.: Data provenance in agriculture. In: Belhajjame, K., Gehani, A., Alper, P. (eds.) IPAW 2018. LNCS, vol. 11017, pp. 257–261. Springer, Cham (2018). https://doi.org/10.1007/978-3-319-98379-0_31
11. Cumpa, J.: A naturalist ontology of instantiation. Ratio **31**(2), 155–164 (2018)
12. da Silveira, F., Lermen, F.H., Amaral, F.G.: An overview of agriculture 4.0 development: systematic review of descriptions, technologies, barriers, advantages, and disadvantages. Comput. Electron. Agric. **189**, 106405 (2021). https://doi.org/10.1016/j.compag.2021.106405, https://www.sciencedirect.com/science/article/pii/S0168169921004221

13. Di Felippo, A., Almeida, G.M.B.: Uma metodologia para o desenvolvimento de wordnets terminológicas em português do Brasil. Tradterm **16**, 365–395 (2010). https://doi.org/10.11606/issn.2317-9511.tradterm.2010.46325, https://www.revistas.usp.br/tradterm/article/view/46325

14. Dooley, D.M., et al.: FoodOn: a harmonized food ontology to increase global food traceability, quality control and data integration. NPJ Sci. Food **2**(1), 23 (2018). https://doi.org/10.1038/s41538-018-0032-6, https://www.nature.com/articles/s41538-018-0032-6

15. Drucker, D., et al.: Implantação da rede temática GO FAIR agro brasil: Primeiros passos. In: Anais do XIII Congresso Brasileiro de Agroinformática, pp. 164–171. SBC, Porto Alegre, RS, Brasil (2021). https://doi.org/10.5753/sbiagro.2021.18387, https://sol.sbc.org.br/index.php/sbiagro/article/view/18387

16. Drury, B., Fernandes, R., Moura, M.F., de Andrade Lopes, A.: A survey of semantic web technology for agriculture. Inf. Process. Agric. **6**(4), 487–501 (2019). https://doi.org/10.1016/j.inpa.2019.02.001, https://www.sciencedirect.com/science/article/pii/S2214317318302580

17. Yeumo, E.D., et al.: Developing data interoperability using standards: a wheat community use case. F1000Research **6**(1843), 10 (2017). https://f1000research.com/articles/6-1843/v2

18. El-Beltagy, S.R., Hazman, M., Rafea, A.: Ontology based annotation of text segments. In: Proceedings of the 2007 ACM Symposium on Applied Computing, p. 1362–1367. ACM, Seoul Korea (2007). https://doi.org/10.1145/1244002.1244296

19. Fileto, R., Liu, L., Pu, C., Assad, E.D., Medeiros, C.B.: Poesia: an ontological workflow approach for composing web services in agriculture. VLDB J. Int. J. Very Large Data Bases **12**(4), 352–367 (2003). https://doi.org/10.1007/s00778-003-0103-3

20. Gnoli, C.: Knowledge Organization Systems (KOSs), p. 71–86. Facet (2020). https://doi.org/10.29085/9781783304677.004

21. Helfer, G.A., Costa, A.B.D., Bavaresco, R.S., Barbosa, J.L.V.: Tellus-onto: uma ontologia para classificação e inferência de solos na agricultura de precisão: tellus-onto: an ontology for soil classification and inference in precision agriculture. In: XVII Brazilian Symposium on Information Systems, pp. 1–7. ACM, Uberlândia Brazil (2021). https://doi.org/10.1145/3466933.3466946, https://dl.acm.org/doi/10.1145/3466933.3466946

22. Hodge, G.: Systems of knowledge organization for digital libraries: beyond traditional authorities files. Technical report, Council on Library and Information Resources, Washington, DC (2000)

23. Jacomy, M., Venturini, T., Heymann, S., Bastian, M.: ForceAtlas2: a continuous graph layout algorithm for handy network visualization designed for the Gephi software. PLoS ONE **9**(6), e98679 (2014)

24. Khan, A.A., Abonyi, J.: Information sharing in supply chains - interoperability in an era of circular economy. Clean. Logist. Supply Chain **5**, 100074 (2022)

25. Lambiotte, R., Delvenne, J.C., Barahona, M.: Laplacian dynamics and multiscale modular structure in networks. arXiv 1 (2008)

26. Letia, I.A., Groza, A.: Developing Hazard Ontology for supporting HACCP systems in food supply chains. In: IEEE 8th International Symposium on Intelligent Systems and Informatics, pp. 57–62 (2010). https://doi.org/10.1109/SISY.2010.5647189

27. Liang, A.C., Salokhe, G., Sini, M., Keizer, J.: Towards an infrastructure for semantic applications: Methodologies for semantic integration of heterogeneous resources. Cataloging Classif. Q. **43**(3–4), 161–189 (2007). https://doi.org/10.1300/J104v43n03_09,http://www.tandfonline.com/doi/abs/10.1300/J104v43n03_09

28. Liu, X., Duan, X., Zhang, H.: Application of ontology in classification of agricultural information. In: 2012 IEEE Symposium on Robotics and Applications (ISRA), pp. 451–454. IEEE (2012)

29. Musker, R., et al.: Towards designing an ontology encompassing the environment-agriculture-food-diet-health knowledge spectrum for food system sustainability and resilience. In: ICBO/BioCreative (2016). https://api.semanticscholar.org/CorpusID:2562803

30. Neethirajan, S., Kemp, B.: Digital livestock farming. Sens. Bio-Sens. Res. **32**, 100408 (2021). https://doi.org/10.1016/j.sbsr.2021.100408, https://www.sciencedirect.com/science/article/pii/S2214180421000131

31. Niloofar, P., et al.: Data-driven decision support in livestock farming for improved animal health, welfare and greenhouse gas emissions: Overview and challenges. Comput.. Electron. Agric. **190**, 106406 (2021). https://doi.org/10.1016/j.compag.2021.106406, https://www.sciencedirect.com/science/article/pii/S0168169921004233

32. Rezayi, S., et al.: Agribert: knowledge-infused agricultural language models for matching food and nutrition. In: Proceedings of the Thirty-First International Joint Conference on Artificial Intelligence, pp. 5150–5156. International Joint Conferences on Artificial Intelligence Organization, Vienna, Austria (2022). https://doi.org/10.24963/ijcai.2022/715, https://www.ijcai.org/proceedings/2022/715

33. Sardinha, T.B.: Linguística de Corpus. Manole, São Paulo (2004)

34. Sharma, V., Tripathi, A.K., Mittal, H.: Technological revolutions in smart farming: current trends, challenges & future directions. Comput. Electron. Agric. **201**, 107217 (2022). https://doi.org/10.1016/j.compag.2022.107217, https://www.sciencedirect.com/science/article/pii/S0168169922005324

35. Shrestha, R., et al.: Bridging the phenotypic and genetic data useful for integrated breeding through a data annotation using the crop ontology developed by the crop communities of practice. Front. Physiol. **3** (2012). https://doi.org/10.3389/fphys.2012.00326, http://journal.frontiersin.org/article/10.3389/fphys.2012.00326/abstract

36. Sivamani, S., Bae, N., Cho, Y.: A smart service model based on ubiquitous sensor networks using vertical farm ontology. Int. J. Distrib. Sens. Netw. **9**(12), 161495 (2013). https://doi.org/10.1155/2013/161495, http://journals.sagepub.com/doi/10.1155/2013/161495

37. Stucky, B.J., Guralnick, R., Deck, J., Denny, E.G., Bolmgren, K., Walls, R.: The plant phenology ontology: a new informatics resource for large-scale integration of plant phenology data. Front. Plant Sci. **9**, 517 (2018). https://doi.org/10.3389/fpls.2018.00517, http://journal.frontiersin.org/article/10.3389/fpls.2018.00517/full

38. Tummers, J., Kassahun, A., Tekinerdogan, B.: Obstacles and features of farm management information systems: a systematic literature review. Comput. Electron. Agric. **157**, 189–204 (2019). https://doi.org/10.1016/j.compag.2018.12.044, https://www.sciencedirect.com/science/article/pii/S0168169918307944

39. Wilkinson, M.D., et al.: The FAIR guiding principles for scientific data management and stewardship. Sci. Data **3**, 1–9 (2016). https://doi.org/10.1038/sdata.2016.18

Design Implications Towards Human-Centric Semantic Recommenders for Sustainable Food Consumption

Gayane Sedrakyan$^{(\boxtimes)}$, Anand Gavai, and Jos van Hillegersberg

Department of Industrial Engineering and Business Information Systems, University of Twente, Enschede, The Netherlands
{g.sedrakyan,a.k.gavai,j.vanhillegersberg}@utwente.nl

Abstract. The significance of food is evident in the myriad challenges confronting contemporary society, including the increasing prevalence of diet-related diseases, food waste with its adverse economic, environmental, and social impacts, and the significant impact of food production on environmental issues, among others. As the negative health and environmental impacts of dietary patterns become more evident, there is a growing demand for personalized and sustainable food recommendations to promote healthier and planet-friendly choices. This study aims to enrich the theoretical underpinnings of food recommender systems with an emphasis on sustainable food consumption, by integrating insights from existing research, behavior change theories, and Industry 5.0 digitization concepts on humanity-centered technologies.

Keywords: food recommender · nutrition consumption behavior · sustainable food consumption · human-centric food recommender · privacy-aware food recommender

1 Introduction

The significance of food is evident in numerous challenges confronting contemporary society. Several health conditions such as obesity, diabetes, hypertension, cancer have been linked to dietary practices [1]. The rising prevalence of diet-related diseases has raised concerns about the challenges people face in achieving and maintaining balanced diets. Additionally, the significant contribution of food production to greenhouse gas emissions [2], deforestation, and biodiversity loss [3] emphasizes the potential role of dietary changes on a global scale in mitigating climate change. Food waste is yet another critical issue that necessitates urgent attention due to its adverse economic, environmental, and social impacts [4], in developed countries, consumers constituting a major source of food waste. Scholars have suggested that modifying individual diets could be a crucial aspect of the solution to climate change [5]. The need to address both nutritional and environmental concerns in recommending diets presents a compelling challenge for researchers and practitioners in this domain. As the negative health and

T. P. Sales et al. (Eds.): ER 2023 Workshops, LNCS 14319, pp. 312–328, 2023.
https://doi.org/10.1007/978-3-031-47112-4_29

environmental impacts of dietary patterns become more evident, there is a growing demand for personalized and sustainable food recommendations to promote healthier and planet-friendly choices.

Existing food recommender systems suffer from various shortcomings. For instance, they often lack personalization, relying on generic choices instead of considering individual preferences, dietary restrictions, and cultural backgrounds among others. On the other hand, some food recommenders heavily rely on user data for personalized recommendations, potentially limiting user exposure to food choices that align with their existing preferences and/or conditions [6, 7], failing to address broader food consumption contexts, support behavior change [8] for healthier lifestyles or conscious choices that align with societal concerns. These systems in addition raise privacy concerns over the use of sensitive user data [9]. Additionally, the underlying algorithms lack transparency and may lead to reduced user trust. Furthermore, inaccuracies and slow adaptation to changing conditions (e.g., physiological status, time, location) can result in outdated and ineffective recommendations.

These various factors collectively form the backdrop for further research and development in the field of food recommender systems. In addition, a prominently emerging area of interest in this domain pertains to sustainability assessment, as evidenced by recent scholarly attention [10]. However, the present landscape comprises only a limited number of tools designed to assess the degree of sustainability associated with individual food items.

Despite its significant importance and complex nature, the field of food recommendation remains in its infancy stage, lacking a dedicated theory exclusively focused on this domain [1]. In this study, we aim to construct a comprehensive conceptual map for developing cutting-edge recommenders to promote design and development of sustainable food consumption. This map will interconnect various essential concepts, including insights from existing research in the field (outlined in Sect. 2), theories related to behavior change (presented in Sect. 3), and the next generation digitization concepts rooted in Industry 5.0 - a humanity-centered digitization paradigm (Sect. 4). By integrating multidisciplinary knowledge and leveraging advances in recommender system techniques that accommodate individual preferences, health aspirations, and sustainability targets within a humanity-centered framework, our goal is contributing to new generation recommenders for more effective and responsible food consumption choices. The research aims to serve a starting platform for researchers and designers for human-centric recommenders for sustainable food consumption.

2 Existing Food Recommenders

In the context of food, recommendation systems aim to enhance user dining experiences, encourage healthier choices, and reduce food waste by empowering users to make informed and responsible food consumption choices. There are several types of techniques employed within recommender systems which include: (1) content-based or constraint-based recommendations that suggest items based on attributes and characteristics of previously liked items or restrictions, e.g. previously liked recipes or recipes that are based on restrictions, e.g. gluten-free; (2) collaborative filtering that utilizes user behavior to identify patterns among like-minded users and recommends items based on their preferences that is especially useful when there is limited item information available; (3) hybrid recommenders combine both content-based and collaborative filtering approaches to leverage the benefits of each for more diverse and accurate suggestions; (4) knowledge-based recommenders that use explicit knowledge about user(s) needs and requirements to generate personalized recommendations, which often use rule-based experts systems. Additionally, advanced methods like (5) matrix factorization and deep learning models are employed to capture intricate user-item interactions to deliver more sophisticated and accurate suggestions.

2.1 Algorithms Employed

Content-based recommenders (also referred to as semantic recommenders) analyze the characteristics of food items, such as ingredients, cuisines, or nutritional profiles, to recommend similar options [11]. In collaborative filtering recommenders [12], when presented with a new user (e.g. a traveler), the system recommends an item by drawing upon the favorably rated selections of comparable users (e.g. preferences of previous tourists in a region). In a hybrid system [13], user behavior data (user-item interaction) can be combined for instance with the textual analysis of ingredient lists and cooking methods to suggest recommendations that not only reflects similar user preferences but also considers the intrinsic attributes of food items. In such systems, if for instance a user has interacted with a variety of Italian pasta recipes (collaborative filtering) and has a preference for dishes containing tomatoes and basil (content-based), the hybrid algorithm might combine these insights. The knowledge-based recommender can rely on a structured database of recipes and attributes, while using user input to filter, match, and rank recipes for personalized recommendations [14]. This approach does not require user interaction with explicit ratings or past behavior, making it suitable for scenarios where user data might be limited or where users are seeking inspiration based on available ingredients and preferences. Collaborative filtering can suffer from the "cold start" problem when new items or users have limited interaction data. Knowledge-based recommenders can handle this better since they don't require historical interaction data.

To achieve their objectives, these systems employ different algorithms. Content-based recommenders use algorithms such as vector-based representations, e.g. cosine similarity to measure the similarity between food items based on their attributes, topic model-based or dependency decision tree representations. Collaborative filtering algorithms, use techniques such as matrix factorization decompose user-item interaction matrices to identify latent features to generate personalized recommendations. Matrix factorization dissects the user-item interaction matrix to uncover latent preferences [11]. For instance, it might reveal that a user prefers vegetarian options and spicy foods, even if they haven't explicitly ordered them before. Knowledge-based algorithms employ knowledge graphs that may integrate various aspects of food such as ingredients, nutritional knowledge and medical conditions, and their relations [14]. Deep learning models, on the other hand, can combine images and descriptions of dishes, e.g. using Convolutional Neural Networks (CNNs) to analyze dish images and Natural Language Processing (NLP) techniques for textual descriptions. By fusing visual and textual features, the platform offers personalized recommendations. For instance, when a user interacts with an image of a visually enticing pasta dish with a flavorful description, the model can combine the extracted visual details (ingredients, colors, textures) and semantics of textual cues (e.g. using Word2Vec to convert words into numerical vectors that capture their semantic relationships) to suggest similar recipes that resonate with the user taste. Recent research indicates that visual cues embedded within online recipes accessed by users have the potential to predict user preferences [15]. Another example of using predictive models within food recommender systems include chatbots. Through chatbots or voice assistants, recommender systems can engage in conversations with users to understand their preferences and provide real-time recommendations. Users can ask questions, receive suggestions, and discuss their culinary desires, which can be used by predictive models to refine further recommendations.

2.2 Intended Users

In the context of food recommenders, existing systems can be classified based on different characteristics, such as user groups. In terms of the users these systems can be classified to the following user categories:

- *Individual (children, elderly)*: These systems delve into the intricacies of personal preferences and contexts such as fitness diet, preventive or beneficial diets for health conditions, cost effectiveness, etc.
- *Family*: These systems encompass the broader canvas of shared meals, navigating the delicate balance of accommodating various preferences of various family members and dietary needs into a culinary mosaic that fosters shared enjoyment, catering to both children cravings and adult dietary considerations, next to budgetary choices.
- *Group of users*: These systems aim to support shared decision making, e.g. what restaurant to attend and what to eat together. For example, as people often eat socially, group recommendation becomes important. In group recommendation situations recommendations are optimised to suit multiple taste profiles with the preferences of different users being traded-off or balanced against each other.
- *Farmers for animals*: Such systems for farmers designed to feed animals are personalized tools that utilize data analysis and machine learning that consider factors such as species, age, weight, and nutritional needs to offer tailored feeding plans. These algorithms prioritize health, sustainability, and efficiency. By recommending optimal combinations of feed resources, e.g. organic and hormone-free options, the systems contribute to animal welfare, reduced environmental impact, and resource efficiency. These systems also promote transparency to consumers about how animals are raised, promoting trust in the quality and ethical production of animal products.
- *Farmers for food*: Food cultivation recommenders guide farmers in sustainable practices such as crop selection, precision farming, water-efficient irrigation, energy-conscious strategies, integrated pest management, soil health enhancement, nutrient optimization, climate adaptation, resource monitoring, and real-time decision support. These tools can promote cost-effective andeco-friendly methods, reduce waste, enhance yields, and provide educational resources for informed, sustainable agriculture that in addition can potentially inform other recommenders above to support sustainable food consumption behavior.

– *Supplier*: Food recommender systems benefit suppliers (like local farmers) often in bi-directional systems by expanding market access, reducing waste, and promoting sustainable practices. For consumers, these systems, besides serving standard recommenders, can encourage local sourcing, seasonal eating, environmentally friendly and cost-effective choices, fostering sustainable food consumption patterns and efficient supply chain.

2.3 Food Profiles

In terms of the food categories recommenders can be classified to the following objectives:

– *Product*: Recommenders can encompass food products such as meats, vegetables, and dairy to suggest recipes centered around favored products by users.
– *Ingredient*: ingredient recommendation include fundamental components used in cooking to create dishes, using e.g. user preferred ingredients, dietary restrictions, and allergies to suggest recipes and meals that align with their culinary preferences.
– *Recipe*: Some recommenders offer recipes, a set of instructions that outlines the steps and quantities required to prepare a specific dish, by analyzing user recipe choices, cooking styles, and preferred flavors to choose/compose recipes that match their tastes.
– *Menu*: Some recommenders offer suggestions for menus which contain multiple predefined dishes and possibly appropriate drinks and snacks.
– *Meal*: Recommenders may offer dishes, beverages, and accompaniments that can be consumed together during a dining occasion by analyzing user historical meal choices, taste preferences, and nutritional goals to suggest well-balanced and satisfying meals catering to specific needs and desires.
– *Meal plan*: Recommenders offer guidance for users seeking a structured approach to their daily diet by recommending menu plans that outline a schedule of meals and snacks over a designated period (sequence of temporal occasions), often designed to meet specific dietary goals, such as weight loss, muscle gain, or overall health improvement.
– *Feeding plan*: Recommenders can assist farmers with livestock nutrition, health and management by analyzing animal data and suggesting balanced diets based on nutritional needs, age, and breed. They customize feeding plans, considering ingredient nutritional profiles and cost efficiency. These tools also promote animal health by preventing imbalances and can adjust plans in real-time. By factoring in sustainability and environmental impact, they help farmers make eco-friendly choices. Record keeping and analysis enable long-term optimization.

2.4 Recommender Types in Food Domain

The applications of food recommender systems span across various domains. In terms of the objectives the food recommender systems can be classified to support:

– *Health*: In health-related contexts, food recommendations go beyond just satisfying user taste preferences by also taking into consideration the healthiness of the recommended food items for individuals who suffer from some kind of disease or seek to prevent illnesses to suggest recovery or preventive diets. The focus here is not only on the immediate appeal of the food but also on its nutritional value and its potential benefits for overall health. Furthermore, while individual meals may be healthy in isolation, recommender algorithms also need to ensure a balanced diet approach when combining them.

– *Recipes*: Recipe recommenders provide personalized recipes based on dietary requirements, cooking skills, and ingredient availability.
– *Cooking*: Cooking-related recommender systems allow users to explore new recipes, offer suggestions on what meals they can prepare based on their available ingredients, cooking skills, and preferred cuisine.
– *Diet*: Diet recommenders design custom diet plans according to individual health goals and nutritional needs. Diet-focused recommenders assist users in achieving specific dietary goals, such as weight loss, muscle gain, or managing medical conditions. These systems analyze the users' dietary requirements and restrictions to suggest suitable meal plans that align with their objectives. The recommendations may emphasize portion control, nutrient balance, and calorie intake, helping users stay on track with their desired diet plans.
– *Grocery*: Grocery recommenders help users create tailored shopping lists and promote healthier food choices.
– *Restaurant*: Restaurant recommenders suggest nearby dining options based on user preferences, reviews, and location.

2.5 Interfaces and Outputs

Recent advancements in the field of food recommender systems have brought about various interfaces for enhancing user experience and satisfaction. These interfaces encompass a diverse range of platforms, including:

– Mobile apps
– Web applications
– Smart plates
– Desktop programs
– Conversational agents

Mobile apps offer on-the-go accessibility, enabling users to receive personalized meal suggestions based on their preferences and dietary restrictions while navigating busy schedules. Web applications extend to larger screens, presenting visually appealing recipe recommendations with interactive features like customizable filters and virtual pantry organization. Smart plates, a cutting-edge interface, merge the physical and digital realms by analyzing the nutritional content of meals placed on them. These plates can suggest healthier alternatives or complementary dishes to balance dietary intakes, fostering mindful eating habits by, for example, analyzing the number of calories in a meal and the time taken to consume it. On the other hand, desktop programs cater to users seeking an immersive browsing experience, allowing them to explore diverse cuisines, cooking techniques, and culinary inspirations through an expansive interface. Conversational agents, powered by natural language processing, introduce a conversational dimension to food recommendations. Users can engage in dialogue with these agents, discussing their tastes, moods, and dietary goals to receive tailored meal ideas and recipe suggestions. These interfaces showcase the transformative potential of AI-driven interactions in shaping culinary preferences and encouraging culinary experimentation.

Food recommenders use a variety of output formats to present recommendations effectively. These may include:

- Text-based descriptions (e.g. recipes)
- Interactive visual elements (e.g. nutrition scores)
- images (e.g. dishes or ingredients)
- Videos (process of cultivating, selecting, harvesting, cooking)
- Nutritional information (e.g. serving size, ingredients, calories, allergen information, glycemic index, health benefits)
- Customizable filters (e.g. search and preference refining features)
- meal plans and grocery lists (e.g. vegetarian breakfast and the ingredients needed for cooking)
- Social sharing (e.g. integration with social media and messengers)

Textual output may contain descriptions or lists of suggested dishes along with relevant details such as ingredients, preparation methods, and flavor profiles. Visual representations of recommended dishes, often accompanied by images or icons, provide users with a quick and visually appealing overview of the options. Users can interact with these elements to explore more details about each dish, such as nutritional content, cooking steps, and variations or find a location of a proposed restaurant. High-quality images of recommended dishes can evoke sensory experiences, helping users get a sense of the final presentation and may prompt them to try recipes that align with their culinary preferences. Recommender systems can also suggest cooking tutorial videos or links to online cooking demonstrations, e.g. with guiding users through the step-by-step process of preparing recommended dishes, thus making it easier for them to follow along and replicate the recipe. Some output formats focus on providing detailed nutritional information for each recommended dish, including calorie counts, macronutrient breakdowns, and dietary information. This allows users to choose options that align with their dietary goals. Users can apply filters based on dietary restrictions, preferred cuisines, cooking times, or specific ingredients. The system then generates recommendations that fit the selected criteria, offering a more tailored and personalized experience. Recommender systems may generate entire meal plans, suggesting combinations of dishes for specific meals throughout the day. Additionally, they can generate grocery lists based on the selected recipes, streamlining the shopping process. Some systems allow users to share their recommended dishes or meal plans on social media platforms, fostering a sense of community and encouraging culinary exploration among peers.

2.6 Challenges with Existing Recommender Algorithms for Food Domain

With regard to the analytics approaches, standard recommendation algorithms perform significantly less well when used for recommending food items than on other problems, such as movies or online purchases in an e-commerce context [1]. Unlike movies or e-commerce products, which may have more universal appeal, food choices are inherently diverse and influenced by various factors that can vary significantly from person to person due to cultural, dietary, and personal factors, making it challenging for standard algorithms to capture such nuances accurately. Context-dependent preferences further complicate the matter. Food choices can be influenced by the time of day, location, or occasion. For instance, users might prefer different types of food for breakfast, lunch, or dinner. Standard algorithms, which typically treat each recommendation as an isolated event, may struggle to incorporate such contextual information, leading to less accurate suggestions. In many recommendation contexts, having a comprehensive and diverse dataset is essential for generating accurate suggestions. However, food data can often be limited to specific cuisines or restaurants, resulting in biased and less varied recommendations. Moreover, the dynamic nature of food menus poses a challenge for recommendation algorithms. Unlike movies or e-commerce products, food menus can change frequently, with new dishes being added or removed regularly. Standard algorithms may not be able to adapt quickly to these dynamic changes, leading to suboptimal recommendations. Additionally, nutritional and dietary constraints need to be considered in food recommendations. Users may have specific dietary preferences or restrictions, such as allergies or cultural dietary practices, which must be taken into account. Furthermore, food choices often involve multiple criteria, such as taste, price, healthiness, and availability. Standard algorithms are typically designed for single-criteria recommendations and may not effectively handle the complexity of multi-criteria food recommendations. Lastly, food recommendations might require additional contextual information other than preferences and historical data, such as financial, seasonal and time constraints, which might not be readily available or challenging to integrate into standard algorithms. For instance, a recipe may be relevant for a user however lack of the time or cooking equipment necessary to prepare the meal can make the recommendation unsuitable [1]. The processing of context information often entails handling data derived from heterogeneous sources with various formats. This can encompass data stream queries, potentially leading to resource consumption that is unaffordable, in addition to data fusion needs [16].

3 Integrating Behavior Change Theories for Sustainable Food Consumption

In the pursuit of addressing pressing global challenges such as climate change and resource depletion, promoting sustainable food consumption has emerged as a critical endeavor. Achieving meaningful shifts towards more eco-friendly dietary choices requires a comprehensive understanding of human behavior and the factors that influence decision-making. This is where behavior change theories come into play. By delving into the intricate interplay of attitudes, motivations, and social dynamics that shape our food choices, these theories can offer valuable insights into how sustainable food consumption can be effectively encouraged and nurtured. While behavior change theories hold significant potential for enhancing the effectiveness of food recommender systems, their extensive application in this context is still relatively limited. While these theories have been widely studied in health promotion and behavior change interventions, their integration into food recommender systems is a developing area of research. Researchers and practitioners have recognized the value of behavior change theories in improving user engagement, adoption of healthier eating habits, and promoting sustainable food choices through recommender systems. However, the comprehensive implementation of these theories, including tailoring recommendations based on attitudes, social influences, and motivations, is not yet mainstream. Some studies have started to explore the integration of behavior change principles into food recommender systems [17], however the full potential of these theories is still being realized. As the field of food recommender systems continues to evolve, there is a growing interest in leveraging behavior change theories to create more impactful and user-centered recommendation strategies. Challenges may include the complexity of adapting theoretical constructs into practical algorithms and the need for personalized data to accurately apply these theories to individual users. Other challenges are related to privacy, trust, and ethical considerations while designing systems that effectively guide users toward healthier and more sustainable eating behaviors.

Integrating behavior change theories within the context of recommenders holds the promise of guiding individuals toward more sustainable food consumption and environmentally conscious dietary habits. When designing food recommender systems, these theories can serve as foundational pillars. *Theory of Planned Behavior* (TPB) [18] can be leveraged to analyze user attitudes towards sustainability and recommend plant-based recipes, aligning with their environmental values and perceived control over food choices. These systems consider user attitudes towards health, social norms related to healthy eating, social influences from eco-conscious communities, sustainability, and their perceived ability and control to make sustainable food decisions.

Similarly, the *theory Social Cognitive Theory* (SCT) [19] supports the notion of humans being influenced by social factors and emphasizes the role of observational learning, social interactions, and modeling in shaping human behavior. This theory can inspire the incorporation of social features within recommender systems by creating a platform where users share their sustainable cooking achievements and engage in discussions by also fostering a sense of community around eco-friendly eating habits. In the context of food choices and dietary behaviors, SCT posits that individuals learn from observing the actions and consequences of others, especially those within their social environment. This includes family members, friends, peers, and role models. People tend to adopt behaviors that they perceive as socially acceptable or that align with the behaviors of those around them. When it comes to sustainable food consumption and using food recommender systems, SCT suggests that individuals are likely to be influenced by the choices and recommendations of their social circles. Recommendations or endorsements from friends, family, or community members can carry significant weight in influencing dietary decisions. Therefore, designing food recommender systems that tap into social influences, provide community-based recommendations, and facilitate social interactions can potentially encourage more sustainable food choices. Practical application of SCT principles into food recommender systems can lead to strategies like showcasing recipes favored by user's social connections, allowing users to share and discuss their meal choices, and highlighting community-endorsed sustainable options. This not only promotes sustainable eating behaviors but also fosters a sense of belonging and social engagement, further enhancing the potential for behavior change.

Yet another theory that can be relevant for behavior change for sustainable food consumption includes the *Health Belief Model* (HBM) [20] which can drive recommendations that emphasize both the health benefits and ecological impact of adopting a sustainable diet. By showcasing the nutritional value and reduced carbon footprint of certain dishes, users are more likely to make informed and planet-conscious choices.

Nudge theory [21], discreetly integrated, can present subtle cues that encourage users to explore sustainable food choices, e.g. meatless or locally sourced options, without imposing strict restrictions. This approach makes sustainable alternatives more salient in their decision-making process. For instance, when searching for protein sources the initial results can showcase legume-based recipes or locally grown vegetables as *default options*. This discreet nudge encourages users to consider sustainable alternatives without explicitly imposing restrictions. When presenting recipe choices, the system can use *visual highlight* or position meatless or sustainable options at the top of the list. By drawing attention to these choices through design elements such as color or placement, users are subtly nudged towards exploring these alternatives first, increasing the likelihood of selecting them. In addition to nutritional information, the recommender system could provide an eco-friendly rating or label for each recipe. For example, recipes with lower carbon footprints or reduced water usage can be marked as environmentally conscious choices. *Community endorsements* are another subtle display of choices that aligns with the principle of social influence and nudge theory, encouraging users to adopt similar behaviors without overtly enforcing restrictions. Nudge techniques can include *time-sensitive notifications* or *alerts* highlighting e.g. special meatless or locally sourced recipes that align with current trends or events. These prompts can gently nudge users to consider sustainable options during specific moments, such as Meatless Mondays or local produce seasons. Recommenders can also analyze user interactions and context information to apply personalized nudges. Additionally, recommenders can incorporate a *progress tracker* that celebrates user adoption of sustainable food choices. Milestones achieved, such as a certain number of meatless meals, could be *rewarded* e.g. with virtual badges or incentives, subtly reinforcing positive behaviors. Incorporating nudge theory into a food recommender system involves strategically applying subtle cues and design elements to guide users towards more sustainable food choices. By discreetly highlighting positive sustainability attributes and making them salient in the decision-making process, the system can encourage users to explore eco-friendly alternatives without imposing strict limitations.

These examples showcase the potential of behavior change theories to shape effective, personalized, and eco-conscious food recommendations.

4 Linking Recommendations with Human-Centric Digitalization Concepts

Food recommenders in the era of Industry 4.0 heavily rely on user data to provide personalized recommendations. However, this can lead to users being only exposed to content that aligns with their existing preferences and conditions, potentially limiting their exploration of new and diverse food choices. Additionally, gathering and analyzing user data for personalized recommendations raises privacy concerns. Users may be hesitant to share their personal information, leading to a lack of data for accurate recommendations or to potential misuse of their data. Some food recommenders may use complex algorithms and machine learning models, making their decision-making processes opaque. Users may not understand why specific recommendations are made, leading to reduced trust in the system. Some food recommenders may not consider the broader context of food consumption, such as cultural preferences, dietary restrictions, or ethical considerations, leading to less relevant or inappropriate recommendations. Among other issues are inaccurate or outdated data that can result in poor recommendations, such as promoting unhealthy or environmentally harmful products. Balancing profitability with such ethical concerns can be challenging for some recommender systems. Last but not least, some food recommenders may struggle to adapt quickly to changing user preferences, conditions or emerging food trends, leading to potentially outdated recommendations.

4.1 Transforming Food Recommenders with Industry 5.0

Industry 5.0 emphasizes the infusion of human values and ethical considerations into technology design. In the context of food recommenders, this means moving beyond a purely algorithmic approach of Industry 4.0 and incorporating human-centric recommendations by considering *broader context* of e.g. cultural, geographical, temporal, ethical, and emotional aspects among others, by incorporating a *semantic approach* instead of solely relying on standard recommender algorithms.. Industry 5.0 leverages psychological insights to drive *behavior change*. Food recommenders can draw from behavior change theories. By thoughtfully integrating behavior change theories into food recommender system design, these enriched algorithms can create user-centric platforms that not only facilitate informed decisions but also empower individuals to embrace and sustain sustainable food consumption practices. As discussed earlier, behavior change theories provide valuable frameworks to align food recommendations with user motivations, beliefs, and social influences. By understanding and applying these theories, researchers and designers can create more effective and user-centered food recommender systems.

As recently suggested by Norman [22] we should talk about *humanity-centered* design of future systems. Under this new paradigm, the value of new technology for humans cannot be prioritized over the impact on the socio-technical and environmental contexts. Therefore, a product that brings high value to people by ineffective use of resources (e.g., consuming too much energy) or by impacting negatively other people cannot be considered humanity centered. Instead of relying purely on data analytics approach, this new "re-humanized" version of digitalization process implies eroding the boundaries between different disciplines. In the domain of food recommenders, this entails crafting conscientious technologies that facilitate human-centric food consumption suggestions. Technological mindfulness also encompasses *involving key stakeholders* in the transformation process, such as *healthcare experts, domain experts e.g. dieticians, privacy and food production policymakers, suppliers, and users* themselves. Engaging relevant stakeholders not only ensures appropriate solutions but also fosters essential values like *trust, privacy, ethics, security, accessibility, usability,* and *transparency* for end-users. Trust and transparency are linked to eXplainable AI (XAI), which employs emerging methods to enhance trust and facilitate *evaluation. Usability,* a determinant of safety (ISO 9241–11) and trust [23], warrants examination in contexts like conversational agents. The predictability [24] of digital agents significantly influences trust, particularly in high-risk decision-making scenarios like health, where trust, comprehension, and explainability are closely interrelated [25].

4.1.1 Explainable Semantic Food Recommenders

In the transition from Industry 4.0 to Industry 5.0, a profound shift in digitization strategies has emerged, one that centers on human values, individual experiences, and the seamless integration of technology within our daily lives. At the heart of this paradigm lies a revolutionary approach to algorithms, that goes beyond mere calculations and enters the realm of semantic understanding. In the context of food recommenders, this transition signifies a departure from the conventional, data-driven algorithms solely based on statistical patterns towards models that are deeply rooted in the semantic understanding of the user preferences and the broader context in which these preferences exist. Semantic algorithms, in contrast to their standard counterparts, possess the capability to comprehend not only what a user likes but also why they like it, by tapping into the intricate interplay of cultural traditions, geographical influences, temporal contexts, ethical considerations, emotional ties, and more. For instance, a semantic algorithm will not just analyze the users' past preferences but can take into account the cultural heritage associated with the event, the regional flavors that resonate with the family background, and even the emotional significance attached to certain ingredients or dishes to craft an experience that is not only delicious but deeply meaningful. The shift towards semantic algorithms can also foster an environment where the food recommender system becomes a trusted culinary companion, one that understands the nuances of individual lifestyles and aspirations, with a capability of offering suggesting local and seasonal ingredients that align with eco-conscious values, and even crafting recipes that evoke nostalgic memories. While this integration of semantic algorithms aligns seamlessly with the principles of humanity-centered digitization, it can also encapsulate the essence of personalization and holistic well-being. By considering the broader semantic context surrounding food choices, these algorithms can empower users to embark on a culinary journey that is not just about sustenance but also about connecting with their heritage, embracing sustainability, and savoring the joys of human experience.

Explainability can enhance user experience [26] by providing clear and transparent insights into why specific recommendations are made, empowering users to make informed choices and fostering a sense of trust; for example, by explaining that a vegan pasta dish is recommended due to its alignment with the user's dietary preference, its rich nutritional profile, and the incorporation of locally sourced ingredients for sustainability, the system can create confidence in user selection and appreciates the system's personalized guidance, e.g."The meal was recommended because it matches your gluten-free options. This recommendation for a quinoa and roasted vegetable bowl also takes into account your nutritional needs with a blend of vitamins and protein, and features locally sourced vegetables to support your sustainability choices".

Semantics can in addition be enhanced by the use of *ontologies*. Within the food domain ontologies can be used to create ontology-based user profiles enabling algorithms to understand user preferences, dietary needs, and culinary context and link with ontologies of nutritional data, ingredient compatibility, culinary techniques and their relationships to provide personalized suggestions with enhanced quality and diversity. Moreover, ontology-based algorithms can facilitate the generation of transparent, explainable and justified recommendations, fostering user trust and understanding. Linked data can provide significant advantages for food recommender algorithms by enabling seamless fusion of data from diverse sources facilitating data integration for more accurate and comprehensive recommendations that leverage a broader spectrum of information from different domains in the context of food consumption and production.

4.1.2 Sustainability Concepts

Sustainability has emerged as a prominent topic in the field of food recommender systems [13]. However, only a limited number of tools are currently available to assess the degree of sustainability of food items. Sustainability recommenders can play a crucial role in promoting environmentally-friendly food choices and reducing food waste. To consider environmental impact in food choices, recommendations must balance user preferences with measures like food miles (distance traveled by food), carbon dioxide emissions associated with food production and transportation, and other environmental/ecological metrics. Recommendations provided by these systems aim to strike a balance between user preferences and environmentally sustainable options, encouraging users to make choices that have a lower ecological footprint.

Eco-assessments are a relatively new concept that quantifies the environmental impact of food. They consider various factors such as greenhouse gas emissions, water use, and land use associated with the production and distribution of each product [27]. Food recommenders use these values to recommend foods with a smaller environmental footprint to promote sustainable consumption habits. These platforms are pioneering the use of alternative ingredient suggestions. They can recommend substitutes for common ingredients that are either healthier, more environmentally friendly, or both. This feature is especially beneficial for people with dietary restrictions or those looking to reduce their environmental impact. In addition, these platforms often can link these suggestions to local farmers to support local agriculture and reduce the carbon footprint associated with food transportation [28].

Some food recommenders also introduce nutritional *health scores* of ingredients by considering various factors such as nutrient composition, dietary guidelines, health impact studies, and user preferences and specific conditions. Nutrition composition assessment takes into account the nutritional content of ingredients, including macronutrients (carbohydrates, proteins, and fats) and micronutrients (vitamins and minerals). Food recommenders often use standardized nutrient databases e.g. the USDA National Nutrient Database or other relevant sources to obtain accurate information on the nutritional composition of each ingredient. Health scores are usually calculated based on dietary guidelines recommended by health organizations like the World Health Organization (WHO) or national dietary guidelines (e.g., USDA's Dietary Guidelines for Americans). These guidelines provide specific recommendations for daily intake levels of various nutrients to maintain a balanced and healthy diet. Many food recommenders in addition incorporate scientific studies and research on the health impacts of different nutrients and food components. These studies help in understanding how certain nutrients can affect health positively or negatively, both in the short and long term. Food recommenders use scoring algorithms that assign weights to different nutrients based on their importance for overall health. These algorithms often take into account factors such as energy density, fiber content, healthy fats, vitamins, and minerals. Some food recommenders allow users to customize their health scores based on personal preferences, dietary restrictions, and health goals. For example, a person following a low-carb diet might prioritize the health score based on lower carbohydrate content.

Different food recommenders, in addition, may employ varying approaches and data sources to compute health scores, and their specific methodologies might not be publicly disclosed due to proprietary reasons, therefore privacy-awareness becomes an important topic in the domain of recommenders in general and food recommender systems in particular.

4.1.3 Privacy-Aware Food Recommender Platforms

In the age of digitalization, privacy-preserving data platforms have emerged as an important tool for promoting sustainable food consumption. By incorporating these mechanisms, privacy-aware novel data platforms for food recommenders strike a delicate balance between delivering personalized recommendations and upholding user privacy. These platforms empower users to make informed food choices without compromising their sensitive information, contributing to a more secure and trustworthy user experience.

Federated learning [29] is a promising privacy-preserving technique that can be employed in food recommendation systems to enhance user privacy while still delivering personalized recommendations. In federated learning, users have full control over their data and its usage. Explicit consent is obtained before any data is collected or utilized for recommendations. The training of machine learning models occurs locally on user devices, and only aggregated insights are shared with the central server. This decentralized approach minimizes the need to transfer raw user data to a central repository, reducing privacy risks. In the context of food recommendation, federated learning operates as follows:

- Data Partitioning: User data remains on their respective devices, ensuring that sensitive information, such as dietary preferences and consumption habits, is not shared externally.
- Local Model Training: Each user device trains a local recommendation model using personal data, considering factors like past food choices, dietary restrictions, and nutritional goals.
- Model Aggregation: The central server aggregates insights from the locally trained model without accessing raw data. This aggregation process captures collective behavioral patterns without compromising individual privacy.
- Global Model Update: The central server updates the global recommendation model based on the aggregated insights. This model is then distributed back to users' devices, ensuring that it benefits from the collective knowledge while preserving privacy.

– Personalized Recommendations: On-device models use the updated global model to generate personalized recommendations for each user, without exposing individual data to the central server.

These platforms, which prioritize user privacy while providing personalized food recommendations, are revolutionizing the way we perceive and consume food. In particular, they stand out for their emphasis on eco-ratings, novel approaches to calculating nutritional value, and suggestions for alternative ingredients that connect with local farmers in a privacy-preserving manner.

In summary, privacy-conscious platforms play a critical role in promoting sustainable food consumption. By highlighting organic ratings, introducing novel methods for calculating nutritional value, and recommending alternative ingredients that connect with local farmers, these platforms enable consumers to make more informed and sustainable food choices. The concepts and interrelationships discussed within this work are additionally depicted via a conceptual framework, as illustrated in Fig. 1.

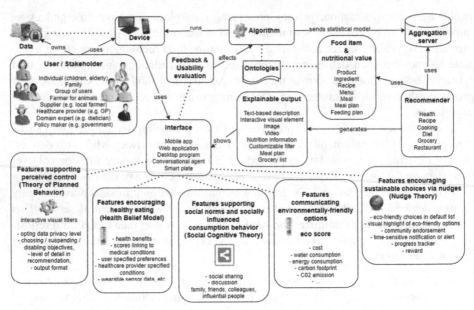

Fig. 1. A conceptual diagram depicting design elements for enhancing recommender systems with privacy awareness and promoting sustainable food consumption.

5   Conclusion

Sustainable food consumption offers diverse benefits, spanning environmental preservation and individual well-being. However, its integration into food recommender systems lacks a unified theoretical foundation, particularly in behaviorally transformative frameworks. While health dimensions are extensively studied, domains like cooking, grocery shopping, and restaurant choices, including menu selections, are relatively underexplored. Algorithms predominantly draw from conventional methodologies, particularly collaborative filtering, often prioritizing individual over group dynamics, which in addition face challenges due to the absence of standardized datasets, compounded by GDPR and copyright limitations on behavioral and recipe data. Notably, a crucial omission is the incorporation of sustainability and environmental considerations, which are increasingly pertinent in modern food consumption paradigms. Initiatives to bridge these gaps and address such critical issues are yet to emerge.

This work enriches the theoretical underpinnings of food recommender systems, emphasizing sustainability by linking the concept of food recommendation with behavior change theories and their applications within the food recommenders. It also situates the food recommenders within human-centric digitization concepts inherent to industry 5.0 such as privacy, trust, explainability, while also connecting relevant stakeholders such as consumers, suppliers, domain experts including healthcare providers, as well as policy makers in the domain. The conceptual map presented in the work offers a foundational platform for the exploration of environmentally friendly human-centric food recommender designs aligned with industry 5.0 principles.

References

1. Elsweiler, D., Hauptmann, H., Trattner, C.: Food recommender Systems. In: Ricci, F., Rokach, L., Shapira, B. (eds.) Recommender Systems Handbook, pp. 871–925. Springer, New York (2012). https://doi.org/10.1007/978-1-0716-2197-4_23
2. Clark, M., Tilman, D.: Comparative analysis of environmental impacts of agricultural production systems, agricultural input efficiency, and food choice. Environ. Res. Lett. **12**(6), 064016 (2017)
3. Stoll-Kleemann, S., Schmidt, U.J.: Reducing meat consumption in developed and transition countries to counter climate change and biodiversity loss: a review of influence factors. Regional Environ. Change **17**, 1261–1277 (2017)
4. Aschemann-Witzel, J., et al.: Consumer-related food waste: causes and potential for action. Sustainability **7**(6), 6457–6477 (2015)
5. Foer, J.S.: We Are the Weather: Saving the Planet Begins at Breakfast. Penguin, UK (2019)
6. Naresh, A., et al.: Understanding user taste preferences for food recommendation. Int. J. Eng. Res. Technol. **9**(6) (2020). ISSN 2278-0181
7. Zeevi, D., et al.: Personalized nutrition by prediction of glycemic responses. Cell **163**(5), 1079–1094 (2015)
8. Agapito, G., et al.: DIETOS: a recommender system for adaptive diet monitoring and personalized food suggestion. In: 2016 IEEE 12th International Conference on Wireless and Mobile Computing, Networking and Communications (WiMob). IEEE (2016)
9. Nurbakova, D., et al.: Adaptive and privacy-aware persuasive strategies to promote healthy eating habits: position paper. In: Adjunct Proceedings of the 31st ACM Conference on User Modeling, Adaptation and Personalization (2023)
10. Asano, Y.M., Biermann, G.: Rising adoption and retention of meat-free diets in online recipe data. Nat. Sustain. **2**(7), 621–627 (2019)
11. Lin, C.-J., Kuo, T.-T., Lin, S.-D.: A content-based matrix factorization model for recipe recommendation. In: Tseng, V.S., Ho, T.B., Zhou, Z.H., Chen, A.L.P., Kao, H.Y. (eds.) PAKDD 2014, Part II. LNCS, vol. 8444, pp. 560–571. Springer, Cham (2014). https://doi.org/10.1007/978-3-319-06605-9_46

12. Rajabpour, N., Naserasadi, A., Estilayee, M.: TFR: a tourist food recommender system based on collaborative filtering. Int. J. Comput. Appl. **975**, 8887 (2018)

13. Freyne, J., Berkovsky, S.: Intelligent food planning: personalized recipe recommendation. In: Proceedings of the 15th International Conference on Intelligent User Interfaces (2010)

14. Haussmann, S., et al.: FoodKG: a semantics-driven knowledge graph for food recommendation. In: Ghidini, C., et al. (eds.) ISWC 2019, Part II. LNCS, vol. 11779, pp. 146–162. Springer, Cham (2019). https://doi.org/10.1007/978-3-030-30796-7_10

15. Yang, L., et al.: Yum-me: a personalized nutrient-based meal recommender system. ACM Trans. Inf. Syst. (TOIS) **36**(1), 1–31 (2017)

16. Rojas Melendez, J.A., et al.: Supporting sustainable publishing and consuming of live linked time series streams. In: Gangemi, A., et al. (eds.) ESWC 2018. LNCS, vol. 11155, pp. 148–152. Springer, Cham (2018). https://doi.org/10.1007/978-3-319-98192-5_28

17. Trang Tran, T.N., et al.: An overview of recommender systems in the healthy food domain. J. Intell. Inf. Syst. **50**, 501–526 (2018)

18. Bosnjak, M., Ajzen, I., Schmidt, P.: The theory of planned behavior: selected recent advances and applications. Eur. J. Psychol. **16**(3), 352 (2020)

19. Bandura, A.: Social cognitive theory of personality. Handb. Pers. **2**, 154–196 (1999)

20. Champion, V.L., Skinner, C.S.: The health belief model. Health Behav. Health Educ.: Theory Res. Pract. **4**, 45–65 (2008)

21. Arno, A., Thomas, S.: The efficacy of nudge theory strategies in influencing adult dietary behaviour: a systematic review and meta-analysis. BMC Public Health **16**(1), 1–11 (2016)

22. Norman, D.A.: Design for a Better World: Meaningful, Sustainable, Humanity Centered. MIT Press, Cambridge (2023)

23. Salanitri, D., et al.: Relationship between trust and usability in virtual environments: an ongoing study. In: Kurosu, M. (ed.) HCI 2015. LNCS, vol. 9169, pp. 49–59. Springer, Cham (2015). https://doi.org/10.1007/978-3-319-20901-2_5

24. Daronnat, S., et al.: Inferring trust from users' behaviours; agents' predictability positively affects trust, task performance and cognitive load in human-agent real-time collaboration. Front. Robot. AI **8**, 194 (2021)

25. Diprose, W.K., et al.: Physician understanding, explainability, and trust in a hypothetical machine learning risk calculator. J. Am. Med. Inform. Assoc. **27**(4), 592–600 (2020)

26. Vultureanu-Albişi, A., Bădică, C.: Recommender systems: an explainable AI perspective. In: 2021 International Conference on INnovations in Intelligent SysTems and Applications (INISTA). IEEE (2021)

27. Sonesson, U., et al.: Protein quality as functional unit–a methodological framework for inclusion in life cycle assessment of food. J. Clean. Prod. **140**, 470–478 (2017)

28. Hansen, P.G., Schilling, M., Malthesen, M.S.: Nudging healthy and sustainable food choices: three randomized controlled field experiments using a vegetarian lunch-default as a normative signal. J. Public Health **43**(2), 392–397 (2021)

29. McMahan, B., et al.: Communication-efficient learning of deep networks from decentralized data. In: Artificial Intelligence and Statistics. PMLR (2017)

CoffeeWKG: A Weather Knowledge Graph for Coffee Regions in Colombia

Cristhian Figueroa[1]([envelope]) [iD], Nadia Yacoubi Ayadi[2] [iD], Nicolas Audoux[3] [iD],
and Catherine Faron[3] [iD]

[1] Department of Telematics, Universidad del Cauca, Popayán 190003, Colombia
cfigmart@unicauca.edu.co
[2] Université de Lyon, CNRS, LIRIS UMR 5205, 6922 Villeurbanne, France
nadia.yacoubi-ayadi@univ-lyon1.fr
[3] Université Côte d'Azur, Inria, CNRS, I3S (UMR 7271), Nice, France
nicolas.audoux@etu.univ-cotedazur.fr, faron@i3s.unice.fr

Abstract. Coffee is one of the major crops produced in Colombia which is the third largest producer of coffee after Brazil and Vietnam. Together, these three countries produce more than 50% of the world's total coffee. One of the main challenges facing coffee producers in Colombia is to determine the effects of climate variability and climate change on their production. This paper presents CoffeeWKG, an RDF knowledge graph focused on weather conditions in the coffee-growing regions of Colombia over 15 years (2006–2020), to facilitate the understanding of climate impacts on coffee crops. CoffeeWKG enables the integration of heterogeneous sensor data collected from different weather stations and the definition of semantic metadata on agro-climatic parameters. This knowledge graph enables coffee growers and experts to explore and query historical weather conditions to establish a correlation between weather data and information on coffee crops, thus revealing the complex interaction between climate and production dynamics. This research is essential to improving the resilience of agriculture and optimizing resources in the face of changing climatic challenges.

Keywords: Coffee crop · Weather Knowledge Graph · Linked Data · Meteorology

1 Introduction

Coffee plays a critical role in Colombia's economy and cultural heritage, serving as the most important agricultural export crop. With a cultivated area of about 884 thousand hectares and 540 families depending economically on coffee production, it significantly contributed to the country's economic growth [10,20]. Colombian coffee is well-marketed in the specialty coffee industry due to its mild and sweet flavor obtained from arabica beans grown in distinctive climatic conditions (precipitation, solar radiation, temperature, and humidity) and due to

T. P. Sales et al. (Eds.): ER 2023 Workshops, LNCS 14319, pp. 329–342, 2023.
https://doi.org/10.1007/978-3-031-47112-4_30

topological and edaphic factors (soil depth, acidity/alkalinity, fertility) in the cultivable areas [10].

However, a significant challenge facing coffee cultivation in Colombia is climate variability, which has become more problematic due to climate change. This variability adversely affects coffee production, resulting in economic losses for farmers. According to the Colombian Federation of Coffee Growers (FNC), in the last 12 months (June 2022 - June 2023), coffee production fell by 13%, from 12.3 million bags to 10.7 million, due to unfavorable climatic variables that affected crucial stages like flowering and grain filling in the coffee-growing regions of the country [14].

In this context, data-driven approaches are increasingly necessary to mitigate the adverse effects of climate change and fostering sustainable economic growth within the coffee farming industry. Access to accurate and comprehensive data, especially meteorological data, becomes a powerful tool for improving coffee production. In this paper, we propose to use the Semantic web models and technologies and adopt the Linked data principles to publish and integrate weather data in coffee-growing regions. We conceived and built the CoffeeWKG knowledge graph for weather conditions in Colombia's central coffee production regions. This knowledge graph aims to help agronomists and farmers understand long-term weather trends and adapt coffee crops to shifting growing conditions, mitigating potential risks and losses.

The rest of this paper is organized as follows. Section 2 presents previous works using semantic web technologies and knowledge graphs for weather and climate conditions in agriculture. Section 3 illustrates the motivation scenario and competency questions that guided the construction of the proposed knowledge graph CoffeeWKG. Section 4 presents the data model in CoffeeWKG. Section 5 describes the data sources and the generation process of CoffeeWKG. Section 6 presents the validation of CoffeeWKG through the implementation of the competency questions as SPARQL queries. Section 7 shows conclusions and future works.

2   State of the Art

The adoption of semantic web models has brought about a ground-breaking approach towards the implementation of intelligent agricultural practices. Knowledge graphs and ontologies integrated with weather and agricultural data offer a novel perspective on understanding, analyzing, and optimizing the intricate interplay between climate, agricultural practices, and crop outcomes [7]. This section presents an overview of the state-of-the-art advancements in applying semantic web technologies to enhance crop management in the general context of Smart Food systems.

2.1   Ontologies and Vocabularies for Agriculture

Semantic web technologies have proven to be instrumental in capturing, sharing, and integrating heterogeneous agricultural data. Ontologies play a central role in semantic representations, acting as formal vocabularies that define domain concepts and their relationships. In coffee crop management, ontologies can capture diverse aspects such as weather conditions, soil properties, disease profiles, and farming practices, enabling comprehensive knowledge representation. Currently, there are many studies describing semantic web technologies for agriculture.

AgroVoc [18] is one of the key controlled vocabularies covering areas of interest to the Food and Agriculture Organization of the United Nations (FAO). It provides concepts and terms in multiple languages to describe agricultural, forestry, fisheries, food, and related domains.

Crop Ontology [11] is a community-based project providing a central place for creating crop-related ontologies. It provides a standardized and controlled vocabulary for describing crop plants' anatomy, morphology, and phenology. Additionally, it helps in data integration and sharing across agricultural research communities.

RustOnt [17] is an ontology for modeling favorable climatic conditions to prevent coffee rust. This ontology gathers relevant concepts and instances of meteorological variables used by coffee rust control systems or models.

BBCH [16] is an ontological framework for the semantic description of plant development stages and for transforming specific scales into RDF vocabularies.

2.2 Knowledge Graphs for Climate Conditions

The AEMET meteorological dataset [2] is the first LOD dataset that made available Spanish meteorological data through a SPARQL endpoint. AEMET dataset is based on a network of ontologies, where each ontology describes a sub-domain involved in the modeling of meteorological measurements. These sub-domains include meteorological parameters such as Measurements, Sensors, Time, and Location.

In a recent study [22], a knowledge graph was created for meteorological observational data in France, referred to as WeKG-MF. This modular ontology-based knowledge graph provides a comprehensive data model that can be customized by weather data providers. WeKG-MF has also been designed to serve different use cases identified in the context of the D2KAB French project.

LinkClimate is a knowledge graph presented in [21] that uses a similar network of ontologies as [22], excluding the GeoSPARQL ontology. The researchers created a knowledge graph that incorporates information from daily climate summaries provided by National Oceanic and Atmospheric Administration (NOAA), and established connections between their graph and external geographic sources in the LOD cloud such as OpenStreetMap and Wikidata.

In our research work, we rely on the study presented in [22] and examine how the WeKG-MF data model can be used to represent weather conditions in Colombia's coffee growing areas. Our first step is to create a list of competency questions that address the needs of coffee producers.

3 Motivating Scenario and Competency Questions

Several research projects have been undertaken to demonstrate the influence of weather conditions on crop growth, encompassing various cultivars and plant species in different spatial and temporal contexts. These studies have shown that farmers can refine their farming practices by taking advantage of historical and real-time weather data. It can enable them to plan planting and harvesting times more effectively, adjust irrigation and fertilisation strategies in response to changing weather conditions, and even predict future harvests more accurately. However, it is essential to provide experts and farmers with access to semantically enriched data presented in an interoperable format, so that it can be queried and reused to meet different requirements and use cases, and integrated with other data sources. Colombian coffee (*Coffea arabica L.*) is grown using arabica beans, known for their light and flowery taste. This crop is farmed in distinctive climatic conditions (precipitation, solar radiation, temperature, and humidity) provided mainly in the intertropical zone between latitude 20° N and 24° S in altitudinal ranges from 700 to 2000 masl[1] [10].

According to [1], the optimum mean temperature range for Coffea arabica per year is about 18–21 °C or up to 26 °C. At temperatures exceeding this range, the maturation and ripening of fruits accelerate, frequently resulting in compromised coffee quality. The optimal rainfall range for this crop is approximately 1200–1800 mm per year, about 120mm per month [6]. Excessive precipitation may damage the flowering of coffee. Moreover, the recommended sun exposure is around 1800 h annually, considering that cultivating coffee crops under the shade of trees is advisable to shield them from excessive solar radiation, particularly during dry seasons. This practice also contributes to organic matter proliferation in the soil and fosters a conducive habitat for beneficial microbial life, bolstering plant nutrition. Consequently, we have elicited a set of competency questions [13, 19] that reflect, amongst all, the specific needs of coffee growing in terms of climatic conditions. Here are some of them:

– *CQ1. What are the meteorological properties measured at stations near coffee-growing areas?*
  Coffee growers and experts can analyze the main climate variables such as air temperature, relative humidity, solar exposure, and cumulative precipitation. By doing so, they can gain insights into the properties measured at the stations near their crops. This can help them identify climate-related risks such as droughts, excessive rainfall, or extreme temperatures and take proactive measures to protect their crops. They can also obtain data about past planting times, which can enable them to choose the best times for planting and harvesting coffee. Furthermore, monitoring favorable weather conditions for plant pests and diseases can allow for early detection and preventive measures to protect the crop, among other benefits.
– *CQ2. What are the closest weather stations to a spatial location of a given coffee farm?*

---

[1] masl is the abbreviation of meters above sea level.

Coffee growers and experts need to locate weather stations close to their crops' location for easy access to weather data. Consequently, a semantic model should offer a spatial module to capture the geographical coordinates of the stations.

– *CQ3. In which months are the highest temperatures during a specific year?* High temperatures are correlated with drought events that produce a water deficit in coffee crops. According to [10], in coffee-growing regions of Colombia, the soil is vulnerable to water deficit, which can affect the fruit load by 25% when it lasts more than 40 days. Coffee growers may benefit of this knowledge to choose the best times for planting and harvesting coffee avoiding extreme temperatures and drought events.

– *CQ4: In which months do the highest rainfalls occur during a specific year?* Excessive rainfalls affect coffee crop productivity, hinder disease control and weeds, increase the lixiviation of nutrients, and elevate the costs of cultivation practices [1,15]. This competency question illustrates the need of farmers and experts to avoid cultivating coffee in months more prone to rainfall.

4   The CoffeeWKG Semantic Model

To publish weather conditions in Colombia's coffee crop regions, we rely and reuse the WeKG semantic model presented in [22]. Accordingly, the CoffeeWKG model comprises semantic representations of weather stations as well as meteorological observations recorded in different coffee-growing areas in Colombia. We first present an overview of the WeKG semantic model and then we detail how weather stations and meteorological observations are represented.

4.1   Overview of the WeKG Semantic Model

WeKG relies on a self-contained semantic model presented in [22] and available on github[2]. It is a modular ontology that reuses and extends standard ontologies, among which the GeoSPARQL ontology for spatial features and their relations [3], the Time ontology [4] for temporal entities, the Sensor, Observation, Sample, and Actuator (SOSA) [9] and Semantic Sensor Network (SSN) ontologies [8] for sensors and observations and the RDF Data Cube ontology [5] for aggregation and multidimensionality features. The WeKG model has several interesting conceptual modeling features, especially the conceptual representation of meteorological observations and the definition of observable properties and features of interest as a SKOS vocabulary.

The WeKG model distinguishes two conceptual levels of meteorological observations, atomic and aggregated ones. Atomic and fine-grained weather observations that can be recorded at different times of the day are modeled by reusing and extending classes from the SOSA/SSN ontologies [8,9]. The WeKG model introduces three additional classes, namely weo:meterologicalObservation,

---

[2] https://github.com/cfigmart/coffeeWKG/blob/main/meteo/WeKG-modular-ontol ogy/ontology.ttl.

weo:WeatherProperty and weo:MeteorologicalFeature, to represent respectively weather observations performed at a specific location and time providing a measurement of a given observable property (e.g., relative air humidity) characterizing some feature of interest (i.e., air). On the other hand, the observations may correspond to aggregated values (for example, the maximum daily air temperature) calculated in most cases on the basis of atomic observations or directly measured by specific instruments, when available at weather stations. In the WeKG model, these types of observations are modeled as multi-measure observations using the RDF Data Cube Vocabulary (DCV) [5], a W3C recommendation to describe multi-dimensional data structures.

Furthermore, the definition of observable properties and features of interest as well as their alignment with existing controlled vocabularies in SOSA/SSN ontologies and W3C RDF Data Cube vocabulary is delegated to the community of interest. The WeKG model includes a formalization of a subset of the World Meteorological Organization (WMO) vocabulary[3] as a comprehensive SKOS vocabulary whose concepts are named individuals of classes and represent the possible observable properties and features of interest in the meteorological domain. They are aligned with concepts from the Climate and Forecast (CF) Metadata Conventions vocabulary[4].

4.2   Modeling Weather Stations with Spatial Features

The WeKG model introduces the weo:WeatherStation class to represent any type of weather station. This class is a subclass of the geosparql:Feature class to capture the spatial location of weather stations. Therefore, each instance of weo:WeatherStation has a geometry embodied as a point with specific coordinates. Following the GeoSPARQL vocabulary, geo-coordinates of a weather station are defined as a Well-Known Text (WKT) literal (e.g., POINT(8.792667 41.918)). The adoption of GeoSPARQL is motivated by the ability to efficiently query spatial data based on a set of spatial functions. It enables to express spatial queries involving meteorological data, e.g., retrieving the cumulative precipitations recorded at the closest station to a given location. The model also reuses latitude, longitude, and altitude datatype properties from the WGS84 vocabulary since WKT literals do not integrate information about the altitude of a station.

4.3   Modeling Meteorological Observations

The CoffeeWKG knowledge graph is based on the available weather data provided by National Center for Coffee Research (Cenicafé) yearbooks which includes daily aggregate values of weather parameters as described further in Sect. 5.1. Thus, we decided to reuse the DCVontology that introduces the class qb:DataSet to identify a collection of observations typed as qb:Observation.

---

[3] https://public.wmo.int/en/resources/language-resources/meteoterm/.
[4] http://cfconventions.org/.

This class corresponds to a set of measurements (values) that share some attributes (e.g., a date) and can be aggregated according to one or more dimensions (e.g., a temporal period and/or a location). The qb:Observation class is reused in order to represent a set of aggregated agro-meteorological parameters calculated based on atomic observations [23].

The WeKG model also reuses the qb:DataStructureDefinition (DSD) class to define a set of dimensions, attributes, and measures included in the dataset along with qualifying information such as the ordering of dimensions and whether attributes are required or optional. Measures, dimensions, and attributes in DCV are represented by RDF properties, instances of the abstract qb:ComponentProperty class. We defined a DSD named :annualTimesSeries available in[5]. According to this DSD, each observation in our dataset includes 6 daily measures: the minimum, maximum and average air temperatures, the cumulative precipitations, the relative air humidity and the sunshine duration. One or more attributes can be attached to the observation instance and thus will apply to the whole measures included, which is the case of :date in the example (Fig. 1). It is also possible to define subsets of observations as an instance of the qb:Slice class, within a qb:Dataset in order to refer to all observations that share the same spatial and temporal dimensions. More specifically, :annualTimesSeries allows to define spatio-temporal slices of observations by fixing the spatial and temporal dimensions: the spatial dimension may refer to the weather station, while the temporal dimension may correspond to a calendar interval.

5   CoffeeWKG: A Weather Knowledge Graph for Coffee Regions in Colombia

In this section we present the source datasets used to generate the CoffeeWKG knowledge graph and the generation process itself.

5.1   Weather Data Sources for Coffee Regions in Colombia

The official Colombian institute responsible for providing information related to weather is the Institute of Hydrology, Meteorology, and Environmental Studies (IDEAM). It deployed weather stations distributed throughout the Colombian territory. However, the FNC also owns around 232 meteorological stations, which, unlike IDEAM stations, are more oriented to collecting meteorological data related to coffee production. These stations cover the entire Colombian coffee regions in a range between 1,000 and 2,000 m of altitude, located on the slopes of the mountain ranges.

The FNC's meteorological stations are classified as major and minor. Major climatic stations are equipped with instruments that measure air temperature,

---

[5] https://github.com/cfigmart/coffeeWKG/blob/main/meteo/dataset-metadata/ DSD-CoffeeWKG.ttl.

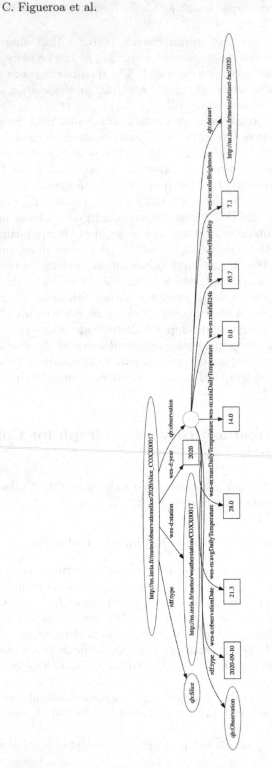

Fig. 1. RDF graph representing an instance of class qb:Observation in CoffeeWKG

relative air humidity, rainfall, and the hours of sunlight, and secondary stations measure a subset of the parameters of the main stations. In this study, we only selected the major climatic stations because these contained most of the variables required for coffee crop studies.

The observations recorded at these stations are centralized by Cenicafé and published as coffee meteorological yearbooks in PDF format. These yearbooks have been available online since 2016 and gather historical agro-climatological information obtained in the stations starting from 2006. These stations are part of the FNC's climatological network, which is strategically located near regions with the most significant influence on coffee crops in Colombia. This network allows coffee growers and experts to permanently analyze the main climate variables such as temperature (max and min), relative humidity, solar brightness, and precipitation.

5.2   Generation of CoffeeWKG

The process for the knowledge graph generation started with downloading the coffee meteorological yearbooks in PDF format from the Cenicafé website[6] from 2006 to 2020. Next, we manually extracted data tables from these pdf files using the Tabula tool[7]. It generated CSV files that were processed to obtain two datasets, one for weather stations and the other one for weather observations. The weather stations CSV file contains data about the stations, such as their code, name, municipality, latitude, longitude, and altitude. The weather observations CSV file contains measurement related to air temperatures (min, max, and mean), relative air humidity, precipitations (night, day, and total), and solar brightness.

To generate the CoffeeWKG knowledge graph from these CSV data, we updated the pipeline for WeKG-MF available on Github[8]. We first defined mapping rules from the available CSV data to RDF triples following the WeKG data model presented in Sect. 4. Then we used these mapping rules to automatically generate CoffeeWKG using the Morph-xR2RML tool[9] [12].

CoffeeWKG is available under an open license at Zenodo[10]. The current version of CoffeeWKG covers a period from 2006 to 2020 with data obtained from Cenicafé yearbooks. The main statistics about CoffeeWKG are shown in Table 1.

6   Querying CoffeeWKG

To validate CoffeeWKG, we implemented the competency questions discussed in Sect. 3 in the SPARQL language. To query CoffeeWKG with them, we installed

---

[6] Cenicafé yearbooks download page: https://www.cenicafe.org/es/index.php/nuest ras_publicaciones/anuarios_meteorologicos.
[7] https://tabula.technology/.
[8] WeKG-MF GitHub repository: https://github.com/Wimmics/weather-kg.
[9] https://github.com/frmichel/morph-xr2rml/.
[10] CoffeeWKG download page: https://zenodo.org/record/8237867.

Table 1. Statistics for the CoffeeWKG knowledge graph

Category	Number of resources
Triples	2.625.361
Weather Stations	62
Observations	786.513
Agro-climatic properties	6

the Apache Jena Fuseki server[11], which allows us to upload our knowledge graph and provides a SPARQL endpoint to execute queries.

6.1 Querying Meteorological Properties

The first competency question (CQ1) is about knowing the meteorological properties available in weather stations located in coffee regions in Colombia. The SPARQL query for this purpose is shown in Listing 1.1.

Listing 1.1. SPARQL query to retrieve the meteorological properties measured by weather stations available in CoffeeWKG (CQ1).

```
SELECT ?measure ?obsproperty
WHERE {
    ?observation qb:observation
            [a qb:Observation; ?measure ?value ]
    ?measure qb:concept ?obsproperty .
}
```

This query helps farmers and experts make informed decisions about their crops by providing information on the capabilities and limitations of nearby weather stations.

6.2 Querying Weather Stations Near a Specific Coffee Farm

The second competency question looks for weather stations near a specific coffee farm, indicated by its latitude and longitude coordinates. For example, the SPARQL query shown in Listing 1.2 finds the closest weather station near a coffee farm at latitude 4° 57′ 1″ and longitude - 74° 24′ 2″.

Listing 1.2. SPARQL query to retrieve the closest station to a specific spatial point (CQ2).

```
SELECT ?stationName ?lat ?long ?coordinates ?distance
WHERE {
    ?station rdfs:label ?stationName ;
        geosparql:hasGeometry [geosparql:asWKT ?coordinates];
```

---

[11] Apache Jena Fuseki Server: https://jena.apache.org/documentation/fuseki2.

```
               geo:lat ?lat; geo:long ?long .
BIND("POINT(-74.40055555555556 4.950277777777778)
          "^^<http://www.opengis.net/ont/geosparql#wktLiteral>
          as ?Currentposition)
BIND (geof:distance(?coordinates,?Currentposition ,
          uom:metre) as ?distance)
}
ORDER BY ?distance
LIMIT 1
```

Understanding the precise locations of meteorological stations in proximity to the crops is instrumental in guiding experts toward decisions on establishing new meteorological stations or devising sophisticated meteorological interpolation systems for data acquisition in these specific zones.

6.3 Querying Highest Temperatures by Month

This competency question identifies the highest monthly temperatures recorded to help farmers identify possible drought events causing a water deficit in coffee crops based on similar past events that look like the current one. Listing 1.3 calculates the number of days per month when the temperature exceeds a threshold (in the sample query 25 °C) as well as the maximum temperature and the average temperature measured at a specific weather station (in the sample query the station is named "Alberto Santos"). It is possible to filter the results for a specific year by adding a FILTER clause.

Listing 1.3. SPARQL query to calculate highest temperatures (CQ3).

```
SELECT ?year ?month
      (SUM(IF(?temp_max>25,1,0)) AS ?nbExtrHeatDays)
      (MAX(?temp_max) AS ?max_T_Month)
      (AVG(?temp_avg ) AS ?month_T_AVG)
WHERE {
    ?station a weo:WeatherStation ;rdfs:label ?stationName .
    ?s a qb:Slice ;
       wes-dimension:station ?station ;
       wes-dimension:year ?year ;
       qb:observation [
               a qb:Observation ;
               wes-attribute:observationDate ?date ;
               wes-measure:maxDailyTemperature ?temp_max ;
               wes-measure:avgDailyTemperature ?temp_avg ].
    BIND (SUBSTR(STR(?date),6,2) AS ?month)
}
GROUP BY ?stationName ?year ?month
ORDER BY DESC(?year) ?month
```

6.4 Querying Highest Rainfalls by Month

This competency question identifies the highest monthly rainfalls to warn farmers about cultivating coffee in months more prone to rainfall. Listing 1.4 shows monthly accumulated rainfall for each year at a specific station identified by its name (in the sample query, the station with a name equal to "Alberto Santos"). It is also possible to filter the results for a specific year by adding a FILTER clause.

Listing 1.4. SPARQL query to calculate cumulative precipitations (CQ4).

```
SELECT ?year ?month (SUM(?rain ) as ?monthlyRain)
WHERE {
    ?station a weo:WeatherStation ;
             rdfs:label ?stationName .
    FILTER(?stationName = "Alberto Santos")
    ?s a qb:Slice ;
       wes-dimension:station ?station ;
       wes-dimension:year ?year ;
       qb:observation [
           a qb:Observation ;
           wes-attribute:observationDate ?date ;
           wes-measure:rainfall24h ?rain ].
    BIND (SUBSTR(STR(?date),6,2) AS ?month)
}
GROUP BY ?year ?stationName ?month
ORDER BY ?year ?month
```

7   Conclusions

Climate variability is a significant constraint in coffee production, which plays a crucial role in Colombia's economy. Therefore, studying and analyzing meteorological data in coffee regions helps farmers and experts fine-tune their agricultural practices based on historical and real-time weather patterns to schedule planting and harvesting periods more effectively, adjust irrigation and fertilization strategies, and even predict future harvests more accurately. In this paper we presented the CoffeeWKG knowledge graph for weather conditions of coffee regions in Colombia constructed with data extracted from FNC yearbooks and processed with an adaptation of the WeKG-MF [22] model initially developed for data from the French weather data provider Météo-France. The CoffeeWKG knowledge graph can be queried using the SPARQL query language to ask for historical weather conditions in coffee regions and correlate this information with coffee crop data to understand how these conditions may favor or hamper coffee production or diseases.

In future work, we intend to complement the meteorological data initially obtained from Cenicafé with satellite data and IDEAM stations located near coffee regions to cover some areas not included in the FNC yearbooks. Furthermore, we are working on integrating other types of data related to coffee production, pest and disease advent, and other factors that may be affected by climate conditions to enrich our knowledge graph. This will enable to meet more complex requirements about weather conditions, coffee production, or diseases. In particular, we plan to link CoffeeWKG with the RustOnt ontology [17] that models data, expressions, and samples related to coffee rust. In the same way, we also plan to link CoffeeWKG with other relevant Linked Open Data (LOD) datasets and to add more weather indicators, e.g., the growing degree days (GDD)[12] described in the Global Change Master Directory (GCMD), which could help farmers to predict plant and pest development rates.

References

1. Ahmed, S., et al.: Climate change and coffee quality: systematic review on the effects of environmental and management variation on secondary metabolites and sensory attributes of Coffea arabica and Coffea canephora. Front. Plant Sci. **12**, 708013 (2021). https://doi.org/10.3389/fpls.2021.708013
2. Atemezing, G., et al.: Transforming meteorological data into linked data. Semant. Web **4**, 285–290 (2013). https://doi.org/10.3233/SW-120089
3. Battle, R., Kolas, D.: Enabling the geospatial semantic web with parliament and GeoSPARQL. Semant. Web **3**, 355–370 (2012). https://doi.org/10.3233/SW-2012-0065
4. Cox, S., Little, C.: Time ontology in OWL. W3C candidate recommendation draft. Technical report, W3C (2022). https://www.w3.org/TR/owl-time/
5. Cyganiak, R., Reynolds, D.: The RDF data cube vocabulary. Technical report, W3C (2014). https://www.w3.org/TR/2014/REC-vocab-data-cube-20140116/
6. DaMatta, F.M., Ronchi, C.P., Maestri, M., Barros, R.S.: Ecophysiology of coffee growth and production. Braz. J. Plant. Physiol. **19**, 485–510 (2007). https://doi.org/10.1590/S1677-04202007000400014
7. Davis, A.P., Gole, T.W., Baena, S., Moat, J.: The impact of climate change on indigenous arabica coffee (Coffea arabica): predicting future trends and identifying priorities. PLoS ONE **7**(11), 1–13 (2012). https://doi.org/10.1371/journal.pone.0047981
8. Haller, A., et al.: The modular SSN ontology: a joint W3C and OGC standard specifying the semantics of sensors, observations, sampling, and actuation. Semant. Web **10**, 9–32 (2018). https://doi.org/10.3233/SW-180320
9. Janowicz, K., Haller, A., Cox, S.J., Phuoc, D.L., Lefrançois, M.: SOSA: a lightweight ontology for sensors, observations, samples, and actuators. J. Web Semant. 56, 1–10 (5 2019). https://doi.org/10.1016/j.websem.2018.06.003
10. León-Burgos, A.F., Unigarro, C., Balaguera-López, H.E.: Can prolonged conditions of water deficit alter photosynthetic performance and water relations of coffee plants in central-west Colombia? South Afr. J. Botany **149**, 366–375 (2022). https://doi.org/10.1016/j.sajb.2022.06.034
11. Matteis, L., et al.: Crop ontology: Vocabulary for crop-related concepts. In: CEUR Workshop Proceedings, vol. 979 (2013)

---

[12] The concept GDD is available at the URI of the GCMD vocabulary https://gcmd.earthdata.nasa.gov/kms/concept/6d808909-ce04-4401-a883-aff4d723d025.

12. Michel, F., Djimenou, L., Faron-Zucker, C., Montagnat, J.: Translation of relational and non-relational databases into RDF with xR2RML. In: Proceedings of the 11th International Conference on Web Information Systems and Technologies, pp. 443–454. SCITEPRESS - Science and and Technology Publications (2015). https://doi.org/10.5220/0005448304430454

13. Peroni, S.: A simplified agile methodology for ontology development. In: Dragoni, M., Poveda-Villalón, M., Jimenez-Ruiz, E. (eds.) OWLED/ORE -2016. LNCS, vol. 10161, pp. 55–69. Springer, Cham (2017). https://doi.org/10.1007/978-3-319-54627-8_5

14. Prensa Federación Nacional de Caficultores de Colombia: en junio importaciones de café disminuyeron un 30 porciento (2023). https://federaciondecafeteros.org/wp/listado-noticias/en-junio-importaciones-de-cafe-disminuyeron-un-30/

15. Ronchi, C.P., Miranda, F.R.: Flowering percentage in arabica coffee crops depends on the water deficit level applied during the pre-flowering stage. Rev. Caatinga **33**, 195–204 (2020). https://doi.org/10.1590/1983-21252020v33n121rc

16. Roussey, C., Delpuech, X., Amardeilh, F., Bernard, S., Jonquet, C.: Semantic description of plant phenological development stages, starting with grapevine. In: Garoufallou, E., Ovalle-Perandones, M.-A. (eds.) MTSR 2020. CCIS, vol. 1355, pp. 257–268. Springer, Cham (2021). https://doi.org/10.1007/978-3-030-71903-6_25

17. Suarez, C., Griol, D., Figueroa, C., Corrales, J.C., Corrales, D.C.: RustOnt: an ontology to explain weather favorable conditions of the coffee rust. Sensors **22**, 9598 (2022). https://doi.org/10.3390/s22249598

18. Subirats-Coll, I., et al.: AGROVOC: the linked data concept hub for food and agriculture. Comput. Electron. Agric. **196**, 105965 (2022). https://doi.org/10.1016/j.compag.2020.105965

19. Uschold, M., Gruninger, M.: Ontologies: principles, methods and applications. Knowl. Eng. Rev. **11**, 93–136 (1996). https://doi.org/10.1017/S0269888900007797

20. Vélez-Vallejo, R.: Informe del gerente - 90 congreso nacional de cafeteros. Technical report, Federación Nacional de Cafeteros de Colombia (2022). https://federaciondecafeteros.org/app/uploads/2022/12/Informe-del-Gerente-D.pdf

21. Wu, J., Orlandi, F., O'Sullivan, D., Dev, S.: LinkClimate: an interoperable knowledge graph platform for climate data. Comput. Geosci. **169**, 105215 (2022). https://doi.org/10.1016/j.cageo.2022.105215

22. Yacoubi Ayadi, N., Faron, C., Michel, F., Gandon, F., Corby, O.: A model for meteorological knowledge graphs: application to Météo-France data. In: ICWE 2022–22nd International Conference on Web Engineering. 22nd International Conference on Web Engineering, ICWE 2022, Bari, Italy (2022). https://doi.org/10.1007/978-3-031-09917-5_19

23. Yacoubi Ayadi, N., Faron, C., Michel, F., Gandon, F., Corby, O.: Computing and visualizing agro-meteorological parameters based on an observational weather knowledge graph. In: Ding, Y., Tang, J., Sequeda, J.F., Aroyo, L., Castillo, C., Houben, G. (eds.) Companion Proceedings of the ACM Web Conference 2023, WWW 2023, Austin, TX, USA, 30 April–4 May 2023, pp. 242–245. ACM (2023). https://doi.org/10.1145/3543873.3587357

Author Index

T. P. Sales et al. (Eds.): ER 2023 Workshops, LNCS 14319, pp. 343–344, 2023.
https://doi.org/10.1007/978-3-031-47112-4

Printed in the United States
by Baker & Taylor Publisher Services